AF710679

★ ★ ★ 2009
RETIRED MILITARY ALMANAC

Ronald S. Hunter - *Publisher*
Debra M. Gordon - *Creative Editor*
Dana L. Smith - *Editor*
Lt Col Sol Gordon, USAF (Ret) - *Editor/Publisher Emeritus*

PRICE—$12.00

(Plus $3.50 P & H = $15.50 via Standard Mail)
(See inside front cover for multiple copy rates)
Call for Government prices.
GSA/Federal Supply Service Contract Number: GS-14F-6023A
(See FSC Group 76, Part 1, Class 7610) Item No. 760-1M

Published Annually by:
UNIFORMED SERVICES ALMANAC, INC.
P.O. BOX 4144, Falls Church, VA 22044-0144
Telephone: (703) 532-1631 Fax: (703) 532-1635
Website: www.militaryalmanac.com

Publishers of:
UNIFORMED SERVICES ALMANAC—for active duty personnel
RESERVE FORCES ALMANAC—for members of the Reserve Forces
NATIONAL GUARD ALMANAC—for members of the National Guard
RETIRED MILITARY ALMANAC—for retired personnel

RETIRED MILITARY ALMANAC

FOREWORD

Recently, while doing some research, I was looking back at some previous editions of the Almanac. Although not the focus of my research, I noticed just how many improvements there have been in the military compensation system since 2001. Here are some of the changes I noticed:

- Basic pay has increased an average of 37.4 percent;
- Retired pay has increased 27 percent;
- Out-of-pocket housing costs have been eliminated;
- Family separation allowance increased from $100 to $250 per month;
- The death gratuity increased from $6,000 to $100,000;
- Extension of the O1E-O3E pay scale to Reserve Component officers;
- Leave expansion to include 21 days leave for adoption and 10 days paternity leave;
- Increase from $10 to $50 per drill for members assigned to high-priority Reserve units, and expansion of the pay to officers;
- Survivor benefits for members who die on active duty with less than 20 years of service;
- Expansion from 10 to 14 years, the period a member can use GI Bill Selected Reserve benefits;
- Expansion of Command Responsibility Pay to Reserve Component officers;
- Enhanced basic pay table to provide longevity increases for some senior members with over 30 years of service;
- Increased the maximum percentage of retired pay from 75 percent to 2.5 percent for each year of service with no maximum;
- Elimination of the Level II Executive pay cap on retired pay;
- Reduced eligibility threshold for a Reserve Component member serving on active duty to receive full housing allowance from 140 to 31 days;
- Reduced the retired pay eligibility age for certain Reserve Component members serving in a contingency operation;
- Reserve Component eligibility to purchase TRICARE health care coverage;
- Improved GI Bill benefits for Reserve Component members serving in a contingency operation;
- Establishment of two special compensation programs for retired members with VA compensable disabilities.

Also, in 1981, the year I joined the Army, a private was paid $501.30 basic pay. Today, a private is paid $1,399.50 or 279 percent more. A second lieutenant was paid $1,224.30 in 1986 – the year I was commissioned. Today, a second lieutenant is paid $2,655.30 or 217 percent more.

The **RETIRED MILITARY ALMANAC** is not an official Government publication of the DoD nor the United States Government. We do, however, rely to a large extent on many DoD and other Government agencies and offices to ensure that the information we provide is as accurate and up-to-date as possible. We therefore express our sincere appreciation to everyone who has worked with us, and particularly acknowledge the assistance and cooperation of the office of the Defense Manpower Data Center and the Office of the Actuary; the Department of Veterans Affairs; the Department of Labor; and several other offices of the military services which have been most generous and cooperative.

Ron Hunter
Publisher

TABLE OF CONTENTS

PUBLICATION DATE — FEBRUARY 2, 2009

PART I MILITARY RETIREMENT

PART II GENERAL INFORMATION

PART III HEALTH CARE, INSURANCE AND MEDICAL FACILITIES

PART IV VETERANS BENEFITS

PART V SBP AND SOCIAL SECURITY

PART VI ARMED FORCES RETIREMENT HOMES, STATISTICS AND ORGANIZATIONS

PART I

MILITARY RETIREMENT BACKGROUND

The military retirement system applies to members of the Army, Navy, Marine Corps, and Air Force. Most of the provisions also apply to retirement systems for members of the Coast Guard (administered by the Department of Homeland Security), officers of the Public Health Service (administered by the Department of Health and Human Services), and officers of the National Oceanic and Atmospheric Administration (administered by the Department of Commerce).

The system is a funded, noncontributory defined benefit plan that includes nondisability retired pay, disability retired pay, retired pay for Reserve service, survivor annuity programs, and special compensation programs for certain disabled retirees. The Service Secretaries may approve immediate nondisability retired pay at any age for members with at least 20 years of active duty service. There is no vesting before retirement. Reserve retirees must have at least 20 years of qualifying service to begin receiving retired pay which normally begins at age 60. However, certain Reservists may begin receiving retired pay earlier due to certain active service performed after January 28, 2008 in support of a contingency operation or other voluntary active service. For every 90 days of qualifying service, retired pay would begin 90 days early, but not earlier than age 50.

Non Disability Retirement From Active Service. There are three distinct nondisability benefit formulas related to three populations within the military retirement system. *Final pay*: Military personnel who first became members of a uniformed service before September 8, 1980, have retired pay equal to final basic pay times a multiplier. The multiplier is equal to 2.5 percent times years of service. Members earn 2.5 percent for each year of active service with no cap. *High-three*: If the retiree first became a member of a uniformed service on or after September 8, 1980, the average of the highest 36 months of basic pay is used instead of final basic pay. *Redux*: Members who first became a member of a uniformed service on or after August 1, 1986 and who elect to receive the Career Status Bonus outlined below are subject to a multiplier penalty if they retire with less than 30 years of service; however, at age 62, their retired pay is recomputed without the penalty. They also have retired pay computed on a base of the average of their highest 36 months of basic pay. Redux members have a choice of (a) receiving High-three benefits or (b) staying under the Redux formula and receiving a $30,000 Career Status Bonus (CSB). Eligible members make their election during the 15th year of service. Those who choose the CSB must remain continuously on active duty until they complete 20 years of active duty service or forfeit a portion of the $30,000.

Retired pay and survivor annuity benefits are automatically adjusted annually to protect the purchasing power of initial retired pay. The benefits associated with members first entering the armed services before August 1, 1986 or those entering on or after that date who do not elect the CSB, have their benefits adjusted annually by the percentage increase in the average Consumer Price Index (CPI). This is commonly referred to as full CPI protection. Benefits associated with members entering on or after August 1, 1986, who elect the CSB are annually increased by the percentage change in the CPI minus one percent, but at the military member's age 62, or when the member would have been age 62 for a survivor annuity, the benefits are restored to the amount that would have been payable had full CPI protection been in effect. This restoration is in combination with that described in the previous paragraph. However, after this restoration, partial indexing (CPI minus one percent) continues for future retired pay and survivor annuity payments.

As of September 30, 2008, there were 1.47 million nondisability active duty retirees receiving retired pay. In FY 2008, nondisability retirees were paid $36.9 billion.

Disability Retirement. A disabled military member is entitled to disability retired pay if the disability is not the result of the member's intentional misconduct or willful neglect, was not incurred during a period of unauthorized absence, and either:

1. the member has at least 20 years of service; or
2. at the time of determination, the disability is at least 30 percent (under a standard schedule of rating disabilities by the Veterans Administration) and one of three additional conditions are met:

1st additional condition. If the disability was incurred:

a. prior to January 28, 2008, it must have been incurred after the member had eight 8 years of active service; or

b. between January 28, 2008 and October 13, 2008, it must have been incurred after the member had six months or more of active military service and the disability must not have been noted at the time of the member's entrance on active duty (unless compelling evidence or medical judgment is such to warrant a finding that the disability existed before the member's entrance on active duty); or

c. effective October 14, 2008, it must have been incurred after the member's entrance on active duty and the disability must not have been noted at the time of the member's entrance on active duty (unless clear and unmistakable evidence demonstrates that the disability existed before the member's entrance on active duty and was not aggravated by active military service).

2nd additional condition. The disability results from active duty.

3rd additional condition. The disability occurred in the line of duty during a time of war or national emergency or after September 14, 1978.

In disability retirement, the member receives retired pay equal to the larger of (a) the accrued nondisability retirement benefit regardless of eligibility to retire or (b) base pay multiplied by the rated percent of disability. The benefit cannot be more than 75 percent of base pay. Only the excess of (a) over (b) is subject to federal income taxes if the member had service on or before September 24, 1975. If not a member of a uniformed service on September 24, 1975, disability retired pay is tax-exempt only for those disabilities that are combat or hazardous duty related. Base pay is equal to final basic pay if the retiree first became a member of a uniformed service before September 8, 1980; otherwise, base pay is equal to the average of the highest 36 months of basic pay.

Members who have disabilities that may not be permanent are placed on a temporary-disability retired list and receive disability retirement pay just as if they were permanently disabled. However, they must be physically examined every 18 months for any change in disability. A final determination must be made within five years. The temporary disability pay is calculated like the permanent disability retired pay, except that it can be no less than 50 percent of base pay.

As of September 30, 2008, there were 86,000 disability retirees receiving retired pay. In FY 2008, disability retirees were paid $1.29 billion.

Reserve Retirement. Reserve Component members may retire after 20 qualifying years of creditable service, the last six of which must be in a Reserve Component. **Note**: The six year requirement is repealed for members who attain 20 creditable years of service on or after April 25, 2005. Reserve retired pay normally begins at age 60, however, a member may begin receiving Reserve Component retired pay three months early for each aggregate 90 days of qualifying active service—performed after January 28, 2008—during a fiscal year period. Qualifying active service is: service in support of a contingency operation and certain voluntary active duty. Retired pay is computed as base pay times 2.5 percent times years of service. If the Reservist was first a member of a uniformed service before September 8, 1980, base pay is defined as the active duty basic pay in effect for the retiree's grade and years of service at the time that retired pay begins. If the Reservist first became a member of the armed services on or after September 8, 1980, base pay is the average basic pay for the member's grade in the highest 36 months computed as if he/she was on active duty for the 36 months immediately preceding age 60. The years of service are determined by using a point system, where 360 points convert to a year of service. Typically, one point is awarded for one day of active duty service (e.g. active duty training) or one period of inactive duty training (IDT). Reservists may perform two IDT periods in one day thereby receiving two retirement points per day. In addition 15 points are awarded for a year's membership in a Reserve Component. A creditable year of service is one in which the member earned at least 50 points. A member cannot retire with less than 20 creditable years, although points earned in non-creditable years are used in the retirement calculation. Inactive duty points are limited in any year to no more than 130. Lesser point limitations have applied in the past.

As of September 30, 2008, there were 329,000 Reserve retirees receiving retired pay. In FY 2008, Reserve retirees were paid $4.2 billion.

Survivor Benefits. Legislation originating in 1953 provided optional survivor benefits. It was later referred to as the Retired Serviceman's Family Protection Plan (RSFPP). The plan proved to be expensive to the participants and inadequate since the survivor annuities were never adjusted for inflation and could not be more than 50 percent of retired pay. RSFPP was designed to be self supporting in the sense that the present value of the reductions in retired pay equaled the present value of the survivor annuities.

On September 21, 1972, the Survivor Benefit Plan (SBP) replaced the RSFPP for new retirees. RSFPP still covers those servicemen retired before 1972 who did not convert to the new plan or who retained RSFPP in conjunction with SBP. RSFPP continues to pay survivor annuities.

Retired pay is reduced, before taxes, for the member's cost of SBP. Total SBP costs are shared by the government and the retiree, so the reductions in retired pay are only a portion of the total cost of the SBP program.

The SBP survivor annuity is 55 percent of the member's base amount. The base amount is elected by the member, but cannot be less than $300 or more than the member's full retired pay. If the member elects the Career Status Bonus with REDUX and is subject to a penalty for service under 30 years in the calculation of retired pay, the maximum base amount is equal to the full retired pay without the penalty.

When the plan started in 1972, benefits for those 62 and older were reduced by the constructed amount of Social Security for which the survivor would be eligible based on the member's military pay. In 1985, that reduction formula was changed so all annuitants 62 and over received a reduced flat rate of 35 percent of the member's base. Beginning October 1, 2005, the age 62 reduced rate was phased out in 5 percent increments. On April 1, 2008, the survivor benefit reduction at age 62 was fully eliminated and the rate of 55 percent of the member's elected base became standard for all survivors, regardless of age.

During FY 1987, the SBP program's treatment of survivor remarriages changed. Prior to the change, a surviving spouse remarrying before age 60 had the survivor annuity suspended. The change lowered the age to 55. If the remarriage ends in divorce or death, the annuity is reinstated.

Members who die on active duty are assumed to have retired with full disability on the day they died and to have elected full SBP coverage for a spouse or former spouse (if applicable), and/or children. Insurable interest elections may be applicable in some cases. These benefits have been improved and expanded over the history of the program.

SBP annuities are reduced by any VA survivor benefits (Dependency and Indemnity Compensation (DIC)) and all premiums relating to the reductions are returned to the survivor. Beginning in October 2008, surviving spouses whose SBP is reduced due to receipt of DIC began receiving a Special Survivor Indemnity Allowance.

SBP annuities and premiums are increased annually by the same cost of living adjustments (COLAs) that were applicable to the retiree. If a member who elected the Career Status Bonus dies before age 62, the survivor is subject to partial COLAs and his/her annuity is increased on what would have been the member's 62nd birthday to the amount that would have been payable had full COLAs been in effect. Partial COLAs continue annually thereafter.

For Reserve retirees, the same set of retired pay reductions applies for survivor coverage after a Reservist turns 60 and begins to receive retired pay. The Reserve Component Survivor Benefit Plan (RCSBP) provides annuities to survivors of Reservists who die after attaining 20 years of qualified service, but before commencement of retired pay. The added cost of this coverage is borne completely by Reservists through deductions from retired pay and survivor annuities.

A paid-up provision eliminating the reduction in retired pay for premiums for SBP and RSFPP coverage became effective October 1, 2008 for participants age 70 or older whose retired pay has been reduced for 30 years or more.

As of September 30, 2008, there were 291,000 survivors of military members receiving annuity payments. In FY 2008, survivors were paid $3.27 billion.

Temporary Early Retirement Authority (TERA). The National Defense Authorization Act for Fiscal Year 1993 (P.L. 102-484) granted temporary authority for the military services to offer early retirements to members with more than 15 but less than 20 years of service. This authority expired on September 1, 2002. The retired pay was calculated in the usual way except that there was a reduction of one percent for every year below 20 years of service. Part or all of this reduction can be restored at age 62 if the retired member works in a qualified public service job during the period from the date of retirement to the date on which the retiree would have completed 20 years of service. Unlike members who leave military service before 20 years with Voluntary Separation Incentives or Special Separation Benefits, these early retirees are generally treated like regular military retirees for the purposes of other retirement benefits.

As of September 30, 2008, there were 56,000 TERA retirees receiving retired pay. In FY 2008, TERA retirees were paid $834 million.

Relationship with Department of Veterans Affairs Benefits. The Department of Veterans Affairs (VA) provides compensation for service-connected and certain non-service-connected disabilities. These VA benefits can be in place of or in combination with DoD retired pay, but through December 31, 2003 were not fully additive. Since VA benefits are exempt from federal income taxes, it is often to the advantage of a member to elect them. Through 2003, retired pay earned from DoD for military service was offset by any payment received from VA for a VA-rated disability. Since 2004, a series of legislative changes have been enacted that increasingly reduces or eliminates the offset to military retired pay due to receipt of VA disability compensation. Members with a combined VA disability rating of 50 percent or greater who have at least 20 years of service will have their offset eliminated under the Concurrent Retirement and Disability Pay (CRDP) program. The CRDP program has a ten-year phase-in schedule that began in 2004; however, the offset is already fully eliminated for members whose disabilities are rated total or make the individual unemployable. Members whose disability meets certain combat-related criteria can elect to receive payments against the offset under the Combat Related Special Compensation (CRSC) program. Under CRSC, members are not subject to a phase-in schedule, are not required to have at least 20 years of service, and are not required to have at least a 50 percent VA disability rating. Although CRSC amounts are calculated based on retired pay lost due to offset and are paid from the Military Retirement Fund, CRSC is not technically considered retired pay. CRSC payments are tax exempt. A member may not participate in both the CRDP and CRSC programs simultaneously, but may change from one to the other during an annual "open season."

VA benefits also overlap survivor benefits through the Dependency and Indemnity Compensation (DIC) program. DIC is payable to survivors of veterans who die from service-connected causes. Although an SBP annuity must be reduced by the amount of any DIC benefit, all SBP premiums relating to the reduction in benefits are returned to the survivor. The National Defense Authorization Act for Fiscal Year 2008 provided a temporary Special Survivor Indemnity Allowance that pays a monthly amount ($50 in FY 2009, increasing to $100 by FY 2014, in $10 increments) to survivors with a DIC offset. The authority for the allowance ends in 2016.

Interrelationship with other Federal Service. For retirement purposes, no credit is given for other federal service, except where cross-service transferability is allowed. Military service is generally creditable toward the federal civilian retirement systems if military retired pay is waived. However, a deposit (equal to a percentage of post-1956 basic pay) must be made to the Civil Service Retirement Fund in order to receive credit. Military service is not generally creditable under both systems (but is for Reservists and certain disability retirees).

SELECTED LEGISLATION

The following is a partial listing of legislation of interest and significance to military personnel. The listing is not all-inclusive, but contains highlights of key legislation passed by Congress and signed into law by the President.

(110th Congress – 2nd Session) DEFENSE APPROPRIATIONS LEGISLATION

P.L. 110-329, September 30, 2009, Consolidated Security, Disaster Assistance, and Continuing Appropriations Act, 2009.

Provides funds for the fiscal year ending September 30, 2009 for the Department of Defense for: (1) members on active duty (except members of the Reserve Components provided for elsewhere) for pay, allowances, individual clothing, subsistence, interest on deposits, gratuities, permanent change of station travel (including temporary duty travel between permanent changes of station), cadets and midshipmen, aviation cadets, members of the Reserve Officers' Training Corps, Social Security taxes, and for payments to the Defense Military Retirement Fund as follows: Army, $36,382,736,000; Navy, $24,037,553,000; Marine Corps, $11,792,974,000; Air Force, $25,103,789,000; and (2) Reserve Component members for pay, allowances, clothing, subsistence, gratuities, travel, and related expenses for Reserve personnel on active duty (under certain provisions of law), or while undergoing Reserve training, or while performing drills or equivalent duty, for expenses related to the Montgomery GI Bill – Selected Reserve, and for payments to the Defense Military Retirement Fund as follows: Army, $3,904,296,000; Navy, $1,855,968,000; Marine Corps, $584,910,000; Air Force, $1,423,676,000; Army National Guard, $6,616,220,000; Air National Guard, $2,741,768,000.

In addition, the Act provides $72,000,000 to pay special pay during FY 2009 to eligible members of the Army, Navy, Air Force, and Marine Corps, including members of their Reserve Components who at any time during FY 2009 serve on active duty while the members' enlistment or period of obligated service is extended, or whose eligibility for retirement is suspended under stop-loss authority. The special pay may not exceed $500 per month for each month or portion of a month during FY 2009 that a member is retained on active duty as a result of application of the stop-loss authority.

DEFENSE AUTHORIZATION LEGISLATION

P.L. 110-417, October 14, 2008, Duncan Hunter National Defense Authorization Act for Fiscal Year 2009.

End Strengths for Active Forces. The Armed Forces are authorized strengths for active duty personnel as of September 30, 2009 as follows: Army, 532,400; Navy, 326,323; Marine Corps, 194,000; Air Force, 317,050.

Revised End Strength Minimum Levels. Adjusts the minimum end strengths for the services on September 30, 2009 as follows: Army, 532,400; Navy, 325,300; Marine Corps, 194,000; Air Force, 317,050.

End Strengths for Selected Reserve. The Armed Forces are authorized strengths for Selected Reserve Personnel of the Reserve Components as of September 30, 2008 as follows: Army Reserve, 205,000; Navy Reserve, 66,700; Marine Corps Reserve, 39,600; Air Force Reserve, 67,400; Coast Guard Reserve, 10,000; Army National Guard, 352,600; Air National Guard, 106,756.

End Strengths for Reserve Component personnel on Active Duty in Support of the Reserve Components. Within the Selected Reserve end strengths, the Reserve Components of the Armed Forces are authorized, as of September 30, 2009, the following number of Reserve or National Guard personnel to be serving on full-time active duty or full-time National Guard duty for organizing, administering, recruiting, instructing, or training the Reserve Components: Army Reserve, 16,170; Navy Reserve, 11,099; Marine Corps Reserve, 2,261; Air Force Reserve, 2,733; Army National Guard, 32,060; Air National Guard, 14,360.

The Act also:

• Increases the rate of basic pay for all military members 3.9 percent effective January 1, 2009.

• Provides that Air Force military technicians who meet dual status requirements may serve beyond a mandatory removal date or maximum years of service until reaching age 60 and becoming eligible for an unreduced annuity.

• Increases the mandatory retirement age to 62 for Reserve Component officers assigned to the Selective Service System or serving as a property and fiscal officer.

• Allows the Secretary of the Army or Air Force to retain to age 66 a Reserve Component Lieutenant General who reaches maximum years of service.

• Provides up to 10 days of leave for a married member whose wife gives birth on or after October 14, 2008.

• Authorizes the *Career Flexibility Pilot Program* allowing a limited number of servicemembers to take a leave of absence from service for up to three years.

• Requires that members entitled to benefits under the Reserve Educational Assistance Program after separation from the Selected Reserve must have been separated under honorable conditions.

• Increases the maximum Temporary Lodging Expense allowance rate from $180 to $290 per day.

• Provides two Family Separation Allowances (one to each member) to married members who are both are simultaneously assigned to duties qualifying both members for the allowance.

• Adds skill incentive pay and proficiency bonus authorities to encourage training in critical foreign languages and cultural studies.

• Adds accession and retention bonus authorities for psychologists.

• Provides that the annual maximum amount of a loan that may be repaid under the Selected Reserve Health Professional Loan Repayment Program shall be the same as the maximum amount in effect for the same year under the Active Duty Health Professional Loan Repayment Program.

• Waives all copayments for certain preventative care services even though the beneficiary has not met the annual TRICARE deductible during the year in which the preventative services are received.

• Establishes a smoking cessation program under TRICARE.

• Directs the Secretary of Defense to establish a preventative health services allowance demonstration project.

• Increases the maximum reimbursement for Extended Health Care Option Benefits from $2,500 per month to $36,000 per year.

• Provides 180 days of transitional health care to a member who is separated from active duty after October 14, 2008 who agrees to become a member of the Selected Reserve.

OTHER LEGISLATION

P.L. 110-317, August 29, 2008, Hubbard Act.

Provides enhanced benefits for members of the Armed Forces who are separated from service after September 11, 2001 under DoD policies relating to sole survivorship.

• Makes mandatory the exemption from recoupment of any portion of a bonus, incentive pay, or similar benefit of a member who receives a sole survivorship discharge. Also provides discretionary exception to the requirement to terminate payment of any unpaid amount of the bonus, incentive pay, or similar benefit.

• Extends separation pay eligibility to a member who receives a sole survivorship discharge even though the member did not serve for six years.

• Provides 180 days of transitional health care benefits and two years transitional commissary and exchange benefits to a member who receives a sole survivorship discharge.

The Act also provides eligibility for the following veteran's and Department of Labor benefits for a member who receives a sole survivorship discharge: housing loan benefits; employment and training benefits; entitlement to GI Bill benefits the member would have otherwise been entitled to; eligibility for unemployment compensation; and veteran's preference in hiring for government jobs.

KEY CONGRESSIONAL COMMITTEES

SENATE COMMITTEE ON ARMED SERVICES

228 Russell Senate Office Bldg.
Washington, D.C. 20510
Phone: (202) 224-3871
Website: http://armed-services.senate.gov/

HOUSE ARMED SERVICES COMMITTEE

2120 Rayburn House Office Bldg.
Washington, D.C. 20515
Phone: (202) 225-4151
Website: http://armedservices.house.gov/

SENATE COMMITTEE ON VETERANS AFFAIRS

412 Russell Senate Bldg.
Washington D.C. 20510
Phone: (202) 224-9126
Website: http://veterans.senate.gov

HOUSE COMMITTEE ON VETERANS AFFAIRS

335 Cannon House Office Bldg.
Washington, D.C. 20515
Phone: (202) 225-9756
Website: http://veterans.house.gov/

MILITARY RETIREMENT

One of the most attractive incentives of a military career is a retirement system which provides monthly retirement income for those who serve a minimum of 20 years, for nondisability retirement, and entitles the retiree to retain many of the benefits of the services such as exchange and commissary privileges, medical care in uniformed services facilities on a space available basis, or through TRICARE, and many other desirable benefits, including provisions to provide income for survivors if desired. (See SURVIVOR BENEFIT PLAN section for details.)

The military retirement system is composed of three separate but interrelated systems: nondisability retirement from active duty; Reserve nondisability retirement; and disability retirement for both regular and Reserve personnel. These three systems support both the active and Reserve forces.

The military retirement systems are funded under an accrual accounting technique. Members make no direct contributions towards military retirement. Retired members are paid from the DoD Military Retirement Fund. The fund receives assets from three sources: contributions from the DoD for the present value of the future retirement benefit for members presently serving; deposits from the Treasury General Fund to amortize the unfunded liability of past service; and interest from investments of fund assets.

The National Oceanic and Atmospheric Administration and the Public Health Service also have retirement systems, under other provisions of the law, based on the military retirement system. In addition, the U.S. Coast Guard has a retirement system under Title 14, United States Code.

TYPES OF RETIREMENT AND RETIRED PAY

Nondisability Retirement

Generally, regular and Reserve commissioned officers, warrant officers, and enlisted members may retire after completing 20 or more years of active service. Tenure is established based on grade. Ordinarily, members may serve a maximum of 30 years prior to mandatory retirement. However, certain members are allowed to serve longer.

Servicemembers are credited for all years of active service. Service performed in a Reserve Component may also be included in the retirement computation.

The retired grade of the member is generally the grade, (whether temporary or permanent), held on the date of retirement. Officers in pay grades O-5 through O-8 must serve in that pay grade for at least three years (may be waived to two years) in order to retire in that pay grade. Officers in pay grade O-4 must have at least six months time in that grade. Warrant officers W-3 through W-5 and enlisted members E-7 through E-9 must have at least two years time in grade to retire at that grade. However, there are some exceptions:

- An enlisted member or warrant officer retiring with less than 30 years of service, who held a temporary higher grade than the grade in which retired, receives retired pay in the grade held at time of retirement until the total active and retired service reaches 30 years. At that time, the retired member is advanced to the higher grade and retired pay is then based on the higher grade. (In the Navy and Marine Corps, the higher grade must have been held as a commissioned officer.)
- Enlisted members and warrant officers holding Reserve commissions who complete 20 years of active service, of which 10 years is commissioned, may retire and receive pay on the basis of the commissioned rank.

• A Reserve officer (O-5 and above) eligible to retire under the Reserve retirement system, who has at least six months, but less than three years time in grade, may retire at the higher grade if transferred from an active status or discharged due to age, years of service, or because of a physical disability that leaves the member not qualified to serve in the Ready Reserve.

All regular officers, unless retired or separated earlier, are retired by law on the first day of the month following the month in which the officer becomes 62 years of age (O-6 and below) and 64 years of age (above O-6). Certain officers assigned to positions at the military academies may serve longer.

In addition, officers serving in a grade above major general or rear admiral may have retirement deferred by the Secretary of Defense up to age 66, and by the President up to age 68. The officer must be retired or separated on the first day of the month following the month in which the officer reaches maximum deferment age.

Three Retirement Systems:

There are three methods for computing nondisability retired pay depending on the date the person first became a member of a uniformed service. The date that an individual first entered the military is the Uniformed Service Initial Entry Calendar Date (DIEUS). The DIEUS determines which retirement system the individual is under. Some individuals have unique circumstances that complicate determining their DIEUS.

MAJOR FEATURES OF THE THREE MILITARY RETIREMENT SYSTEMS			
Plan	**Eligible**	**Annuity Formula**	**Cost of Living Adjustments**
Final Basic Pay	Servicemembers with a DIEUS prior to Sept. 8, 1980.	50 percent of Final Basic Pay after 20 years of service; 2.5 percent for each additional year. A 40-year retiree receives 100 percent of Final Basic Pay.	Annual COLAs based on changes in the Consumer Price Index.
High-three	Servicemembers with a DIEUS between Sept. 8, 1980, and July 31, 1986.	Same as above except the 2.5 percent is applied to average basic pay for the highest 36 months rather than Final Basic Pay.	Annual COLAs based on changes in the Consumer Price Index.
REDUX	Servicemembers with a DIEUS after July 31, 1986, who accept a mid-career bonus and agree to remain on active duty for at least 20 years.	40 percent of the average of highest 36 months basic pay (applies to the first 20 years); 3.5 percent for each additional year up to 30 years, then 2.5 percent for each year after 30. When the retiree turns 62, annuities increase to match the High-three formula of 2.5 percent for each year served.	Annual COLAs set one percentage point below the Consumer Price Index. Retired pay is adjusted at age 62 to add full COLAs since retirement. But partial COLAs continue after the one-time catch-up.

Here are a few examples: An academy graduate who entered the academy prior to, but graduated after, August 1986 is under the High-three System; An ROTC graduate who began a ROTC scholarship program or enlisted as a Reserve in the Senior ROTC program prior to August 1986 is also under the High-three System; An officer or an enlisted member who initially entered the military prior to August 1986, separated, and after a break in service, rejoined the military, is under the High-three System, even if their pay entry basic date was adjusted to a date of August 1, 1986 or later; A member who enlisted under the delayed entry program is under the system in effect when they entered the delayed entry program, not when they initially reported for duty.

Servicemembers are responsible for ensuring their personnel records reflect an accurate DIEUS. If you have unusual circumstances and are unsure of when your DIEUS date is, or believe your records show an incorrect DIEUS date, contact your personnel office to discuss your particular situation.

Final Basic Pay System. Members with a DIEUS prior to September 8, 1980 retire under the Final Basic Pay System, which is 50 percent of basic pay after 20 years of military service plus 2.5 percent for each additional year of active service. This equals, but is not limited to, 100 percent for 40 years. Cost of Living Adjustments (COLAs) are added annually based on the increase in the Consumer Price Index (CPI), a measure of inflation. Under the Final Basic Pay System, the annual COLA is equal to CPI.

RETIREMENT PAY PERCENT OF BASIC PAY FOR MEMBERS WHO HAVE A DIEUS DATE

PRIOR TO AUGUST 1, 1986				ON OR AFTER AUGUST 1, 1986			
Years of Service	Percent	Years of Service	Percent	Years of Service	Percent	Years of Service	Percent
20	50.0	31	77.5	20	40.0	31	77.5
21	52.5	32	80.0	21	43.5	32	80.0
22	55.0	33	82.5	22	47.0	33	82.5
23	57.5	34	85.0	23	50.5	34	85.0
24	60.0	35	87.5	24	54.0	35	87.5
25	62.5	36	90.0	25	57.5	36	90.0
26	65.0	37	92.5	26	61.0	37	92.5
27	67.5	38	95.0	27	64.5	38	95.0
28	70.0	39	97.5	28	68.0	39	97.5
29	72.5	40	100	29	71.5	40	100
30	75.0			30	75.0		

High-three System. Members with a DIEUS between September 8, 1980 and July 31, 1986 retire under the High-three System. In addition, this plan will also apply to an individual who has a DIEUS on or after August 1, 1986 who does not elect the REDUX retirement system with the career status bonus at their 15th year of service. Retired pay is computed using 50 percent of the average of the "High-three" (36 months) of basic pay for 20 years with 2.5 percent for each additional year of active service. This multiplier is applied against the average basic pay for the highest 36 months of the individual's career. This typically, though not always, equals the average basic pay for the final three years of service. Like the Final Basic Pay System, under the High-three System, the annual COLA is equal to CPI.

Military Retirement Reform Act. Members with a DIEUS on or after August 1, 1986 have a choice of two retirement options: (1) Retire under the High-three System; or (2) Retire under provisions of the Military Retirement Reform Act (MRRA), also known as REDUX, and elect to receive a $30,000 Career Status Bonus.

Note: Servicemembers with a DIEUS on or after August 1, 1986, are the only ones who may choose between the High-three System and the REDUX retirement system that comes with a $30,000 Career Status Bonus.

If the member chooses to retire under REDUX, retired pay is computed using 40 percent of the "high-three" years of service for 20 years of service with an additional 3.5 percent for each additional year up to 75 percent for 30 years of service and 2.5 percent for each year after 30 years, equaling 100 percent for 40 years of service. In addition, annual cost of living (COLA) adjustments for retirees are one percentage point less than inflation, as measured by the CPI. A feature unique to REDUX is a recomputation of retirement pay at age 62 when two adjustments are made. The first adjusts the multiplier to what it would have been under the High-three System. This adjustment does not apply to a member who retires with 30 or more years. For example, a 20-year retiree's new multiplier becomes 50 percent, a 24-year retiree's multiplier becomes 60 percent, but a member who retires with 30 or more years continues to receive the multiplier applied at retirement.

This new multiplier, if applicable, is applied against the individual's original average basic pay for his or her highest 36 months. Then the second adjustment is applied. Full CPI for every retirement year is applied to this amount to compute a new base retirement salary. At age 62, the REDUX and High-three retirement salaries are equal. However, REDUX COLAs in later years will again be set at CPI minus one percent.

Those members who elect the REDUX retirement system at their 15th year of service receive a $30,000 Career Status Bonus. To receive this bonus, the member must agree to complete a twenty-year active duty career. The member, of course, may continue beyond twenty years, but the obligation is only to that point. The entire $30,000 bonus may be paid in a lump sum, or the member may elect two, three, four, or five installments. An advantage of electing multiple payments is the opportunity to use the bonus as a tax deferred contribution to the Thrift Savings Plan. If the member does not complete the obligation of the twenty-year career, the member must repay a prorated share of the bonus.

Choosing the REDUX retirement system and Career Status Bonus is a "package deal." The combination of these two items can be advantageous to many individuals. The REDUX portion determines retirement income (the longer one's career, the higher that income) and the $30,000 career status bonus provides current cash available for investing, major purchases, or setting up a business after retirement.

Under REDUX, the longer an individual stays on active duty the closer the multiplier is to what it would have been under High-three up to the 30-year point where the multipliers are equal.

Disability Retirement

The military disability retirement system is for servicemembers that are considered to have a permanent disability that is not the result of intentional misconduct or willful neglect, and not incurred during a period of unauthorized absence. Members must have either 20 years of service creditable toward retirement; or a disability rating that is at least 30 percent in accordance with the standard schedule of rating disabilities in use by the Department of Veterans Affairs at the time of determination.

The system applies to members in a regular component of the Armed Forces and members on active duty for more than 30 days. However, a member with less that 20 years of service for retirement must also meet one of the following four requirements: (1) the disability was not noted at the time of the member's entrance on active duty (unless clear and unmistakable evidence demonstrates that the disability existed before the member's entrance on active duty and was not aggravated by active military service); (2) have a disability that is the result of performing active duty; (3) have a disability that was incurred in the line of duty in time of war or national emergency; or (4) have a disability that was incurred in the line of duty after September 14, 1978.

The disability retirement system also applies to cadets attending the United States Military Academy, United States Air Force Academy, or United States Coast Guard Academy, and midshipmen attending the United States Naval Academy with respect to a physical disability incurred after October 28, 2004.

Members on active duty for less than 30 days and Reserve Component members performing inactive-duty training or funeral honors duty may also be eligible for disability retirement benefits. The disability must have been the result of an injury, illness, or disease incurred or aggravated while: (1) performing active duty, inactive duty for training, or funeral honors duty; (2) traveling directly to or from the place of performing the duty; or (3) while remaining overnight: (a) immediately before performing the inactive duty training or funeral honors duty, or (b) between successive periods of such duty at or in the vicinity of the training or duty site.

A member with a disability described above that is not considered permanent, may be placed on the Temporary Disability Retired List (TDRL). The member is subject to a physical examination not less than once every 18 months. After five years, the member must be retired for permanent disability or, if the disability is less than 30 percent and the member has less than 20 years of service, returned to either active duty or Reserve service, or separated.

Disability retirement pay is computed by multiplying the basic pay of the member's retired grade by the greater of: the percentage of the disability; or 2.5 percent multiplied by the number of years of service. For a permanent disability retirement, the minimum benefit is 30 percent of basic pay with a maximum benefit of 75 percent. While on the TDRL, the minimum benefit is 50 percent of basic pay. For members with a DIEUS on or after September 8, 1980, the "high-three" average of basic pay is used in the computation. **Note:** The monthly amount of basic pay for a cadet or midshipman is used in the retired pay computation for disability retired pay for a member in one of those categories.

Disability Severance Pay. A member of the Armed Forces with fewer than 20 years of service who is found physically unfit for military service, and who has less than a 30 percent disability under the VA schedule for rating disabilities, is entitled to disability severance pay when discharged. The computation depends on whether the disability is or is not combat related.

SPECIAL COMPENSATION PROGRAMS FOR CERTAIN RETIREES RECEIVING VA DISABILITY COMPENSATION

The Department of Defense has two programs for certain disabled retirees designed to reduce the reduction in retired pay due to receipt of VA compensation. Both programs have specific eligibility requirements.

Concurrent Retirement and Disability Payments

This program (CRDP) was implemented January 1, 2004 in accordance with a schedule provided by law. It provides for a 10 year phase-in of restored retired pay to an eligible member who is receiving VA compensation for either a service-connected disability or combination of service-connected disabilities rated 50 percent or higher. In the case of a disability retiree, the amount of retired pay is still subject to offset to the extent that the disability retired pay may exceed the amount of retired pay to which the member would have received for years of service.

Eligibility. The Program automatically applies to the following: (1) active duty members who are retired for length of service; (2) Reserve retirees who are retired for length of service, upon commencement of retired pay; and (3) those retired for disability as long as they have at least 20 years of service.

Phase-in Period. The phase-in began with a fixed dollar rate in 2004 based on percentage of VA disability. In each year, 2005 through 2013, a graduated percentage of the remaining offset is added to the amount the member receives in monthly retired pay. The reduction in retired pay for members rated between 50 and 90 percent will be fully eliminated by 2014.

Members who are rated by the VA as 100 percent service-connected disabled, or 100 percent due to being rated unemployable by the VA are not subject to the phase-in period and receive 100 percent of their retired pay.

Computation. The formula for computing the amount of retired pay to be restored under CRDP during the phase-in period is:

(GRP – CBO) + ((CBO – 2004 Base) x Phase-in %) + 2004 Base

GRP = Gross retired pay before any offset for VA disability compensation.

CBO = Current Baseline Offset which is the lesser of ARP and VADC.

ARP = Applicable Retired Pay which is that portion earned for years of service.

VADC = VA Disability Compensation (the monthly disability compensation received).

VAPD = VA Disability Percentage currently assigned by the VA.

Phase-in % = The cumulative percentage for a specific calendar year during the phase-in.

2004 Base = The dollar value assigned by law for a specific VA disability percentage.

2004 BASE RATES		PHASE-IN %			
100%	$750	2004	N/A	2009	84.88%
90%	$500	2005	10.00%	2010	93.95%
80%	$350	2006	28.00%	2011	98.18%
70%	$250	2007	49.60%	2012	99.64%
60%	$125	2008	69.76%	2013	99.96%
50%	$100				

The finance center computes CRDP payments monthly since there are many factors that may cause retired pay and VA disability compensation payments to change. It is also important to note that Concurrent Retirement and Disability Payments are taxable as a restoration of earned longevity retired pay.

Combat Related Special Compensation

This program (CRSC) provides special compensation for military retirees who have a qualifying combat-related disability and who have their military retired pay reduced in order to receive VA disability compensation for those qualified disabilities.

Eligibility. With one exception, all military retirees entitled to retired pay, including those retired under the Temporary Early Retirement Authority (TERA) and disability retirees with less than 20 years of service are eligible for CRSC). ***Exception*:** Reserve Component members retired with more than 15 but less than 20 years of service due to physical disability not incurred in the line of duty (10 USC 12731b) are not eligible for CRSC. In addition, those meeting the basic eligibility requirements must have a disability that is compensated by the VA and was incurred: (1) as a direct result of armed conflict; (2) while engaged in hazardous service; (3) while performing duty simulating war; or (4) through an instrumentality of war.

The amount of compensation paid to a member under this program is equal to the amount of retired pay offset due to VA disability compensation that is attributable to a qualifying disability listed above. Members must apply for benefits under this program through their military service. The DFAS website, www.dfas.mil, has a link to the application and service points of contact. Payments under this program are not considered retired pay and are not taxable. In the case of a disability retiree, the maximum amount of CRSC may not exceed the amount of retired pay the member would have received if the member's retired pay had been calculated based on years of service.

Annual Open Season

A member who is eligible for both Concurrent Retirement and Disability Payments and Combat Related Special Compensation must choose one program over the other. Members may not receive compensation from both programs. In addition, in no case may payments from any combination of DoD sources exceed the amount of full retired pay. There may be advantages to selecting one program over another. For example, a member with a 50 percent combat related disability may choose Combat Related Special Compensation ($770 in 2009) over Concurrent Retirement and Disability Payments ($668 in 2009). Not only is the payment higher, it is also nontaxable. The member will have a chance each year, in an open season, to move from one program to the other. Each year, prior to the open season, DFAS will mail each eligible retiree a letter that provides an explanation and instructions along with an election form that must be returned to DFAS. Once an election is made, it may not be changed until the next open season.

ADJUSTMENTS TO RETIRED PAY

Prior to June 1, 1958, each time military basic pay was increased, pay for a retired member was recomputed on the basis of those newly established pay rates. That practice was discontinued on June 1, 1958 when the pay of retired personnel was increased by six percent, which was approximately the increase in the cost of living since 1955 when retired pay was last increased. In 1963, a revised system of increasing retired pay, based on a formula reflecting increases in the cost of living, was adopted.

Under the revised system, whenever the monthly Consumer Price Index (CPI) reached a level of three percent above the Base Index which was the basis for the most recent increase in retired pay, and remained at or above that level for three consecutive months, military retired pay was increased by the highest percentage above that Base Index attained by the CPI in any month during that three month period. The effective date of the increase was the first day of the third month beginning after that three month period.

On December 16, 1967, P.L. 90-207 further refined the retired pay adjustment formula with a "look-back" provision to ensure that (1) members with the same grade, length of service, and basic pay at the time of retirement receive the same retired pay; and (2) members retired after an increase in the basic pay rate would, at the first ensuing retired pay increase, receive an increase equal to the percent of increase in the CPI between the month preceding the basic pay increase and the month which was the basis for the retired pay increase (Base Index Month). Prior to this legislation, individuals retiring before CPI adjustments received full adjustments, while members retiring the day of the adjustment or after received no adjustment. The "look-back" provision was not used extensively until several years after enactment when the rate of inflation was significantly greater than the annual increases in basic pay.

P.L. 91-179, effective November 1, 1969, added one percent to retired pay adjustments under the CPI formula in order to offset the erosion of retired pay resulting from the time delay in the adjustment formula.

On October 7, 1975, P.L. 94-106 modified the computation of retired pay so that members initially entitled to receive retired pay on or after January 1, 1971 would not receive less pay than had they retired earlier in their career. The purpose of this legislation was to prevent retired pay inversions that could occur whenever retired pay increases exceed active duty pay increases. Each retirement eligible member would have his or her retired pay computed, on an individual comparative basis, using one of the following options:

- Pay at present grade and years of service at time of retirement.
- One-year "look-back" using the prior pay scale at present grade and years of service at time of retirement plus cumulative CPI increases in retired pay from the prior year's pay scale.
- Any pay scale on or after July 1, 1971 at the grade and years of service a member would have had at that time (if retirement eligible under those pay rates) plus cumulative CPI increase accruing from that earlier rate.

Legislation enacted in October 1976 changed the formula for computing retiree pay raises. That formula eliminated the one percent "add-on" and provided for automatic COLA raises every six months based on the actual CPI increase. These semi-annual increases were effective on March 1 and September 1 and were based on the percentage increase, rounded to the nearest tenth of one percent, between the CPI increases from June to December and December to June respectively. Repeal of the previous CPI adjustment formula was necessary because, in the high inflation rates of the early 1970s, retired pay was overcorrected for cost of living.

In July 1981, Congress eliminated the semi-annual cost of living increases in favor of a single annual increase. This action was designed to save the Government hundreds of millions of dollars each year and provided for retiree cost of living increases based on CPI changes over a full year, from December to December, with the increase payable the following March. The delay in applying the adjustment was to slow down the rate of retired pay increases and provide substantial savings to the Government.

The Omnibus Budget Reconciliation Act of 1982, P.L. 97-253 further modified the retirement pay formula. For FY 83, 84, and 85, nondisabled military retirees under age 62 were to receive only 50 percent of the projected CPI, plus the full difference of the actual CPI over those projections. Disabled retirees and those over age 62 were to receive full COLA increases. Additionally, these COLA increases were delayed one month in FY 83, FY 84 and FY 85. The FY 83 increase was effective April 1, 1983 instead of March 1. The 98th Congress delayed the FY 84 COLA until December 1, 1984 with payment scheduled for the first day of the following month instead of the last day of the current month. This delay caused a one-time reduction in taxable income for retirees in 1984 since the December 1984 increase was paid in January 1985.

Another significant feature of the Act provided some relief for those military retirees who combine their military service with civil service, waive military retired pay, and retire under the civil service retirement system for a single annuity. Since these individuals received credit for Social Security (after December 31, 1956), the law required the civil service annuity to be recomputed at age 62 to eliminate credit for military service after 1956 to preclude coverage under both systems for the same period of service. This generally meant a significant reduction in retirement benefits. The Act provides that members who are employed in the civil service after September

1982 may pay into the civil service retirement fund a sum equal to seven percent of their military pay for active service after 1956 or three percent, if employed by the civil service as of January 1, 1984 or later. Such a deposit entitles the member to both the civil service annuity and Social Security benefits.

The National Defense Authorization Act for Fiscal Year 1984 further modified retired pay as follows:

• Repealed the one-year "look-back" provision. Any member who first became eligible to retire after September 24, 1983 could no longer use the one-year "look-back" in computing retired pay which allowed retirees to base their annuities on the basic pay scales in effect when they retired or those in effect the previous year plus any CPI adjustments. **Note:** This repeal did not affect the provisions of the "Tower" amendment which provided that a member's retired pay will not be less than it would have been if he or she retired on an earlier eligible date.

• Required that monthly retired or retainer pay be rounded down to the next lower full dollar. This same rounding down applied to survivor annuities.

• Six month rounding rule. Considers any full month of service as one-twelfth of a year, whether or not the member has more or less than six full months for retired pay computation.

The 98th Congress also passed several other measures which impacted military retired pay. P.L. 98-270 changed the basis for COLAs and delayed the FY 84 increase from May to December and changed the basis for the COLA to the change in the Consumer Price Index (CPI) from the average for the third quarter (July, August, and September) of one year to the next. This action resulted in the permanent loss of a 2.6 percent adjustment which would have occurred if the change in CPI from the last adjustment (April 1983) to the beginning of the new CPI base figure (average for the third quarter 1983) was included in the calculation.

The Second Supplemental Appropriations Act, FY 84, eliminated the "half COLA" provision for non-disability retirees under age 62 which had been in effect starting in FY 83.

The Military Retirement Reform Act (REDUX), P.L. 99-348, July 1, 1986, significantly changed the computation for retired pay for military service. All members with a Uniformed Service Initial Entry Calendar Date (DIEUS) on or after August 1, 1986 are affected. Members who entered the service prior to the effective date, including service academy cadets and midshipmen, delayed entry program members, ROTC students under contract, Reserve Component members, and members already retired are not affected.

Under REDUX, the retired pay formula is the product of 2.5 multiplied by the years of creditable service minus one percentage point for each year less than 30 years. Annual COLA increases are calculated at CPI minus one percentage point. At age 62, the amount of retired pay is recalculated on the 2.5 percent multiplied by years of service formula and full COLA increases. That amount becomes the new retired pay base amount. However, COLA increases continue at CPI minus one percentage point.

The National Defense Authorization Act for Fiscal Year 2000 contained a provision that made REDUX optional and required members first entering the service after July 31, 1986 to make a retirement option choice during their fifteenth year of service. The choices include remaining under the REDUX retirement system and receiving a career status bonus, or converting to the High-three retirement system where the member would receive 50 percent of their High-three years of service and full COLAs.

The National Defense Authorization Act for Fiscal Year 1996 established the COLA payment date for military retirees to conform to the payment date established for federal civilian retirees by making the military retired pay COLA for FY 96 and FY 97 effective the first day of March 1996 and the first day of December 1996, respectively, and that the effective date for the FY 98 and subsequent COLA increases will conform to the date prescribed for federal civilian retirees.

The National Defense Authorization Act for Fiscal Year 2007 changed nondisability retired pay as follows:

- Extends the pay table, effective April 1, 2007, to 40 years of service for pay. Increases rates for over 26 years of service for pay in grades E-8; E-9; W-4; W-5; and O-6 thru O-10.
- Provides that retired pay for general and flag officers who retire on or after October 1, 2006 will be computed using the uncapped rates of basic pay. Previously, the retired pay was subject to being computed using the maximum pay for Level II of the Executive Schedule.
- Extends, beyond 75 percent, the retired pay multiplier for the retired pay of members retiring on or after January 1, 2007 with more than 30 years of active service. The increased multiplier increases at the rate of 2.5 percent for each additional year of creditable service (for active duty members), and service credited under 10 USC 12733 (for Reserve Component members).

The National Defense Authorization Act for Fiscal Year 2008 contained a provision that allows a member to begin receiving Reserve Component retired pay three months early for each aggregate 90 days of qualifying active service—performed after January 28, 2008—during a fiscal year period.

RETIRED PAY COST OF LIVING ADJUSTMENTS (COLA)

The following Cost of Living Adjustments (COLAs) are effective December 1, 2008, based on the increase in the Consumer Price Index for Urban Wage Earners and Clerical Workers (CPI-W) through September 2008. The term "retired pay" includes retainer pay and the term "members" includes former members.

1. The retired pay COLA for those who first became a member of a uniformed service before September 8, 1980, is specified according to the effective date of the active duty basic pay rate used to compute their retired pay as follows:

Retired Pay Based on Rates of Pay Effective	Percent Increase
Before January 1, 2008	5.8 percent
January 1 – December, 2008	5.0 percent

2. The retired pay COLA for those who first became a member of a uniformed service on or after September 8, 1980, is specified according to the effective date of their retirement as follows:

Retirement Effective	Percent Increase
Before January 1, 2008	5.8 percent
January 1 - March 31, 2008	5.0 percent
April 1 - June 30, 2008	3.8 percent
July 1 - September 30, 2008	1.2 percent
October 1 - December 31, 2008	0.0 percent

Note: The potential exists for the provisions of Title 10, United States Code, Section 1401(f)(1) to apply to retirements that would otherwise be computed using the basic pay rates effective January 1, 2009. The new basic pay rate reflects a 3.9 percent increase over the rates for 2008. Some members may be entitled to more retired pay using the 2008 pay rates increased by a partial COLA instead of using the increased pay rates with more service credit. Under the above provisions, allowing computation of retired pay based on retirement at an earlier date prevents inversions to retired pay entitlements. In such computations, the grade and service credit must be those applicable on the earlier date being used. Also, the retiree must meet the retirement eligibility requirements as of the earlier date. DFAS will compute the retired pay in both cases and apply the most favorable computation to the member's retired pay.

3. The retired pay COLA for those who first became a member of a Uniformed Service on or after August 1, 1986, who elected to receive a career status bonus under the provisions of Section 354 of Title 37, United States Code, is specified according to the date of retirement as follows:

Retirement Effective	Percent Increase
Before January 1, 2008	4.8 percent
January 1 - March 31, 2008	4.2 percent
April 1 - June 30, 2008	3.3 percent
July 1 - September 30, 2008	1.0 percent
October 1 - December 31, 2008	0.0 percent

The percentage increase in the CPI-W determines the percentage increase in retired pay and not the change in index points. Under current procedures, the percentage adjustment is determined by the percentage increase in the average CPI-W for the same three-month period of the preceding year. The average CPI-W for the third quarter of 2008 was 215.495 (July 216.304, August 215.247, September 214.935). Subtracting the average third quarter CPI-W for 2007, 203.596, results in a point difference of 11.899. Dividing 11.899 by the 2007 third quarter average of 203.596, results in the 5.8 percent increase effective December 1, 2008.

Any questions regarding the amount of retired pay should be directed to the member's finance and accounting service center.

CPI-W SUMMARY JANUARY 2006 THROUGH NOVEMBER 2008

Month	CPI	Month	CPI	Month	CPI
Jan 06	194.0	Jan 07	197.559	Jan 08	206.744
Feb 06	194.2	Feb 07	198.544	Feb 08	207.254
Mar 06	195.3	Mar 07	200.612	Mar 08	209.147
Apr 06	197.2	Apr 07	202.130	Apr 08	210.698
May 06	198.2	May 07	203.661	May 08	212.788
Jun 06	198.6	Jun 07	203.906	Jun 08	215.223
Jul 06	199.2	Jul 07	203.700	Jul 08	216.304
Aug 06	199.6	Aug 07	203.199	Aug 08	215.247
Sept 06	198.4	Sept 07	203.889	Sept 08	214.935
Oct 06	197.0	Oct 07	204.338	Oct 08	212.182
Nov 06	196.8	Nov 07	205.891	Nov 08	207.296
Dec 06	197.2	Dec 07	205.777		

RETIRED PAY INCREASE TABLE

Retired Pay Increases									
Year	Month	Actual %	Cumulative % (over base of 100)	Reverse Cumulative % Increase	Year	Month	Actual %	Cumulative % (over base of 100)	Reverse Cumulative% Increase
1958	Jun	6.0%	6.0%	713.1%	1983	Apr	3.3%	291.4%	114.6%
1963	Oct	5.0%	11.3%	667.1%	1984	Dec	3.5%	305.1%	107.8%
1965	Sep	4.4%	16.2%	630.6%	1986	Dec	1.3%	310.3%	100.7%
1966	Dec	3.7%	20.5%	599.8%	1987	Dec	4.2%	327.6%	98.2%
1968	Apr	3.9%	25.2%	574.8%	1988	Dec	4.0%	344.7%	90.2%
1969	Feb	4.0%	30.2%	549.5%	1989	Dec	4.7%	365.6%	82.9%
1969	Nov	5.3%	37.1%	524.5%	1990	Dec	5.4%	390.7%	74.6%
1970	Aug	5.6%	44.8%	493.1%	1991	Dec	3.7%	408.9%	65.7%
1971	Jun	4.5%	51.3%	461.6%	1992	Dec	3.0%	424.2%	59.8%
1972	Jun	4.8%	58.6%	437.4%	1993	Dec	2.6%	437.8%	55.1%
1973	Jul	6.1%	68.2%	412.8%	1994	Dec ††	2.8%	452.8%	51.2%
1974	Jan	5.5%	77.5%	383.3%	1995	Dec xx	2.6%	467.2%	47.1%
1974	Jul	6.3%	88.7%	358.1%	1996	Dec	2.9%	483.7%	43.4%
1975	Jan	7.3%	102.4%	331.0%	1997	Dec	2.1%	495.9%	39.3%
1975	Aug	5.1%	112.8%	301.7%	1998	Dec	1.3%	503.7%	36.4%
1976	Mar	5.4%	124.3%	282.2%	1999	Dec	2.5%	518.8%	34.7%
1977	Mar	4.8%	135.0%	262.6%	2000	Dec	3.5%	540.4%	31.4%
1977	Sep	4.3%	145.1%	246.0%	2001	Dec	2.6%	557.1%	27.0%
1978	Mar	2.4%	151.0%	231.7%	2002	Dec	1.4%	566.3%	23.8%
1978	Sep	4.9%	163.3%	223.9%	2003	Dec	2.1%	580.3%	22.0%
1979	Mar	3.9%	173.6%	208.8%	2004	Dec	2.7%	598.6%	19.5%
1979	Sep	6.9%	192.4%	197.2%	2005	Dec	4.1%	627.3%	16.4%
1980	Mar	6.0%	210.0%	178.0%	2006	Dec	3.3%	651.3%	11.8%
1980	Sep	7.7%	233.9%	162.3%	2007	Dec	2.3%	668.5%	8.2%
1981	Mar	4.4%	248.6%	143.5%	2008	Dec	5.8%	713.1%	5.8%
1982	Mar	8.7%	278.9%	133.3%					

1994: ††Delayed to March 1, 1995 by FY95 DoD Authorization Act.
1995: xxDelayed to March 1, 1996.

NOTE: The Retired Pay Increase of 3.1% for December 1985 was cancelled.

RETIRED PAY COMPUTATION

For members entering the service prior to September 8, 1980, retired pay is computed by multiplying the basic pay rate at the time of retirement by 2.5 percent for each year of service. Members entering the service between September 8, 1980 and July 31, 1986 receive retired pay computed on 50 percent of the "High-three" years of basic pay. The law provides for adjustments to retired pay to reflect cost of living increases. These adjustments are based on increases in the Consumer Price Index (CPI). Personnel entering the service on or after August 1, 1986 choose between receiving retired pay computed at 50 percent of their "High-three" years of basic pay or receiving a career status bonus (currently $30,000) and 40 percent of basic pay upon completion of 20 years of service (computed at 2.5 percent times years of service minus one percent for each year less than 30 years) up to 75 percent of basic pay and 2.5 percent for each year over 30.

Although active duty pay increases are also, to some degree, affected by cost of living, basic pay increases did not go up as frequently or by as much as the retired pay increases in the early 1970s. A retired pay "inversion" was created which resulted in certain retirement eligible members facing a possible loss of retirement income by remaining on active duty.

The law protects retired members from receiving less than they would receive by serving on active duty for a longer period. Under the "Tower Amendment," a member initially entitled to receive retired pay on or after January 1, 1971, may not receive less pay than he or she would have received had he or she retired earlier in his or her career. However, there are still differences among members with similar grades. For example, as a result of previous pay inversions and because members entering service on or after September 8, 1980 have retired pay computed using the high-three average of basic pay, an O-5 who retired in 1972 with 20 years of service is receiving $527 more per month than an O-5 who retired in January 2009 with the same amount of service. Due to the various methods used in computing retired pay, the application of only partial CPI increases in some situations, and depending on when the member retires, the member's finance and accounting service center will make all calculations and automatically provide retirees with the benefit of the most advantageous retirement formula on an individual basis.

COMPARATIVE 20 YEAR RETIREMENT*
(Figures rounded to lowest whole dollar)

PAY SCALE (CPI ADJUSTED) THROUGH JANUARY 1, 2009									
Grade	2009	2008	2007	2006	2005	2000	1990	1980 L	1972 N
O-7	5,115	5,020	5,167	5,087	5,112	5,204	4,958	4,607L	5,857N
O-6	3,919	3,846	3,959	3,896	3,916	3,985	3,789	3,521L	4,479N
O-5	3,524	3,460	3,560	3,504	3,521	3,604	3,426	3,186L	4,051N
O-4	3,179	3,121	3,212	3,162	3,177	3,116	2,965	2,755L	3,504N
O-3	2,733	2,682	2,760	2,716	2,730	2,694	2,563	2,380L	3,027N
W-4	2,668	2,603	2,658	2,607	2,602	2,589	2,458	2,285L	2,905N
W-3	2,421	2,353	2,384	2,340	2,329	2,275	2,160	2,005L	2,556N
E-9	2,323	2,281	2,346	2,302	2,299	2,187	2,080	1,886L	2,397N
E-8	2,032	1,994	2,052	2,015	2,014	1,918	1,820	1,651L	2,102N
E-7	1,850	1,803	1,838	1,804	1,802	1,690	1,606	1,452L	1,849N
E-6	1,586	1,546	1,575	1,547	1,553	1,480	1,405	1,279L	1,627N

*This table illustrates the gap in nondisability retirement pay for individuals of the same grade and years of service due to the retired pay inversion. Assumes years of service for pay is same as year of active service, and a DIEUS date prior to September 8, 1980 for years prior to 2005, and a DIEUS date of September 8, 1980 or later for years 2005-2009.

NOTE: The suffix attached to some of the rates indicate that a new retiree was eligible for a one-year "look-back" if a higher amount resulted. (L = Look-back, N = not)

The Retirement Pay Tables later in this section are provided as general guides for retirees whose pay was based on pay scales for years indicated. These tables include all CPI increases through January 1, 2009. For example, an officer, grade O-5, retired in 1990 with 22 years of service can refer to the appropriate table using the 1990 pay scale and find that his or her retired pay is $3,908. Similarly, the individual who retired in 1998 would refer to the 1998 pay scale for his or her grade to determine the amount of retirement pay for that period.

OTHER RETIRED PAY INFORMATION

Retired Paychecks

Retirees are responsible for arranging for receipt and deposit of retired pay and to notify their servicing finance center to forward the payments to a financial institution. Retirees who reside outside the United States or who have checks sent to an address outside the United States must submit a semi-annual Report of Existence to DFAS or their finance service center. The legal representative of retirees whose checks are made payable to a guardian, trustee, fiduciary, or other legal representative must submit the semi-annual Report of Existence for the retiree. The report must be signed by the retiree or designated legal representative and submitted to ensure uninterrupted receipt of retired pay. For further information concerning requirements for Reports of Existence, contact the member's finance and accounting service center.

myPay

myPay is a tool to help retirees manage their pay. The self-service tool lets retirees make changes to their pay account information via the Internet at https://mypay.dfas.mil/mypay.aspx. The system is available 24 hour a day, seven days a week. The member accesses the system using their Social Security Number and Personal Identification Number (PIN). In addition, the system is available via an Interactive Voice Response System using a touch telephone to call toll free: 1-877-363-3677.

Using myPay, retirees can: see their last two Retiree Account Statements; see their last IRS 1099-R; change their Federal and state tax withholding status and exemptions; start, stop, or change allotments; start, stop, or change bonds; change or update correspondence address; submit a semiannual Report of Existence (if required); and update bank information for deposit of retired pay. Changes are usually available for review on myPay within three days.

Allotments from Retired Pay

Retirees are authorized up to six discretionary allotments from retired pay. These allotments may be carried over from active duty into retirement.

Examples of discretionary allotments include, but are not limited to, savings allotment, home loans, dependency allotment, life insurance, and premiums for health insurance. In addition, nondiscretionary allotments allowed include: purchase of U.S. Savings Bonds; payment of delinquent federal and state taxes; court ordered garnishments; and repayment of loans to the Navy/Marine Corps Relief Society, Army Emergency Relief, Air Force Aid Society, and American Red Cross.

A properly executed form, written request, or an automated data exchange (from specific organizations) may be used to establish, discontinue, or change an allotment. In addition, DFAS will accept telephone requests from members without written documentation, but only after the member's identity has been validated. Requests for allotment actions should be sent to the DFAS Cleveland Center or the retiree's servicing finance center.

Waiver of Retired Pay

A member may not waive his or her right to retired pay. Even if retired pay is not being sent to a member on a monthly basis (for example, the member's whereabouts is unknown), the retired pay continues to accrue. Since, for income tax purposes, such retired pay is deemed to have been received by the member once it is made available and can be drawn upon, it is subject to the withholding and reporting requirements of ordinary wages.

However, eligible retired personnel may elect to waive all or part of their retired pay to receive VA compensation or pension, or to credit active military service under the civil service retirement system. A member electing to receive disability compensation or pension from the VA may subsequently elect to have the military retired pay reinstated. Retired military pay that is waived to establish eligibility for civil service retirement may also be reinstated under certain conditions.

Waiving military retired pay does not divest the retiree, eligible family members, or survivors of the privileges to which they are otherwise entitled. Continued use of ID and privilege cards is authorized.

Suspension of Retired Pay

Retired military pay may be suspended (placed in a non-payment status) for a definite or indefinite period of time. The suspension may be removed when the basis for the suspension no longer apply.

The most common reasons that a retiree's pay is suspended is when the member: is recalled to active duty; requests waiver of retired pay to receive VA disability compensation or pension payments; or the military service is being used for purposes of qualifying for a civil service retirement annuity.

Other reasons for suspension include: the member has completed five years on the Temporary Disabled Retired List (TDRL); failure to report for a required physical examination while on the TDRL or while a member of the Fleet Reserve/Fleet Marine Corps Reserve; employment by a foreign government without applicable congressional or secretarial approvals; the member is reported or found to be mentally incapable of managing his or her personal affairs, and no guardian, trustee, or other legal representative has been appointed; or the member is missing, and there is no information concerning his or her whereabouts. In addition, the Secretary concerned may also order the suspension of a retiree's pay under certain circumstances.

Garnishment of Retired Pay

Title 42, United States Code, Section 659 authorizes the garnishment (or attachment) of active duty pay, retired pay, VA disability pay, and survivor benefits to enforce obligations of alimony and child support.

The Consumer Credit Protection Act, Title 15, United States Code, Section 1673, sets limits on the amount that can be garnished or attached. The limits are based on the individuals' aggregate disposable earnings, which would include all pay, pensions, disability compensation, Social Security benefits, etc., received from any agency of the Federal Government (allowances, such as BAH, are not subject to garnishment for child support). The limits allow garnishment/attachment of Government earnings of no more than:

a. 50 percent of disposable earnings if the individual is supporting a second family.

b. 60 percent of disposable earnings if the individual is not supporting a second family.

c. An additional five percent in each of the above if the individual is in arrears for more than 12 weeks.

A valid court order or similar legal process must be issued prior to the garnishment of active duty or retired pay. The order should name DFAS as the garnishee and state on its face that it is to enforce an obligation to provide child support or to make alimony payments. It must include the debtor's full legal name and Social Security number. Court costs and attorney's fees can also be deducted if the court order directs such action. An individual attempting to have such a court order reversed must assume any costs involved in such an action.

Garnishment orders involving military personnel, either active or retired, should be sent to: Defense Finance and Accounting Service-Cleveland, DFAS-HGA/CL, P.O. Box 998002, Cleveland, OH 44199-8002. The request can also be sent by FAX to: (216) 522-6960. When submitting correspondence regarding garnishment, ensure that you include the member's Social Security number on documents. Also ensure that your return mailing address and phone number is included on the correspondence and not just on the mailing envelope.

For more information, contact the Garnishment Operations Customer Service Center toll-free at 1-888-332-7411, Option 3. The hours of operation are 7:30 a.m. - 6:00 p.m. on workdays or visit: www.dfas.mil/garnishment.html.

Title 5, United States Code, Section 5520a permits creditors to obtain an involuntary allotment from the wages of active duty members who are delinquent on paying bills. Under Title 15, United States Code, Section 1673(a)(1), collection for commercial debt is limited to a maximum of 25 percent of the employee's disposable earnings. However, state limitations prevail when they provide a maximum collection percentage that is less than 25 percent.

An employee's disposable earnings are defined as pay subject to garnishment such as wages, salary, bonuses and incentive pay, less authorized deductions. A complete list of the payments included in the definition of "disposable earnings" is found the Federal Regulation 5 C.F.R., Section 582.103. Section 582.103(a) specifies the amounts that are excluded from the amount subject to garnishment. These include amounts owed to the United States, federal, state and local income tax withholding, Social Security withholding (FICA) and deductions for health insurance premiums.

The Office of Personnel Management (OPM) published an "*Application for Federal Employee Commercial Garnishment*," Optional Form 311 (available at www.opm.gov/Forms). This form is designed for use by creditors to facilitate the Federal Government's compliance with commercial garnishment orders.

JAG and financial counseling offices may assist members who receive notices of wage garnishment, but military lawyers generally cannot represent, in court, a member seeking to have a court ordered debt modified or reversed.

RETIRED PAY CENTERS

The Defense Finance and Accounting Service (DFAS) operates the Defense Retiree and Annuitant Pay System (DRAS). Military retiree and annuitant accounts are administered at the DFAS center in Cleveland, OH.

When dealing with DFAS, the following points will help reduce problems:

• Contact the retired and annuity pay contact center by calling toll-free, 1-800-321-1080. Representatives are available from 7:00 a.m. to 7:00 p.m. Eastern time, Monday - Friday. In addition, the DFAS website www.dfas.mil has information for annuitants and a link to send email questions. A toll-free FAX line, 1-800-469-6559 is also available. Remember to include your complete name and Social Security number, as well as your signature on any correspondence, including faxes.

• A current address ensures timely receipt of all mailings from the center, including your annual tax statement. Filing an address change will also update your official personnel file.

FINANCE AND ACCOUNTING SERVICE CENTERS

Army, Air Force, Navy and Marine Corps

Defense Finance and Accounting Service, U.S. Military Retired Pay, P.O. Box 7130, London, KY 40742-7130, 1-800-321-1080, (216) 522-5955.

Coast Guard/National Oceanic & Atmospheric Administration

Commanding Officer (RAS), U.S. Coast Guard Personnel Service Center, 444 S.E. Quincy St., Topeka, KS 66683, 1-866-772-8724.

United States Public Health Service

USPHS, Compensation Branch, Parklawn Bldg., 5600 Fishers Ln., Rm. 4-50, Rockville, MD 20857, 1-800-638-8744, (301) 594-2963.

MULTI-YEAR MILITARY NONDISABILITY RETIREMENT PAY TABLES

The figures reflect CPI and pay increases prior to and including January 1, 2009. As a result of certain provisions for advancement on the retired rolls or because of the "Tower Amendment," some individuals would receive differing amounts. Years of service and pay are assumed to be equal. Pay scale indicates rates in effect for active duty personnel at time of retirement. The suffix attached to some of the rates indicate that a new retiree was eligible for a one-year "look-back" if a higher amount resulted (L= Look back, N = Not. For example, 10/78L and 10/72N).

These tables apply only to members retiring from active duty with the grade and years of service indicated and under the specific pay scale shown. The table on page 42 contains retired pay amounts for those retiring after January 1, 2007 with over 30 years of service. Because retirement pay must be computed on an individual basis, it is not possible to reflect in a simple table the exact amount received by all members, even with same grade and years of service. These tables should be considered only as a guide to retirement pay for most individuals retiring in the time frame reflected. (See preceding text for additional information pertaining to RETIRED PAY COMPUTATION). For specific information regarding your retirement pay, contact the financing and accounting service center for your branch of service.

MONTHLY AMOUNT OF MILITARY NONDISABILITY RETIREMENT PAY (For individuals retiring on or after January 1, 2009)

PAY GRADE	YEARS OF ACTIVE SERVICE																				
	Over 20	Over 21	Over 22	Over 23	Over 24	Over 25	Over 26	Over 27	Over 28	Over 29	Over 30	Over 31	Over 32	Over 33	Over 34	Over 35	Over 36	Over 37	Over 38	Over 39	Over 40
COMMISSIONED OFFICERS																					
O-10	5,488	6,350	7,167	8,029	8,392	8,818	9,234	9,770	10,250	11,311	12,286	12,696	13,105	13,515	14,620	15,050	15,480	15,911	17,158	17,609	18,061
O-9	5,488	6,034	6,516	7,045	7,387	7,785	8,151	8,623	9,045	9,980	10,840	11,202	11,563	11,925	12,900	13,280	13,659	14,039	15,139	15,537	15,936
O-8	5,488	5,918	6,278	6,699	7,047	7,398	7,694	7,990	8,286	9,042	9,588	9,907	10,227	10,547	11,138	11,466	11,793	12,121	12,449	12,776	13,104
O-7	5,115	5,484	5,745	6,007	6,268	6,529	6,790	7,064	7,338	8,020	8,463	8,745	9,027	9,309	9,591	9,873	10,156	10,438	10,720	11,002	11,284
O-6	3,919	4,250	4,523	4,844	5,098	5,403	5,667	6,034	6,357	7,044	7,433	7,680	7,928	8,176	8,424	8,671	8,919	9,167	9,415	9,663	9,910
O-5	3,524	3,768	3,983	4,244	4,472	4,703	4,891	5,079	5,267	5,748	5,946	6,144	6,342	6,541	6,739	6,937	7,135	7,334	7,532	7,730	7,928
O-4	3,179	3,350	3,509	3,669	3,828	3,988	4,147	4,307	4,466	4,874	5,042	5,210	5,378	5,546	5,714	5,882	6,050	6,219	6,387	6,555	6,723
O-3	2,733	2,869	3,006	3,143	3,279	3,416	3,553	3,689	3,826	4,175	4,319	4,463	4,607	4,751	4,895	5,039	5,183	5,327	5,471	5,615	5,759
O-2	2,008	2,109	2,209	2,310	2,410	2,511	2,611	2,712	2,812	3,069	3,174	3,280	3,386	3,492	3,598	3,704	3,809	3,915	4,021	4,127	4,233
O-1	1,585	1,664	1,743	1,823	1,902	1,981	2,060	2,140	2,219	2,421	2,505	2,588	2,672	2,755	2,839	2,922	3,006	3,089	3,173	3,256	3,340
COMMISSIONED OFFICERS WITH: 1,460 POINTS TOWARDS A RESERVE RETIREMENT; OR 4 YEARS OF ACTIVE SERVICE, AS AN ENLISTED MEMBER AND/OR WARRANT OFFICER																					
O-3E	2,890	3,062	3,208	3,354	3,500	3,646	3,792	3,937	4,083	4,456	4,610	4,763	4,917	5,071	5,224	5,378	5,532	5,685	5,839	5,993	6,146
O-2E	2,326	2,442	2,559	2,675	2,791	2,908	3,024	3,140	3,257	3,554	3,676	3,799	3,921	4,044	4,166	4,289	4,412	4,534	4,657	4,779	4,902
O-1E	1,968	2,066	2,165	2,263	2,362	2,460	2,559	2,657	2,755	3,007	3,111	3,214	3,318	3,422	3,525	3,629	3,733	3,836	3,940	4,044	4,148
WARRANT OFFICERS																					
W-5	2,668	2,969	3,249	3,583	3,801	4,057	4,269	4,540	4,768	5,331	5,791	5,984	6,177	6,370	6,891	7,094	7,296	7,499	8,087	8,300	8,513
W-4	2,668	2,866	3,036	3,260	3,456	3,689	3,884	4,134	4,345	4,844	5,111	5,281	5,452	5,622	5,793	5,963	6,133	6,304	6,474	6,645	6,815
W-3	2,421	2,611	2,772	2,941	3,093	3,266	3,424	3,629	3,803	4,249	4,395	4,542	4,688	4,835	4,981	5,128	5,274	5,421	5,567	5,714	5,860
W-2	2,143	2,290	2,426	2,581	2,712	2,868	2,999	3,142	3,259	3,577	3,701	3,824	3,948	4,071	4,194	4,318	4,441	4,564	4,688	4,811	4,935
W-1	1,999	2,148	2,278	2,402	2,507	2,611	2,716	2,820	2,925	3,251	3,363	3,475	3,587	3,699	3,811	3,923	4,035	4,148	4,260	4,372	4,484
ENLISTED MEMBERS																					
E-9	2,323	2,505	2,666	2,867	3,030	3,239	3,413	3,660	3,868	4,298	4,668	4,824	4,979	5,135	5,555	5,719	5,882	6,045	6,519	6,691	6,863
E-8	2,032	2,191	2,316	2,479	2,625	2,795	2,930	3,126	3,302	3,668	3,870	3,999	4,128	4,257	4,386	4,515	4,644	4,773	4,902	5,031	5,160
E-7	1,850	1,966	2,067	2,197	2,321	2,467	2,582	2,770	2,939	3,277	3,390	3,503	3,616	3,729	3,842	3,955	4,068	4,181	4,294	4,407	4,521
E-6	1,586	1,670	1,749	1,829	1,908	1,988	2,067	2,147	2,226	2,443	2,527	2,611	2,695	2,780	2,864	2,948	3,032	3,117	3,201	3,285	3,369
E-5	1,336	1,403	1,469	1,536	1,603	1,670	1,737	1,804	1,870	2,050	2,121	2,192	2,262	2,333	2,404	2,474	2,545	2,616	2,686	2,757	2,828
E-4	1,052	1,105	1,157	1,210	1,263	1,315	1,368	1,421	1,473	1,608	1,663	1,719	1,774	1,830	1,885	1,941	1,996	2,052	2,107	2,163	2,218
E-3	882	926	970	1,014	1,058	1,103	1,147	1,191	1,235	1,348	1,394	1,441	1,487	1,534	1,580	1,627	1,673	1,720	1,766	1,813	1,859
E-2	744	781	818	856	893	930	967	1,005	1,042	1,137	1,176	1,215	1,254	1,294	1,333	1,372	1,411	1,451	1,490	1,529	1,568
E-1	664	697	730	763	796	830	863	896	929	1,014	1,049	1,084	1,119	1,154	1,189	1,224	1,259	1,294	1,329	1,364	1,399

NOTES: Assumes current year retirees with less than 29 completed years of active duty service entered the military after September 7, 1980. These retirees are subject to the high-three formula. Figures assume final three years at the same grade in which retired. Also assumes retiree declined the Career Status Bonus.

Assumes current year retirees with 29 or more completed years of active duty service entered the military before September 8, 1980. Retired pay is based on the final pay formula. Figures are based on the member serving the exact number of years in the column heading and final pay being the active duty pay for a member with that amount of years of service.

Figures shown assume that active service and pay service are equal.

Pay Grade	Pay Scale Mo/Yr	Over 20 50%	Over 21 52 1/2%	Over 22 55%	Over 23 57 1/2%	Over 24 60%	Over 25 62 1/2%	Over 26 65%	Over 27 67 1/2%	Over 28 70%	Over 29 72 1/2%	Over 30 75%
COMMISSIONED OFFICERS												
O-8	Oct-72	6,473 N	6,800 N	7,411 N	7,750 N	8,087 N	8,428 N	8,762 N	9,099 N	9,495 N	9,837 N	10,176 N
	Oct-78	5,483 L	5,755 L	6,277 L	6,562 L	6,848 L	7,134 L	7,421 L	7,704 L	7,987 L	8,274 L	8,560 L
	Oct-79	5,341 L	5,612 L	6,121 L	6,398 L	6,673 L	6,955 L	7,233 L	7,512 L	7,790 L	8,070 L	8,350 L
	Oct-80	5,099 L	5,347 L	5,835 L	6,102 L	6,367 L	6,631 L	6,894 L	7,163 L	7,427 L	7,699 L	7,963 L
	Jan-85	5,444	5,721	6,234	6,518	6,802	7,088	7,366	7,654	7,937	8,222	8,507
	Jan-90	5,482	5,753	6,178	6,460	6,743	7,029	7,306	7,586	7,872	8,154	8,431
	Jan-95	5,469	5,743	6,169	6,447	6,729	7,007	7,288	7,571	7,850	8,130	8,414
	Jan-96	5,437	5,712	6,132	6,409	6,690	6,967	7,248	7,526	7,805	8,085	8,361
	Jan-98	5,520	5,795	6,223	6,505	6,790	7,072	7,357	7,639	7,923	8,205	8,489
	Jan-99	5,640	5,921	6,358	6,650	6,937	7,227	7,514	7,805	8,092	8,384	8,674
	Jan-00	5,757	6,044	6,488	6,786	7,079	7,375	7,672	7,968	8,261	8,556	8,854
	Jan-01	5,075	6,064	6,510	6,805	7,101	7,398	7,693	7,988	8,285	8,582	8,877
	Jan-02	5,242	5,653	6,701	7,005	7,310	7,616	7,920	8,224	8,529	8,834	9,141
	Jan-03	5,353	5,774	6,125	7,165	7,475	7,786	8,102	8,412	8,725	9,036	9,346
	Jan-04	5,430	5,857	6,212	6,629	7,626	7,942	8,259	8,578	8,895	9,213	9,531
	Jan-05	5,484	5,915	6,276	6,696	7,044	7,949	8,269	8,587	8,905	9,224	9,541
	Jan-06	5,459	5,885	6,243	6,662	7,008	7,358	8,205	8,522	8,837	9,152	9,469
	Jan-07	5,543	5,978	6,341	6,766	7,119	7,473	7,773	8,472	8,785	9,099	9,414
	Jan-08	5,387	5,809	6,163	6,575	6,918	7,263	7,553	7,844	8,822	9,138	9,689
	Jan-09	5,488	5,918	6,278	6,699	7,047	7,398	7,694	7,990	8,286	9,042	9,588
O-7	Oct-72	5,857 N	6,143 N	6,444 N	6,737 N	7,033 N	7,323 N	7,616 N	7,911 N	8,253 N	8,548 N	8,843 N
	Oct-78	4,958 L	5,203 L	5,456 L	5,703 L	5,951 L	6,200 L	6,452 L	6,697 L	6,945 L	7,193 L	7,446 L
	Oct-79	4,835 L	5,070 L	5,316 L	5,560 L	5,799 L	6,044 L	6,285 L	6,528 L	6,771 L	7,017 L	7,259 L
	Oct-80	4,607 L	4,840 L	5,068 L	5,300 L	5,529 L	5,765 L	5,994 L	6,229 L	6,458 L	6,691 L	6,922 L
	Jan-85	4,920	5,170	5,422	5,667	5,911	6,160	6,406	6,654	6,898	7,143	7,398
	Jan-90	4,958	5,203	5,455	5,700	5,950	6,200	6,449	6,695	6,942	7,192	7,444
	Jan-95	4,945	5,194	5,440	5,687	5,937	6,184	6,430	6,678	6,927	7,176	7,422
	Jan-96	4,916	5,163	5,409	5,652	5,902	6,145	6,395	6,645	6,887	7,134	7,379
	Jan-98	4,991	5,241	5,493	5,742	5,990	6,239	6,487	6,742	6,991	7,239	7,490
	Jan-99	5,099	5,356	5,609	5,867	6,122	6,377	6,631	6,887	7,141	7,399	7,651
	Jan-00	5,204	5,467	5,725	5,987	6,249	6,509	6,768	7,030	7,289	7,550	7,813
	Jan-01	4,730	5,482	5,744	6,005	6,265	6,527	6,825	7,086	7,348	7,611	7,873
	Jan-02	4,885	5,239	5,913	6,181	6,452	6,720	7,025	7,295	7,565	7,836	8,107
	Jan-03	4,990	5,349	5,605	6,322	6,598	6,874	7,184	7,461	7,737	8,015	8,290
	Jan-04	5,062	5,427	5,684	5,944	6,729	7,008	7,326	7,608	7,890	8,171	8,454
	Jan-05	5,112	5,481	5,742	6,005	6,265	7,017	7,334	7,616	7,899	8,180	8,461
	Jan-06	5,087	5,453	5,714	5,973	6,233	6,494	7,279	7,558	7,838	8,119	8,398
	Jan-07	5,167	5,540	5,804	6,067	6,332	6,596	6,860	7,514	7,793	8,071	8,349
	Jan-08	5,020	5,384	5,640	5,896	6,153	6,410	6,667	6,935	7,825	8,104	8,552
	Jan-09	5,115	5,484	5,745	6,007	6,268	6,529	6,790	7,064	7,338	8,020	8,463

NOTE: Pay scales are in effect for active duty personnel at time of retirement. See detailed footnotes on page following tables.

Pay Grade	Pay Scale Mo/Yr	Over 20 50%	Over 21 52 1/2%	Over 22 55%	Over 23 57 1/2%	Over 24 60%	Over 25 62 1/2%	Over 26 65%	Over 27 67 1/2%	Over 28 70%	Over 29 72 1/2%	Over 30 75%
COMMISSIONED OFFICERS												
O-6	Oct-72	4,479 N	4,705 N	5,211 N	5,456 N	5,694 N	5,933 N	6,691 N	6,945 N	7,248 N	7,506 N	7,768 N
	Oct-78	3,789 L	3,981 L	4,417 L	4,619 L	4,814 L	5,015 L	5,665 L	5,883 L	6,102 L	6,318 L	6,535 L
	Oct-79	3,695 L	3,882 L	4,306 L	4,498 L	4,696 L	4,894 L	5,519 L	5,735 L	5,947 L	6,167 L	6,372 L
	Oct-80	3,521 L	3,699 L	4,106 L	4,290 L	4,481 L	4,665 L	5,266 L	5,467 L	5,670 L	5,877 L	6,080 L
	Jan-85	3,765	3,955	4,385	4,586	4,788	4,986	5,628	5,844	6,058	6,278	6,500
	Jan-90	3,789	3,983	4,415	4,616	4,813	5,017	5,665	5,882	6,099	6,318	6,534
	Jan-95	3,783	3,971	4,406	4,607	4,970	5,181	5,648	5,867	6,084	6,304	6,520
	Jan-96	3,763	3,953	4,380	4,580	4,942	5,148	5,615	5,835	6,050	6,265	6,480
	Jan-98	3,820	4,011	4,446	4,649	5,014	5,227	5,702	5,919	6,140	6,359	6,579
	Jan-99	3,907	4,101	4,545	4,750	5,126	5,339	5,827	6,050	6,276	6,500	6,722
	Jan-00	3,985	4,184	4,637	4,849	5,233	5,449	5,947	6,175	6,405	6,632	6,862
	Jan-01	3,579	4,382	4,712	4,924	5,272	5,493	5,993	6,224	6,454	6,685	6,916
	Jan-02	3,724	4,037	4,849	5,069	5,428	5,655	6,170	6,407	6,646	6,882	7,119
	Jan-03	3,822	4,146	4,411	5,186	5,550	5,781	6,310	6,553	6,795	7,038	7,280
	Jan-04	3,877	4,205	4,476	4,792	5,661	5,898	6,433	6,682	6,929	7,176	7,425
	Jan-05	3,916	4,247	4,520	4,840	5,095	5,903	6,442	6,689	6,937	7,185	7,432
	Jan-06	3,896	4,226	4,497	4,817	5,069	5,373	6,391	6,638	6,884	7,129	7,375
	Jan-07	3,959	4,293	4,569	4,892	5,150	5,457	5,724	6,600	6,844	7,088	7,332
	Jan-08	3,846	4,172	4,440	4,755	5,005	5,303	5,562	5,923	6,873	7,119	7,511
	Jan-09	3,919	4,250	4,523	4,844	5,098	5,403	5,667	6,034	6,357	7,044	7,433
O-5	Oct-72	4,051 N	4,253 N	4,617 N	4,825 N	5,037 N	5,244 N	5,456 N	5,667 N	5,909 N	6,123 N	6,336 N
	Oct-78	3,426 L	3,596 L	3,908 L	4,079 L	4,262 L	4,441 L	4,620 L	4,798 L	4,974 L	5,152 L	5,333 L
	Oct-79	3,339 L	3,509 L	3,809 L	3,981 L	4,156 L	4,329 L	4,500 L	4,673 L	4,850 L	5,020 L	5,197 L
	Oct-80	3,186 L	3,345 L	3,629 L	3,796 L	3,961 L	4,127 L	4,294 L	4,460 L	4,623 L	4,790 L	4,958 L
	Jan-85	3,405	3,580	3,878	4,055	4,230	4,408	4,588	4,761	4,944	5,119	5,296
	Jan-90	3,426	3,601	3,908	4,080	4,262	4,442	4,617	4,796	4,972	5,152	5,331
	Jan-95	3,422	3,594	3,898	4,075	4,254	4,431	4,609	4,787	4,963	5,142	5,316
	Jan-96	3,401	3,572	3,876	4,053	4,225	4,406	4,581	4,756	4,934	5,112	5,287
	Jan-98	3,456	3,627	3,932	4,112	4,291	4,472	4,654	4,829	5,008	5,189	5,369
	Jan-99	3,530	3,705	4,021	4,205	4,386	4,570	4,753	4,934	5,118	5,302	5,484
	Jan-00	3,604	3,784	4,105	4,290	4,478	4,664	4,850	5,039	5,227	5,412	5,599
	Jan-01	3,201	3,832	4,135	4,322	4,511	4,699	4,889	5,076	5,264	5,453	5,641
	Jan-02	3,324	3,584	4,259	4,452	4,644	4,839	5,031	5,229	5,422	5,613	5,809
	Jan-03	3,438	3,676	3,883	4,552	4,751	4,950	5,148	5,345	5,543	5,742	5,941
	Jan-04	3,486	3,729	3,941	4,198	4,845	5,048	5,248	5,451	5,653	5,856	6,057
	Jan-05	3,521	3,767	3,981	4,242	4,468	5,054	5,256	5,459	5,661	5,863	6,065
	Jan-06	3,504	3,747	3,961	4,220	4,447	4,676	5,215	5,416	5,617	5,817	6,017
	Jan-07	3,560	3,806	4,023	4,287	4,518	4,750	4,941	5,385	5,585	5,784	5,984
	Jan-08	3,460	3,699	3,910	4,166	4,390	4,616	4,801	4,986	5,608	5,808	6,009
	Jan-09	3,524	3,768	3,983	4,244	4,472	4,703	4,891	5,079	5,267	5,748	5,946

NOTE: Pay scales are in effect for active duty personnel at time of retirement. See detailed footnotes on page following tables.

Pay Grade	Pay Scale Mo/Yr	Over 20 50%	Over 21 52 1/2%	Over 22 55%	Over 23 57 1/2%	Over 24 60%	Over 25 62 1/2%	Over 26 65%	Over 27 67 1/2%	Over 28 70%	Over 29 72 1/2%	Over 30 75%
COMMISSIONED OFFICERS												
O-4	Oct-72	3,504 N	3,679 N	3,854 N	4,029 N	4,209 N	4,384 N	4,557 N	4,732 N	4,942 N	5,113 N	5,293 N
	Oct-78	2,965 L	3,113 L	3,264 L	3,413 L	3,562 L	3,710 L	3,857 L	4,010 L	4,157 L	4,310 L	4,452 L
	Oct-79	2,888 L	3,038 L	3,180 L	3,325 L	3,470 L	3,618 L	3,762 L	3,910 L	4,053 L	4,205 L	4,346 L
	Oct-80	2,755 L	2,890 L	3,033 L	3,170 L	3,311 L	3,453 L	3,587 L	3,729 L	3,862 L	4,006 L	4,139 L
	Jan-85	2,945	3,093	3,240	3,393	3,536	3,686	3,834	3,983	4,128	4,275	4,429
	Jan-90	2,965	3,113	3,264	3,413	3,562	3,709	3,856	4,009	4,157	4,308	4,452
	Jan-95	2,958	3,109	3,258	3,403	3,553	3,700	3,850	4,001	4,148	4,296	4,446
	Jan-96	2,943	3,091	3,237	3,388	3,534	3,681	3,829	3,980	4,126	4,272	4,419
	Jan-98	2,988	3,139	3,288	3,438	3,587	3,736	3,886	4,037	4,185	4,334	4,486
	Jan-99	3,054	3,207	3,360	3,512	3,665	3,820	3,971	4,126	4,277	4,433	4,583
	Jan-00	3,116	3,275	3,430	3,586	3,742	3,899	4,054	4,211	4,366	4,524	4,678
	Jan-01	2,872	3,299	3,457	3,614	3,770	3,928	4,087	4,244	4,400	4,558	4,716
	Jan-02	2,981	3,157	3,609	3,774	3,938	4,103	4,266	4,433	4,595	4,759	4,923
	Jan-03	3,079	3,243	3,399	3,862	4,029	4,196	4,364	4,532	4,699	4,869	5,037
	Jan-04	3,138	3,306	3,463	3,620	4,108	4,279	4,451	4,622	4,793	4,965	5,137
	Jan-05	3,177	3,347	3,507	3,667	3,825	4,285	4,456	4,628	4,800	4,970	5,142
	Jan-06	3,162	3,330	3,490	3,649	3,805	3,965	4,423	4,592	4,763	4,933	5,103
	Jan-07	3,212	3,383	3,545	3,706	3,868	4,028	4,190	4,567	4,735	4,904	5,074
	Jan-08	3,121	3,289	3,444	3,601	3,758	3,915	4,071	4,228	4,755	4,925	5,095
	Jan-09	3,179	3,350	3,509	3,669	3,828	3,988	4,147	4,307	4,466	4,874	5,042
O-3	Oct-72	3,027 N	3,179 N	3,330 N	3,480 N	3,634 N	3,787 N	3,939 N	4,094 N	4,269 N	4,425 N	4,573 N
	Oct-78	2,562 L	2,689 L	2,819 L	2,947 L	3,079 L	3,204 L	3,333 L	3,466 L	3,589 L	3,727 L	3,851 L
	Oct-79	2,498 L	2,621 L	2,749 L	2,874 L	2,998 L	3,126 L	3,249 L	3,378 L	3,503 L	3,627 L	3,752 L
	Oct-80	2,380 L	2,501 L	2,620 L	2,739 L	2,859 L	2,983 L	3,102 L	3,221 L	3,337 L	3,462 L	3,581 L
	Jan-85	2,545	2,676	2,801	2,934	3,055	3,184	3,312	3,439	3,572	3,697	3,823
	Jan-90	2,563	2,691	2,820	2,948	3,079	3,205	3,333	3,467	3,591	3,727	3,851
	Jan-95	2,561	2,688	2,815	2,943	3,071	3,201	3,329	3,459	3,587	3,718	3,842
	Jan-96	2,545	2,672	2,800	2,930	3,056	3,182	3,311	3,439	3,566	3,693	3,821
	Jan-98	2,585	2,713	2,841	2,972	3,102	3,231	3,361	3,492	3,618	3,747	3,878
	Jan-99	2,640	2,770	2,905	3,038	3,169	3,303	3,436	3,567	3,698	3,834	3,965
	Jan-00	2,694	2,832	2,966	3,100	3,236	3,371	3,506	3,641	3,778	3,912	4,044
	Jan-01	2,503	2,840	2,975	3,110	3,244	3,381	3,514	3,651	3,787	3,923	4,058
	Jan-02	2,588	2,718	3,093	3,232	3,373	3,513	3,654	3,795	3,935	4,078	4,218
	Jan-03	2,652	2,785	2,917	3,307	3,451	3,596	3,738	3,881	4,026	4,170	4,315
	Jan-04	2,698	2,834	2,969	3,103	3,518	3,665	3,814	3,960	4,106	4,253	4,400
	Jan-05	2,730	2,867	3,003	3,140	3,276	3,672	3,818	3,965	4,111	4,258	4,405
	Jan-06	2,716	2,853	2,988	3,125	3,260	3,397	3,788	3,934	4,079	4,225	4,371
	Jan-07	2,760	2,898	3,036	3,175	3,312	3,451	3,589	3,912	4,057	4,202	4,346
	Jan-08	2,682	2,817	2,950	3,085	3,219	3,353	3,488	3,621	4,074	4,218	4,364
	Jan-09	2,733	2,869	3,006	3,143	3,279	3,416	3,553	3,689	3,826	4,175	4,319

NOTE: Pay scales are in effect for active duty personnel at time of retirement. See detailed footnotes on page following tables.

Pay Grade	Pay Scale Mo/Yr	Over 20 50%	Over 21 52 1/2%	Over 22 55%	Over 23 57 1/2%	Over 24 60%	Over 25 62 1/2%	Over 26 65%	Over 27 67 1/2%	Over 28 70%	Over 29 72 1/2%	Over 30 75%
COMMISSIONED OFFICERS												
O-2	Oct-72	2,239 N	2,352 N	2,465 N	2,579 N	2,692 N	2,803 N	2,922 N	3,033 N	3,162 N	3,277 N	3,394 N
	Oct-78	1,893 L	1,994 L	2,087 L	2,183 L	2,276 L	2,373 L	2,468 L	2,564 L	2,661 L	2,756 L	2,849 L
	Oct-79	1,846 L	1,940 L	2,036 L	2,129 L	2,222 L	2,319 L	2,408 L	2,501 L	2,594 L	2,687 L	2,780 L
	Oct-80	1,761 L	1,849 L	1,939 L	2,033 L	2,113 L	2,208 L	2,294 L	2,385 L	2,469 L	2,563 L	2,653 L
	Jan-85	1,884	1,978	2,073	2,172	2,263	2,357	2,456	2,548	2,643	2,739	2,834
	Jan-90	1,896	1,995	2,087	2,185	2,281	2,375	2,469	2,567	2,664	2,759	2,858
	Jan-95	1,894	1,993	2,085	2,183	2,277	2,372	2,466	2,563	2,659	2,755	2,851
	Jan-96	1,886	1,979	2,075	2,171	2,265	2,359	2,454	2,549	2,646	2,739	2,834
	Jan-98	1,917	2,010	2,107	2,202	2,301	2,395	2,491	2,591	2,685	2,780	2,876
	Jan-99	1,959	2,055	2,154	2,251	2,350	2,450	2,546	2,646	2,746	2,841	2,939
	Jan-00	1,999	2,101	2,198	2,300	2,400	2,501	2,599	2,701	2,800	2,899	2,999
	Jan-01	1,854	2,106	2,206	2,308	2,408	2,508	2,606	2,707	2,808	2,908	3,010
	Jan-02	1,916	2,011	2,273	2,374	2,479	2,583	2,686	2,791	2,891	2,995	3,100
	Jan-03	1,958	2,054	2,153	2,429	2,534	2,640	2,747	2,852	2,958	3,063	3,169
	Jan-04	1,986	2,085	2,185	2,284	2,585	2,693	2,801	2,908	3,018	3,126	3,233
	Jan-05	2,005	2,107	2,208	2,308	2,409	2,697	2,804	2,911	3,021	3,129	3,237
	Jan-06	1,997	2,096	2,196	2,295	2,396	2,496	2,784	2,891	2,998	3,106	3,213
	Jan-07	2,029	2,130	2,231	2,332	2,434	2,536	2,638	2,875	2,981	3,088	3,194
	Jan-08	1,971	2,070	2,168	2,267	2,366	2,465	2,563	2,662	2,994	3,100	3,207
	Jan-09	2,008	2,109	2,209	2,310	2,410	2,511	2,611	2,712	2,812	3,069	3,174
O-3E	Oct-72	3,077 N	3,230 N	3,384 N	3,536 N	3,691 N	3,846 N	4,004 N	4,156 N	4,333 N	4,490 N	4,649 N
	Oct-78	2,600 L	2,727 L	2,861 L	2,994 L	3,126 L	3,257 L	3,385 L	3,515 L	3,647 L	3,780 L	3,910 L
	Oct-79	2,532 L	2,662 L	2,791 L	2,921 L	3,044 L	3,170 L	3,302 L	3,426 L	3,556 L	3,682 L	3,813 L
	Oct-80	2,417 L	2,537 L	2,661 L	2,782 L	2,899 L	3,027 L	3,145 L	3,272 L	3,394 L	3,512 L	3,633 L
	Jan-85	2,582	2,713	2,841	2,972	3,104	3,232	3,363	3,492	3,623	3,751	3,882
	Jan-90	2,600	2,731	2,866	2,993	3,125	3,256	3,384	3,515	3,645	3,780	3,910
	Jan-95	2,597	2,726	2,859	2,988	3,118	3,249	3,381	3,509	3,639	3,770	3,901
	Jan-96	2,582	2,712	2,841	2,971	3,100	3,230	3,361	3,491	3,617	3,747	3,878
	Jan-98	2,621	2,756	2,885	3,017	3,150	3,280	3,409	3,544	3,676	3,806	3,936
	Jan-99	2,680	2,815	2,948	3,084	3,216	3,351	3,486	3,620	3,753	3,888	4,023
	Jan-00	2,735	2,872	3,010	3,146	3,284	3,421	3,559	3,695	3,834	3,969	4,107
	Jan-01	2,573	3,031	3,174	3,321	3,464	3,608	3,752	3,898	4,042	4,186	4,331
	Jan-02	2,719	2,855	3,300	3,451	3,601	3,750	3,901	4,051	4,202	4,352	4,500
	Jan-03	2,804	2,971	3,114	3,530	3,682	3,837	3,989	4,144	4,298	4,451	4,604
	Jan-04	2,854	3,024	3,169	3,312	3,754	3,912	4,069	4,224	4,383	4,537	4,695
	Jan-05	2,888	3,060	3,205	3,351	3,498	3,917	4,073	4,229	4,388	4,544	4,700
	Jan-06	2,873	3,044	3,189	3,334	3,479	3,625	4,042	4,199	4,354	4,509	4,665
	Jan-07	2,920	3,093	3,240	3,388	3,535	3,682	3,829	4,174	4,330	4,484	4,638
	Jan-08	2,836	3,006	3,149	3,293	3,435	3,579	3,722	3,865	4,348	4,503	4,658
	Jan-09	2,890	3,062	3,208	3,354	3,500	3,646	3,792	3,937	4,083	4,456	4,610

NOTE: Pay scales are in effect for active duty personnel at time of retirement. See detailed footnotes on page following tables.

Pay Grade	Pay Scale Mo/Yr	Over 20 50%	Over 21 52 1/2%	Over 22 55%	Over 23 57 1/2%	Over 24 60%	Over 25 62 1/2%	Over 26 65%	Over 27 67 1/2%	Over 28 70%	Over 29 72 1/2%	Over 30 75%
COMMISSIONED OFFICERS												
O-2E	Oct-72	2,599 N	2,727 N	2,861 N	2,992 N	3,125 N	3,256 N	3,385 N	3,514 N	3,669 N	3,798 N	3,932 N
	Oct-78	2,197 L	2,307 L	2,420 L	2,530 L	2,642 L	2,752 L	2,867 L	2,974 L	3,085 L	3,196 L	3,307 L
	Oct-79	2,145 L	2,250 L	2,357 L	2,465 L	2,571 L	2,684 L	2,793 L	2,895 L	3,004 L	3,113 L	3,222 L
	Oct-80	2,044 L	2,147 L	2,247 L	2,350 L	2,456 L	2,561 L	2,661 L	2,764 L	2,869 L	2,971 L	3,071 L
	Jan-85	2,183	2,290	2,402	2,518	2,621	2,731	2,841	2,952	3,062	3,175	3,286
	Jan-90	2,199	2,310	2,420	2,530	2,642	2,753	2,866	2,977	3,085	3,197	3,308
	Jan-95	2,197	2,309	2,419	2,529	2,639	2,750	2,860	2,970	3,080	3,191	3,298
	Jan-96	2,185	2,294	2,408	2,514	2,624	2,733	2,843	2,952	3,062	3,170	3,282
	Jan-98	2,220	2,330	2,443	2,551	2,664	2,774	2,888	2,996	3,108	3,220	3,331
	Jan-99	2,267	2,380	2,497	2,609	2,720	2,836	2,949	3,062	3,176	3,292	3,402
	Jan-00	2,315	2,430	2,546	2,662	2,778	2,895	3,012	3,128	3,242	3,360	3,475
	Jan-01	2,148	2,440	2,556	2,672	2,789	2,904	3,022	3,136	3,254	3,370	3,486
	Jan-02	2,220	2,330	2,630	2,752	2,871	2,990	3,111	3,230	3,349	3,470	3,589
	Jan-03	2,267	2,379	2,494	2,814	2,937	3,059	3,181	3,304	3,425	3,549	3,672
	Jan-04	2,299	2,416	2,530	2,646	2,995	3,120	3,244	3,369	3,496	3,618	3,744
	Jan-05	2,323	2,440	2,556	2,672	2,789	3,124	3,249	3,373	3,498	3,623	3,748
	Jan-06	2,312	2,428	2,544	2,659	2,775	2,890	3,224	3,348	3,472	3,597	3,720
	Jan-07	2,348	2,467	2,583	2,702	2,819	2,937	3,054	3,329	3,453	3,577	3,699
	Jan-08	2,284	2,397	2,511	2,626	2,740	2,854	2,969	3,082	3,467	3,591	3,714
	Jan-09	2,326	2,442	2,559	2,675	2,791	2,908	3,024	3,140	3,257	3,554	3,676
W-5	Jan-95	2,734	2,872	3,125	3,269	3,508	3,654	3,961	4,113	4,266	4,418	4,571
	Jan-96	2,720	2,858	3,106	3,249	3,488	3,634	3,938	4,091	4,243	4,393	4,548
	Jan-98	2,762	2,899	3,156	3,297	3,543	3,689	3,998	4,153	4,307	4,459	4,617
	Jan-99	2,821	2,966	3,223	3,371	3,618	3,770	4,087	4,244	4,401	4,558	4,717
	Jan-00	2,881	3,027	3,290	3,439	3,694	3,846	4,172	4,332	4,493	4,654	4,813
	Jan-01	2,313	3,070	3,328	3,478	3,750	3,908	4,196	4,358	4,520	4,681	4,841
	Jan-02	2,409	2,684	3,492	3,650	3,935	4,100	4,402	4,573	4,743	4,910	5,079
	Jan-03	2,483	2,769	3,039	3,734	4,024	4,192	4,502	4,676	4,849	5,022	5,196
	Jan-04	2,547	2,831	3,103	3,423	4,106	4,276	4,591	4,769	4,945	5,121	5,298
	Jan-05	2,602	2,879	3,144	3,468	3,660	4,280	4,597	4,773	4,951	5,127	5,304
	Jan-06	2,607	2,884	3,139	3,452	3,642	3,879	4,562	4,737	4,913	5,090	5,264
	Jan-07	2,658	2,940	3,198	3,506	3,699	3,939	4,143	4,711	4,885	5,060	5,234
	Jan-08	2,603	2,897	3,156	3,478	3,675	3,918	4,119	4,380	5,201	5,387	5,851
	Jan-09	2,668	2,969	3,249	3,583	3,801	4,057	4,269	4,540	4,768	5,331	5,791

NOTE: Pay scales are in effect for active duty personnel at time of retirement. See detailed footnotes on page following tables.

Pay Grade	Pay Scale Mo/Yr	Over 20 50%	Over 21 52 1/2%	Over 22 55%	Over 23 57 1/2%	Over 24 60%	Over 25 62 1/2%	Over 26 65%	Over 27 67 1/2%	Over 28 70%	Over 29 72 1/2%	Over 30 75%
COMMISSIONED OFFICERS												
W-4	Oct-72	2,905 N	3,053 N	3,307 N	3,460 N	3,610 N	3,760 N	4,215 N	4,378 N	4,563 N	4,728 N	4,894 N
	Oct-78	2,458 L	2,582 L	2,797 L	2,928 L	3,053 L	3,180 L	3,570 L	3,705 L	3,845 L	3,981 L	4,119 L
	Oct-79	2,395 L	2,519 L	2,725 L	2,849 L	2,977 L	3,102 L	3,477 L	3,613 L	3,744 L	3,881 L	4,015 L
	Oct-80	2,285 L	2,400 L	2,600 L	2,720 L	2,838 L	2,956 L	3,317 L	3,441 L	3,578 L	3,699 L	3,826 L
	Jan-85	2,443	2,564	2,780	2,903	3,033	3,161	3,543	3,681	3,817	3,955	4,095
	Jan-90	2,458	2,585	2,797	2,928	3,055	3,182	3,570	3,704	3,844	3,983	4,120
	Jan-95	2,455	2,580	2,794	2,924	3,144	3,276	3,562	3,698	3,836	3,971	4,111
	Jan-96	2,443	2,564	2,778	2,904	3,126	3,257	3,543	3,679	3,816	3,953	4,089
	Jan-98	2,482	2,603	2,819	2,948	3,172	3,306	3,595	3,734	3,873	4,011	4,150
	Jan-99	2,534	2,661	2,884	3,016	3,242	3,380	3,674	3,817	3,959	4,100	4,242
	Jan-00	2,589	2,716	2,942	3,077	3,311	3,449	3,749	3,896	4,038	4,183	4,327
	Jan-01	2,313	2,758	2,987	3,123	3,365	3,505	3,762	3,907	4,051	4,196	4,343
	Jan-02	2,409	2,582	3,147	3,292	3,548	3,694	3,965	4,118	4,271	4,425	4,576
	Jan-03	2,483	2,667	2,825	3,367	3,626	3,779	4,055	4,211	4,367	4,525	4,680
	Jan-04	2,547	2,737	2,899	3,098	3,782	3,939	4,226	4,390	4,553	4,717	4,879
	Jan-05	2,602	2,795	2,961	3,166	3,341	3,944	4,232	4,395	4,558	4,721	4,884
	Jan-06	2,607	2,801	2,968	3,171	3,347	3,562	4,201	4,362	4,524	4,684	4,848
	Jan-07	2,658	2,855	3,024	3,233	3,412	3,632	3,818	4,337	4,498	4,659	4,820
	Jan-08	2,603	2,797	2,963	3,181	3,360	3,584	3,768	4,010	4,726	4,895	5,164
	Jan-09	2,668	2,866	3,036	3,260	3,456	3,689	3,884	4,134	4,345	4,844	5,111
W-3	Oct-72	2,556 N	2,680 N	2,911 N	3,045 N	3,179 N	3,311 N	3,570 N	3,705 N	3,868 N	4,008 N	4,139 N
	Oct-78	2,159 L	2,268 L	2,464 L	2,575 L	2,689 L	2,802 L	3,018 L	3,136 L	3,256 L	3,369 L	3,486 L
	Oct-79	2,105 L	2,215 L	2,401 L	2,515 L	2,621 L	2,727 L	2,944 L	3,059 L	3,169 L	3,286 L	3,400 L
	Oct-80	2,005 L	2,107 L	2,287 L	2,394 L	2,501 L	2,603 L	2,803 L	2,919 L	3,023 L	3,131 L	3,240 L
	Jan-85	2,147	2,253	2,450	2,562	2,676	2,785	2,998	3,113	3,231	3,345	3,466
	Jan-90	2,160	2,273	2,465	2,579	2,691	2,802	3,019	3,136	3,256	3,370	3,486
	Jan-95	2,158	2,271	2,463	2,576	2,688	2,800	3,016	3,129	3,248	3,366	3,478
	Jan-96	2,148	2,256	2,450	2,561	2,672	2,785	2,998	3,113	3,230	3,343	3,462
	Jan-98	2,181	2,288	2,487	2,600	2,713	2,826	3,043	3,162	3,279	3,396	3,512
	Jan-99	2,229	2,342	2,542	2,657	2,770	2,890	3,110	3,232	3,350	3,470	3,589
	Jan-00	2,275	2,392	2,595	2,712	2,832	2,948	3,175	3,297	3,419	3,544	3,665
	Jan-01	2,019	2,425	2,629	2,749	2,965	3,086	3,313	3,441	3,568	3,696	3,823
	Jan-02	2,112	2,268	2,785	2,910	3,138	3,270	3,509	3,645	3,781	3,916	4,050
	Jan-03	2,194	2,359	2,501	2,979	3,211	3,344	3,591	3,729	3,868	4,006	4,144
	Jan-04	2,267	2,430	2,575	2,737	3,275	3,412	3,661	3,801	3,943	4,083	4,224
	Jan-05	2,329	2,487	2,628	2,793	2,929	3,415	3,664	3,805	3,947	4,088	4,228
	Jan-06	2,340	2,498	2,630	2,787	2,924	3,094	3,638	3,778	3,917	4,058	4,198
	Jan-07	2,384	2,546	2,679	2,831	2,969	3,143	3,305	3,755	3,895	4,034	4,173
	Jan-08	2,353	2,534	2,673	2,831	2,971	3,137	3,295	3,495	4,145	4,293	4,441
	Jan-09	2,421	2,611	2,772	2,941	3,093	3,266	3,424	3,629	3,803	4,249	4,395

NOTE: Pay scales are in effect for active duty personnel at time of retirement. See detailed footnotes on page following tables.

Pay Grade	Pay Scale Mo/Yr	Over 20 50%	Over 21 52 1/2%	Over 22 55%	Over 23 57 1/2%	Over 24 60%	Over 25 62 1/2%	Over 26 65%	Over 27 67 1/2%	Over 28 70%	Over 29 72 1/2%	Over 30 75%
COMMISSIONED OFFICERS												
W-2	Oct-72	2,287 N	2,402 N	2,621 N	2,746 N	2,867 N	2,985 N	3,106 N	3,222 N	3,363 N	3,480 N	3,602 N
	Oct-78	1,938 L	2,036 L	2,222 L	2,326 L	2,422 L	2,525 L	2,624 L	2,725 L	2,830 L	2,934 L	3,033 L
	Oct-79	1,887 L	1,982 L	2,162 L	2,262 L	2,361 L	2,461 L	2,561 L	2,659 L	2,756 L	2,858 L	2,954 L
	Oct-80	1,804 L	1,890 L	2,063 L	2,158 L	2,251 L	2,346 L	2,442 L	2,532 L	2,630 L	2,722 L	2,819 L
	Jan-85	1,923	2,021	2,206	2,306	2,408	2,505	2,609	2,711	2,812	2,911	3,010
	Jan-90	1,938	2,038	2,221	2,326	2,424	2,525	2,625	2,726	2,832	2,933	3,033
	Jan-95	1,937	2,035	2,220	2,319	2,421	2,523	2,623	2,724	2,825	2,929	3,027
	Jan-96	1,925	2,023	2,206	2,309	2,410	2,507	2,606	2,708	2,808	2,909	3,010
	Jan-98	1,957	2,054	2,239	2,342	2,447	2,545	2,649	2,752	2,853	2,953	3,056
	Jan-99	2,000	2,102	2,288	2,394	2,500	2,601	2,707	2,812	2,914	3,021	3,125
	Jan-00	2,041	2,143	2,337	2,445	2,549	2,657	2,763	2,868	2,976	3,083	3,189
	Jan-01	1,820	2,171	2,349	2,457	2,650	2,759	2,869	2,981	3,091	3,201	3,311
	Jan-02	1,894	2,037	2,502	2,616	2,817	2,935	3,053	3,169	3,288	3,405	3,523
	Jan-03	1,960	2,107	2,232	2,673	2,884	3,002	3,124	3,243	3,365	3,483	3,604
	Jan-04	2,009	2,159	2,288	2,449	2,940	3,062	3,186	3,307	3,430	3,553	3,676
	Jan-05	2,040	2,193	2,324	2,485	2,622	3,067	3,188	3,311	3,436	3,558	3,680
	Jan-06	2,036	2,182	2,312	2,473	2,610	2,779	3,165	3,286	3,408	3,530	3,652
	Jan-07	2,074	2,217	2,349	2,512	2,651	2,823	2,968	3,268	3,389	3,510	3,632
	Jan-08	2,066	2,207	2,338	2,489	2,624	2,777	2,915	3,059	3,490	3,615	3,740
	Jan-09	2,143	2,290	2,426	2,581	2,712	2,868	2,999	3,142	3,259	3,577	3,701
E-9	Oct-72	2,397 N	2,520 N	2,780 N	2,903 N	3,036 N	3,161 N	3,612 N	3,747 N	3,888 N	4,024 N	4,169 N
	Oct-78	2,033 L	2,131 L	2,350 L	2,458 L	2,566 L	2,677 L	3,053 L	3,170 L	3,292 L	3,409 L	3,527 L
	Oct-79	1,977 L	2,077 L	2,288 L	2,395 L	2,503 L	2,605 L	2,981 L	3,093 L	3,204 L	3,323 L	3,435 L
	Oct-80	1,886 L	1,978 L	2,186 L	2,285 L	2,386 L	2,487 L	2,838 L	2,947 L	3,059 L	3,167 L	3,278 L
	Jan-85	2,063	2,168	2,393	2,501	2,612	2,720	3,108	3,227	3,345	3,467	3,586
	Jan-90	2,080	2,183	2,410	2,520	2,628	2,740	3,127	3,249	3,370	3,491	3,612
	Jan-95	2,076	2,181	2,408	2,518	2,726	2,841	3,123	3,241	3,364	3,480	3,605
	Jan-96	2,064	2,169	2,393	2,501	2,712	2,825	3,104	3,223	3,342	3,466	3,585
	Jan-98	2,100	2,201	2,427	2,538	2,756	2,868	3,154	3,275	3,396	3,514	3,639
	Jan-99	2,143	2,250	2,483	2,596	2,815	2,933	3,222	3,345	3,470	3,592	3,720
	Jan-00	2,187	2,297	2,532	2,650	2,872	2,993	3,288	3,415	3,543	3,668	3,793
	Jan-01	1,991	2,380	2,594	2,712	2,935	3,057	3,327	3,456	3,582	3,710	3,839
	Jan-02	2,091	2,235	2,785	2,910	3,153	3,284	3,588	3,729	3,865	4,004	4,142
	Jan-03	2,178	2,340	2,476	3,032	3,281	3,419	3,754	3,900	4,044	4,189	4,333
	Jan-04	2,247	2,422	2,570	2,750	3,422	3,565	3,924	4,075	4,225	4,378	4,529
	Jan-05	2,299	2,479	2,638	2,830	2,992	3,569	3,928	4,080	4,230	4,382	4,533
	Jan-06	2,302	2,482	2,642	2,842	3,003	3,211	3,899	4,048	4,199	4,349	4,499
	Jan-07	2,346	2,529	2,692	2,895	3,061	3,272	3,448	4,026	4,174	4,324	4,474
	Jan-08	2,281	2,459	2,617	2,814	2,975	3,179	3,350	3,593	4,192	4,342	4,717
	Jan-09	2,323	2,505	2,666	2,867	3,030	3,239	3,413	3,660	3,868	4,298	4,668

NOTE: Pay scales are in effect for active duty personnel at time of retirement. See detailed footnotes on page following tables.

Pay Grade	Pay Scale Mo/Yr	Over 20 50%	Over 21 52 1/2%	Over 22 55%	Over 23 57 1/2%	Over 24 60%	Over 25 62 1/2%	Over 26 65%	Over 27 67 1/2%	Over 28 70%	Over 29 72 1/2%	Over 30 75%
COMMISSIONED OFFICERS												
E-8	Oct-72	2,102 N	2,208 N	2,450 N	2,562 N	2,676 N	2,785 N	3,221 N	3,345 N	3,470 N	3,594 N	3,725 N
	Oct-78	1,779 L	1,866 L	2,072 L	2,165 L	2,260 L	2,354 L	2,725 L	2,833 L	2,939 L	3,041 L	3,146 L
	Oct-79	1,730 L	1,817 L	2,019 L	2,109 L	2,206 L	2,294 L	2,657 L	2,759 L	2,867 L	2,966 L	3,067 L
	Oct-80	1,651 L	1,735 L	1,924 L	2,014 L	2,103 L	2,189 L	2,531 L	2,632 L	2,726 L	2,829 L	2,928 L
	Jan-85	1,808	1,896	2,105	2,203	2,300	2,395	2,770	2,880	2,988	3,094	3,200
	Jan-90	1,820	1,917	2,123	2,220	2,320	2,415	2,791	2,897	3,004	3,114	3,222
	Jan-95	1,820	1,911	2,121	2,219	2,418	2,520	2,789	2,892	3,001	3,110	3,216
	Jan-96	1,810	1,898	2,108	2,203	2,403	2,503	2,769	2,879	2,987	3,093	3,200
	Jan-98	1,836	1,929	2,141	2,238	2,442	2,542	2,814	2,924	3,032	3,141	3,248
	Jan-99	1,879	1,971	2,187	2,287	2,494	2,598	2,876	2,987	3,098	3,208	3,322
	Jan-00	1,918	2,012	2,233	2,335	2,545	2,653	2,937	3,049	3,161	3,276	3,389
	Jan-01	1,734	2,074	2,272	2,373	2,584	2,690	2,958	3,072	3,187	3,302	3,415
	Jan-02	1,818	1,943	2,427	2,537	2,761	2,877	3,163	3,286	3,407	3,527	3,650
	Jan-03	1,894	2,030	2,149	2,645	2,824	2,942	3,236	3,360	3,483	3,608	3,734
	Jan-04	1,967	2,106	2,227	2,388	2,925	3,044	3,349	3,477	3,607	3,735	3,865
	Jan-05	2,014	2,165	2,289	2,453	2,598	3,050	3,353	3,482	3,610	3,741	3,869
	Jan-06	2,015	2,173	2,296	2,459	2,604	2,773	3,327	3,455	3,584	3,711	3,839
	Jan-07	2,052	2,213	2,339	2,504	2,651	2,823	2,959	3,436	3,563	3,691	3,818
	Jan-08	1,994	2,150	2,272	2,433	2,576	2,744	2,876	3,068	3,578	3,706	3,911
	Jan-09	2,032	2,191	2,316	2,479	2,625	2,795	2,930	3,126	3,302	3,668	3,870
E-7	Oct-72	1,849 N	1,942 N	2,176 N	2,274 N	2,373 N	2,476 N	2,891 N	3,005 N	3,123 N	3,232 N	3,344 N
	Oct-78	1,563 L	1,643 L	1,840 L	1,924 L	2,008 L	2,095 L	2,452 L	2,545 L	2,642 L	2,734 L	2,830 L
	Oct-79	1,527 L	1,604 L	1,793 L	1,877 L	1,954 L	2,041 L	2,388 L	2,482 L	2,571 L	2,665 L	2,759 L
	Oct-80	1,452 L	1,528 L	1,710 L	1,790 L	1,867 L	1,943 L	2,276 L	2,367 L	2,456 L	2,544 L	2,632 L
	Jan-85	1,594	1,671	1,874	1,953	2,044	2,130	2,491	2,589	2,685	2,782	2,879
	Jan-90	1,606	1,686	1,886	1,972	2,056	2,145	2,513	2,604	2,705	2,800	2,897
	Jan-95	1,604	1,687	1,885	1,969	2,158	2,250	2,505	2,601	2,701	2,797	2,892
	Jan-96	1,596	1,676	1,875	1,960	2,147	2,237	2,491	2,591	2,687	2,783	2,879
	Jan-98	1,619	1,703	1,905	1,991	2,181	2,272	2,530	2,628	2,725	2,823	2,922
	Jan-99	1,656	1,739	1,944	2,036	2,229	2,320	2,588	2,688	2,788	2,888	2,986
	Jan-00	1,690	1,776	1,984	2,076	2,275	2,368	2,640	2,744	2,847	2,945	3,047
	Jan-01	1,541	1,822	2,001	2,093	2,291	2,390	2,661	2,764	2,867	2,970	3,071
	Jan-02	1,622	1,737	2,175	2,274	2,441	2,543	2,833	2,941	3,050	3,160	3,270
	Jan-03	1,698	1,811	1,916	2,324	2,495	2,600	2,897	3,010	3,120	3,232	3,343
	Jan-04	1,757	1,872	1,975	2,116	2,578	2,686	2,992	3,108	3,222	3,336	3,453
	Jan-05	1,802	1,916	2,016	2,157	2,286	2,690	2,996	3,111	3,226	3,342	3,457
	Jan-06	1,804	1,919	2,020	2,155	2,284	2,438	2,972	3,087	3,202	3,316	3,431
	Jan-07	1,838	1,954	2,056	2,194	2,325	2,482	2,605	3,070	3,183	3,298	3,412
	Jan-08	1,803	1,915	2,015	2,143	2,268	2,414	2,533	2,719	3,197	3,311	3,426
	Jan-09	1,850	1,966	2,067	2,197	2,321	2,467	2,582	2,770	2,939	3,277	3,390

NOTE: Pay scales are in effect for active duty personnel at time of retirement. See detailed footnotes on page following tables.

Pay Grade	Pay Scale Mo/Yr	Over 20 50%	Over 21 52 1/2%	Over 22 55%	Over 23 57 1/2%	Over 24 60%	Over 25 62 1/2%	Over 26 65%	Over 27 67 1/2%	Over 28 70%	Over 29 72 1/2%	Over 30 75%
COMMISSIONED OFFICERS												
E-6	Oct-72	1,627 N	1,709 N	1,792 N	1,875 N	1,952 N	2,038 N	2,116 N	2,203 N	2,283 N	2,367 N	2,450 N
	Oct-78	1,374 L	1,440 L	1,516 L	1,581 L	1,652 L	1,722 L	1,793 L	1,864 L	1,932 L	2,001 L	2,069 L
	Oct-79	1,338 L	1,406 L	1,478 L	1,544 L	1,612 L	1,677 L	1,748 L	1,815 L	1,884 L	1,946 L	2,018 L
	Oct-80	1,279 L	1,341 L	1,405 L	1,472 L	1,536 L	1,602 L	1,664 L	1,730 L	1,794 L	1,861 L	1,923 L
	Jan-85	1,393	1,465	1,530	1,604	1,672	1,747	1,815	1,886	1,952	2,027	2,100
	Jan-90	1,405	1,476	1,544	1,617	1,690	1,757	1,829	1,898	1,972	2,044	2,111
	Jan-95	1,403	1,476	1,543	1,615	1,688	1,755	1,827	1,896	1,968	2,040	2,109
	Jan-96	1,397	1,468	1,537	1,606	1,677	1,746	1,818	1,887	1,959	2,027	2,101
	Jan-98	1,419	1,487	1,561	1,632	1,704	1,775	1,846	1,918	1,987	2,058	2,130
	Jan-99	1,448	1,522	1,595	1,667	1,742	1,815	1,887	1,960	2,033	2,106	2,178
	Jan-00	1,480	1,554	1,630	1,704	1,778	1,852	1,925	2,001	2,074	2,149	2,223
	Jan-01	1,372	1,564	1,640	1,715	1,791	1,866	1,939	2,012	2,090	2,164	2,238
	Jan-02	1,436	1,518	1,767	1,847	1,926	2,008	2,088	2,171	2,249	2,331	2,413
	Jan-03	1,491	1,576	1,651	1,890	1,972	2,054	2,138	2,220	2,303	2,384	2,466
	Jan-04	1,530	1,617	1,694	1,771	2,011	2,095	2,180	2,264	2,347	2,431	2,515
	Jan-05	1,553	1,639	1,718	1,797	1,873	2,098	2,183	2,267	2,350	2,434	2,520
	Jan-06	1,547	1,632	1,709	1,785	1,865	1,942	2,167	2,249	2,332	2,416	2,500
	Jan-07	1,575	1,657	1,736	1,815	1,894	1,973	2,052	2,237	2,321	2,403	2,486
	Jan-08	1,546	1,627	1,704	1,782	1,860	1,936	2,014	2,091	2,383	2,468	2,553
	Jan-09	1,586	1,670	1,749	1,829	1,908	1,988	2,067	2,147	2,226	2,443	2,527
E-5	Oct-72	1,379 N	1,446 N	1,520 N	1,588 N	1,654 N	1,725 N	1,798 N	1,867 N	1,937 N	2,004 N	2,076 N
	Oct-78	1,165 L	1,224 L	1,284 L	1,341 L	1,400 L	1,458 L	1,520 L	1,578 L	1,636 L	1,700 L	1,754 L
	Oct-79	1,137 L	1,192 L	1,250 L	1,305 L	1,366 L	1,425 L	1,482 L	1,537 L	1,595 L	1,652 L	1,710 L
	Oct-80	1,086 L	1,138 L	1,192 L	1,246 L	1,299 L	1,355 L	1,413 L	1,469 L	1,521 L	1,577 L	1,629 L
	Jan-85	1,180	1,241	1,298	1,357	1,418	1,479	1,537	1,597	1,654	1,713	1,779
	Jan-90	1,190	1,250	1,308	1,370	1,432	1,488	1,547	1,612	1,668	1,730	1,791
	Jan-95	1,190	1,249	1,308	1,369	1,430	1,487	1,548	1,611	1,667	1,730	1,791
	Jan-96	1,182	1,243	1,301	1,364	1,425	1,481	1,540	1,601	1,661	1,719	1,778
	Jan-98	1,201	1,262	1,323	1,382	1,442	1,506	1,565	1,628	1,687	1,746	1,808
	Jan-99	1,230	1,292	1,353	1,416	1,478	1,539	1,600	1,662	1,722	1,783	1,847
	Jan-00	1,256	1,318	1,380	1,444	1,508	1,571	1,634	1,698	1,759	1,823	1,886
	Jan-01	1,169	1,327	1,389	1,452	1,517	1,578	1,643	1,707	1,771	1,832	1,894
	Jan-02	1,220	1,283	1,487	1,556	1,622	1,691	1,759	1,827	1,894	1,964	2,031
	Jan-03	1,265	1,329	1,392	1,593	1,662	1,733	1,801	1,870	1,940	2,009	2,080
	Jan-04	1,297	1,362	1,427	1,491	1,695	1,765	1,837	1,908	1,978	2,049	2,119
	Jan-05	1,315	1,381	1,446	1,512	1,578	1,768	1,839	1,910	1,981	2,051	2,123
	Jan-06	1,308	1,375	1,439	1,506	1,571	1,636	1,825	1,895	1,965	2,036	2,106
	Jan-07	1,330	1,396	1,464	1,529	1,596	1,663	1,729	1,885	1,955	2,026	2,094
	Jan-08	1,303	1,368	1,434	1,498	1,564	1,629	1,695	1,759	2,000	2,071	2,143
	Jan-09	1,336	1,403	1,469	1,536	1,603	1,670	1,737	1,804	1,870	2,050	2,121

NOTE: Pay scales are in effect for active duty personnel at time of retirement. See detailed footnotes on page following tables.

Pay Grade	Pay Scale Mo/Yr	Over 31 77 1/2%	Over 32 80%	Over 33 82 1/2%	Over 34 85%	Over 35 87 1/2%	Over 36 90%	Over 37 92 1/2%	Over 38 95%	Over 39 97 1/2%	Over 40 100%
COMMISSIONED OFFICERS											
0-8	Jan-07	9,727	10,041	10,355	10,668	10,983	11,297	11,610	11,924	12,238	12,552
	Jan-08	10,011	10,335	10,658	11,256	11,586	11,918	12,249	12,580	12,910	13,242
	Jan-09	9,907	10,227	10,547	11,138	11,466	11,793	12,121	12,449	12,776	13,104
0-7	Jan-07	8,629	8,907	9,185	9,463	9,742	10,020	10,298	10,576	10,855	11,133
	Jan-08	8,837	9,122	9,408	9,692	9,978	10,262	10,548	10,832	11,118	11,403
	Jan-09	8,745	9,027	9,309	9,591	9,873	10,156	10,438	10,720	11,002	11,284
0-6	Jan-07	7,578	7,822	8,067	8,311	8,556	8,801	9,045	9,289	9,533	9,778
	Jan-08	7,761	8,012	8,262	8,512	8,763	9,013	9,264	9,514	9,765	10,014
	Jan-09	7,680	7,928	8,176	8,424	8,671	8,919	9,167	9,415	9,663	9,910
0-5	Jan-07	6,184	6,382	6,581	6,782	6,981	7,180	7,380	7,580	7,779	7,979
	Jan-08	6,209	6,409	6,609	6,810	7,010	7,210	7,410	7,611	7,812	8,012
	Jan-09	6,144	6,342	6,541	6,739	6,937	7,135	7,334	7,532	7,730	7,928
W-5	Jan-07	5,408	5,584	5,758	5,933	6,106	6,282	6,456	6,630	6,805	6,980
	Jan-08	6,046	6,242	6,437	6,963	7,168	7,373	7,578	8,172	8,387	8,602
	Jan-09	5,984	6,177	6,370	6,891	7,094	7,296	7,499	8,087	8,300	8,513
W-4	Jan-07	4,980	5,141	5,301	5,462	5,623	5,783	5,944	6,104	6,266	6,426
	Jan-08	5,337	5,509	5,681	5,853	6,025	6,198	6,370	6,542	6,714	6,886
	Jan-09	5,281	5,452	5,622	5,793	5,963	6,133	6,304	6,474	6,645	6,815
E-9	Jan-07	4,623	4,772	4,921	5,069	5,219	5,369	5,518	5,667	5,815	5,965
	Jan-08	4,875	5,031	5,189	5,614	5,779	5,944	6,108	6,588	6,762	6,935
	Jan-09	4,824	4,979	5,135	5,555	5,719	5,882	6,045	6,519	6,691	6,863
E-8	Jan-07	3,945	4,072	4,200	4,327	4,454	4,582	4,710	4,836	4,964	5,091
	Jan-08	4,041	4,171	4,301	4,433	4,563	4,693	4,823	4,953	5,084	5,215
	Jan-09	3,999	4,128	4,257	4,386	4,515	4,644	4,773	4,902	5,031	5,160
E-7	Jan-07	3,525	3,639	3,752	3,866	3,980	4,094	4,207	4,320	4,436	4,549
	Jan-08	3,540	3,654	3,768	3,882	3,997	4,111	4,225	4,339	4,454	4,568
	Jan-09	3,503	3,616	3,729	3,842	3,955	4,068	4,181	4,294	4,407	4,521

NOTES: Figures assume current year retirees with less than 29 completed years of active duty service entered the military after September 7, 1980. Same is true for previous year retirees with less than 28 years; two year ago retirees with less than 27 years and so on. These retirees are subject to the high-three formula. Figures assume retiree spent the final three years at the same grade in which retired. Figures also assume retiree did not receive any increases during the last three years of active service other than annual pay increases, declined the Career Status Bonus, and retired mid-year (June).

Figures assume current year retirees with 29 or more completed years of active duty service entered the military before September 8, 1980 and their retired pay is based on the final pay formula.

Figures are based on the member serving the exact number of years in the column heading and final pay being the active duty pay for a member with that amount of years of service.

Because retirees received different increases in the last three years and did not receive the same partial COLAs and may have been affected by other various amendments, some individuals receive differing amounts than those shown in the tables.

Pay Scale indicates pay scale in effect for active duty personnel at time of retirement.

Figures shown assume that active service and pay service are equal.

Military retired pay reflects CPI increases where required prior to and including January 1, 2009. For members who retired on or after January 1, 2007, retired pay is computed without considering the active pay cap.

UNIFORMED SERVICES FORMER SPOUSES PROTECTION ACT (USFSPA)

The following is a general discussion of the legislative provisions of the USFSPA that is codified in Title 10, United States Code, Section 1408. This discussion is not designed to answer detailed questions concerning individual cases; rather, it serves as general information for use by Retirement Services Officers, Military Personnel Officers, active duty and retired servicemembers, spouses, and former spouses. It does not provide legal or judicial interpretation of enacted laws and does not deal with case law. Individuals seeking legal assistance are encouraged to contact military legal assistance or retain civilian counsel. **This explanation is not a legal brief nor does it outline a legal position. Do not use it as evidence of intent, interpretation, or precedent in any legal action.**

Background. Prior to the 1981 McCarty vs. McCarty case, state courts disagreed on whether they were authorized or constrained by federal legal precedent in dividing military retired pay in divorce-related property settlements. On June 26, 1981, the U.S. Supreme Court ruled in the McCarty vs. McCarty case that military retired pay could not be treated as community property in divorce cases. In response, Congress enacted the Uniformed Services Former Spouses Protection Act (USFSPA), which decreed that state courts could treat military retired pay as property in divorce cases if they so chose. The USFSPA was enacted by Public Law 97-252, September 8, 1982. It has subsequently been amended by Public Laws 98-94, September 24, 1983; 98-525, September 27, 1984; 99-145, November 8, 1985; 99-661, November 14, 1986; 100-180, December 4, 1987; 101-510, November 5, 1990; 102-484, October 23, 1992; and 104-201, September 23, 1996.

Division of Retired Pay as Property. The USFSPA allows state courts to consider military retired pay as divisible property in divorce settlements after June 25, 1981. It also establishes procedures by which a former spouse could receive all or a portion of the court settlement as a direct payment from the finance service center. Some state courts continued to divide property in divorce cases adjudicated prior to the June 1981 date, and the 101st Congress took action to correct this. The National Defense Authorization Act for Fiscal Year 1991, November 5, 1990, permitted cases so adjudicated to continue payments from November 1990 for two years and no longer. In addition, the Act provided that no other pre-June 25, 1981 cases could be reopened.

Understanding the distinction between the provision that authorizes a court to divide retired pay and the provision that allows direct payment of divided retired pay to a former spouse is essential. Specifically, the law does not direct state courts to divide retired pay; it permits them to. The law stipulates that only "disposable" pay may be divided. Disposable pay is defined as that pay (other than the percentage of pay) remaining after the following deductions:

- Debts owed the U.S. Government
- Federal, state, or local income taxes (for divorces prior to February 5, 1991 only)
- Amounts of retired pay waived for the receipt of VA disability compensation
- National Service Life Insurance (NSLI) premiums
- SBP costs for a former spouse receiving USFSPA payments.

The law does not confer an entitlement to a portion of the retired pay to a former spouse as a result of length of marriage or number of years overlap in the marriage and service. However, once a court has awarded a former spouse a portion of retired pay as property, the former spouse may apply to the finance service center to receive that pay as a direct payment. To qualify for a direct payment, the law requires a former spouse to have been married to the member during at least 10 years of the member's service that is creditable for retired pay.

The law further stipulates that finance service centers may not send more than 50 percent of the member's disposable retired pay as a direct payment unless there are additional garnishments for alimony or child support under Title 42, United States Code, Section 659. In such cases, up to 65 percent of the disposable pay may be sent to the former spouse as a direct payment.

In addition, the law does not allow a court to consider military retired pay in a divorce-related property settlement unless the court has jurisdiction over the military servicemember or retiree by reason of his or her: residence other than by reason of military assignment in the territorial jurisdiction of the court; domicile in the territorial jurisdiction of the court; or consent to the jurisdiction of the court.

For active duty members, the law also requires compliance with the provisions of the Servicemembers Civil Relief Act.

When more than one former spouse has been awarded a division of a servicemember's retired pay, payments will be handled on a first-come, first-served basis. When conflicting court orders exist, the law instructs the service concerned to send the amount specified in the lower of the two conflicting orders (not to exceed 50 percent of disposable pay), and retain the difference until the matter is resolved.

The finance center that services the member's account must begin direct payment to the former spouse within 90 days of receipt of a valid court order. If a servicemember is not retired at the time of the court order, payments must begin no later than 90 days after the servicemember retires. The law does not authorize a court to order a member to apply for retirement or to retire at a specified time in order to start payment.

Remarriage of a former spouse does not stop the direct payment of retired pay as property unless the court so orders.

A former spouse can apply for direct payment by sending to the finance service center concerned a signed DD Form 2293, *"Application for Former Spouse Payments from Retired Pay"* or a statement that includes:

Notice to make a direct payment to the former spouse from the servicemember's retired pay.

- A copy of the court order and other documents dividing retired pay as property. These must be certified by an official of the court that issued them.
- A statement from a lawyer that the court order has not been amended, superseded, or set aside.
- Sufficient identifying information about the member to process the application. The identification should give the member's full name, Social Security number, and uniformed service.
- An agreement by the former spouse that any future overpayments are recoverable and subject to involuntary collection from the former spouse or the estate of the former spouse.
- An agreement by the former spouse to promptly notify the service concerned if the court order upon which payment is based is vacated, modified, or set aside. This shall include notice of the former spouse's remarriage or notice of a change in eligibility for child support payment, (e.g., death,

emancipation, adoption, or attainment of majority) of a child whose support is provided by a former spouse through direct payments from retired pay.

A request for direct payment is not considered complete until the finance service center receives all the required information. Court orders, garnishments, applications, or notifications pertaining to the division of retired pay may be sent by FAX, email, or by mail, to the appropriate finance service center. (See list of addresses at the end of this section.) Within 30 days of receiving the completed DD Form 2293, the finance service center will contact the servicemember with a notification that includes:

- A copy of the court order and accompanying documentation.
- An explanation of the limitations affecting direct payment to a former spouse from a servicemember's retired pay.
- A request that the servicemember submit notification to the designated agent if the court order has been amended, superseded, or set aside. The servicemember must provide an authenticated or certified copy of the operative court documents when there are conflicting court orders.
- The amount or percentage that will be deducted if the member fails to respond to the notification.
- The tentative effective date of direct payments to the former spouse.
- Notice to the servicemember that failure to respond within 30 days of the date the notice was mailed may result in the division of retired pay as provided in the notification.
- Notice that if the servicemember submits information in response to the notification, the servicemember thereby consents to the disclosure of such information to the former spouse or the former spouse's agent.

The finance service center will not honor the court order if it is defective, modified, superseded, or set aside.

Payments to the former spouse will begin within 90 days after the finance service center receives the required documentation. Payments conform with the normal pay and disbursement cycle of the servicemember's retired pay. Payments that are a percentage of retired pay as property will change in direct proportion to and on the effective date of future cost-of-living adjustments to retired pay, unless the court order directs otherwise. Payments will stop if the servicemember dies, the former spouse dies, or as stated in the court order, whichever occurs first.

In 1996, the Act was modified to strengthen the provisions concerning jurisdiction in the case of court orders involving the modification of original (or subsequent) divorce decrees. Specifically, the Service Secretary concerned may not accept service of a court order that is an out-of-state modification, or comply with the provisions of such a court order, unless the court issuing that order has jurisdiction over both the military member and the spouse or former spouse.

A court order is considered to be an out-of-state modification if the order:

1. Modifies a previous order upon which payments are already being made; and
2. Is issued by a court of a state other than the court of the state that issued a previous court order.

The FY 85 DoD Authorization Act, Public Law 98-525, October 19, 1984, expanded medical, commissary, and base exchange privileges previously authorized by Public Law 97-252. Effective January 1, 1985, a former spouse married for 20 years to a person who had 20 years of service creditable toward retirement and who had 15 years overlap between the length of marriage and years of service (20/20/15 rule) will, under certain circumstances, be provided with military medical coverage. Being divorced after September 30, 1988, unremarried former spouses are eligible for medical benefits for one year from date of divorce, followed by the right to convert to the Continued Health Care Benefit Program (CHCBP). The program is intended to provide benefits similar to TRICARE for a specific period of time (18-36 months) to former servicemembers and their family members, some unremarried former military spouses, and emancipated children, who enroll and pay quarterly premiums. Benefits are similar to those in TRICARE Standard.

If a former spouse is no longer eligible for an ID card, the retiree sponsor must notify the Defense Enrollment Eligibility Reporting System (DEERS) at 1-800-538-9552. A retiree who fails to notify DEERS that a former spouse is no longer eligible for care may be liable for the former spouse's medical care costs paid by TRICARE during a period of ineligibility. Medical benefits for an unremarried former spouse whose marriage covered at least 20 years of military service will be granted regardless of the date of divorce, providing the spouse is not covered by employer-sponsored medical insurance. Commissary and exchange privileges may be reinstated to a 20/20/20 former spouse whose remarriage ends in death, divorce, or annulment. Medical benefits are not restored. For further information, contact any ID card issuing facility.

Retirees may also name a former spouse as a beneficiary under the Survivor Benefit Plan (SBP) at the time of his or her retirement, even if he or she has remarried. Retirees who have SBP coverage for a spouse, and later divorce, have one year from the date of the divorce to elect former spouse coverage. For divorces that occurred prior to November 14, 1986, coverage for a former spouse could not be court ordered, but must have been voluntarily elected by the member. However, former spouse coverage can be included as a part of the final divorce decree with the consent of the member. If the member fails to make that election, the former spouse has one year from the date of the court order to request a deemed SBP election. Public Law 99-661 permits state courts to order a member to provide SBP to a former spouse (for divorces finalized on or after November 14, 1986). Only one SBP coverage can be elected. Therefore, if a former spouse is covered by SBP, the member's current or later acquired spouse cannot be covered.

The National Defense Authorization Act for Fiscal Year 1991, Public Law 101-510, November 5, 1990, reemphasized that disability pay is not part of disposable retired pay. It also clarified payment of taxes on court divided retired pay and requires finance service centers to issue two IRS Form 1099, one to each recipient which requires both to pay taxes on their respective shares.

Servicemembers, regardless of whether still on active duty or retired, may also be required by a court to provide coverage for a former spouse under the Survivor Benefit Plan (SBP). There is no length-of-marriage service requirement as there is with the division of retired pay that must be satisfied before such a court order is enforceable. If a member becomes divorced while on active duty and is required by court order to elect former spouse SBP coverage upon retirement, the member must make the election (or be held in contempt of court). However, if the member does not make the election, and the former spouse did not request, in writing, within one year of the date of the court order, that the election be deemed on the member's behalf, the finance center will not enforce the court ordered election. (See the SBP section for more details.)

All correspondence between a former spouse and the finance service center must include the servicemember's Social Security number. Section (h) of the USFSPA, added in 1992 by Section 653 of Public Law 102-484, provides that the spouse or former spouse is eligible to receive payments of the retired pay of a military member whose entitlement to retired pay is terminated as a result of misconduct involving abuse of a spouse or dependent child while a member. Payments are made by the Government and continue as long as the former spouse remains unmarried. This means that, within the same Act, an apparent dichotomy exists in that payments (made by the Government) to an abused spouse cease upon remarriage while payments (made by the military member) to an ex-spouse who was not abused continue upon remarriage.

In 1996, the law was amended to provide for the protection of former spouses awarded military retirement pay under provisions of Title 10, United States Code in cases where the military member accepts civil service employment after retirement from the military. In such cases, military retirement pay may be waived in order that the military service for which retirement pay is being received may be credited for purposes of civil service retirement.

This modification to Title 5, United States Code specifies that a condition of acceptance of civil service employment is an agreement that the Director of the Office of Personnel Management may deduct and withhold from the annuity payable to the former military member an amount equal to that awarded under Title 10, United States Code for payment to the former spouse. This modification is applicable to all military members receiving retired pay who accept civil service employment on or after January 1, 1997.

MILITARY RETIREMENT FOR RESERVE COMPONENT MEMBERS

One of the many benefits of serving in the National Guard or Reserve is the opportunity to qualify for a nonregular (or Reserve) retirement. The basic requirements to qualify for a Reserve retirement are: (1) attaining 20 qualifying years of service (a qualifying year of service is credited for each year in which the member has earned at least 50 points during his or her anniversary year) and (2) serving the last six years in a Reserve Component. **Note:** The six year requirement is repealed for members who attain 20 creditable years of service on or after April 25, 2005.

RESERVE RETIREMENT SYSTEM

Reserve officers, warrant officers and enlisted members who complete a minimum of 20 years of satisfactory service (creditable retirement years), of which at least six years are as a member of a Reserve Component, become entitled to receive Reserve retired pay at age 60 (or a lower age if eligible). The six year requirement is repealed for members who attain 20 creditable years of service on or after April 25, 2005. Although entitled, the member must submit an application, in accordance with service regulations, prior to receiving such pay.

While entitlement to receive retired pay will normally begin at age 60, a member may begin receiving Reserve Component retired pay three months early for each aggregate 90 days of qualifying active service—performed after January 28, 2008—during a fiscal year period. Qualifying active service is: service in support of a contingency operation; voluntary active duty under 10 USC 12301(d)—but not active Guard and Reserve duty; and full time National Guard duty under 32 USC 502(f) authorized by the President or Secretary of

Defense for the purpose of responding to a national emergency declared by the President or supported by federal funds.

The eligibility age can not go below age 50 and, regardless of the age that a member begins receiving retired pay, health care benefits are not authorized until the member reaches age 60.

A retirement year is defined as the 365 or 366 day period starting when the member first joins the military. Example: the member first joins the military on July 17. The retirement year is from July 17 to July 16 the following year. The retirement year does not change unless the member has a break in service.

A creditable retirement year, under this system, is a year in which the Reservist earns at least 50 retirement points during their retirement year. Inactive point credit is earned for inactive duty training, Reserve membership, equivalent instruction, and correspondence courses. By law, members may receive credit for up to 60 inactive points for retirement years that ended before September 23, 1996, up to 75 inactive points for retirement years ending on or after September 23, 1996 and before October 30, 2000, up to 90 points in the retirement year ending on or after October 30, 2000 and before October 30, 2007, and up to 130 points in the retirement year that includes October 30, 2007 and in any subsequent year of service. Points from these sources may be added to points earned from active duty and active duty for training for a maximum total of 365 or 366 points per retirement year. Points are credited on the following basis:

- One point for each day of active service (active duty or active duty for training).
- 15 points for each year of membership in a Reserve Component.
- One point for each period of inactive duty training.
- One point for each day of full time State active duty performed by a National Guardmember on or after September 11, 2001 and before October 1, 2002 in the following counties to support a Federal declaration of emergency following the terrorist attacks on the United States on September 11, 2001: In New York: Bronx; Kings; New York (boroughs of Brooklyn and Manhattan); Queens; Richmond; Delaware; Dutchess; Nassau; Orange; Putnam; Rockland; Suffolk; Sullivan; Ulster; and Westchester. In New Jersey: Bergen, Hudson, Union, and Middlesex. In Virginia: Arlington.
- One point for each day in which a member is in a funeral honors duty status.
- Satisfactory completion of accredited correspondence courses at one point for each three credit hours earned.

The Secretary of the military department concerned (Secretary of Homeland Security for the Coast Guard) notifies, in writing, members of the Reserve Forces who have completed the eligibility requirements for retirement and receipt of retired pay at age 60 (or earlier age, if eligible). Notice is sent to the member within one year of reaching eligibility. Reserve Component members generally have three options upon receiving notice of eligibility:

1. Remain in the Ready Reserve and continue to perform inactive duty training, annual training and active duty for training depending on their training and pay category, or remain on the active status list of the Standby Reserve and continue to perform unpaid training for the purpose of accumulating retirement points.
2. Transfer to the Retired Reserve.
3. Request discharge from the Reserve Components. Regardless of the option chosen, the member is entitled to receive retired pay at age 60 (or earlier age, if eligible), but must apply for it.

STATUTORY REQUIREMENTS FOR MANDATORY RETIREMENT OR DISCHARGE FOR OFFICERS IN A RESERVE COMPONENT

Reserve Grade	Age	Service or Failure of Selection for Promotion
General (Chief of the National Guard Bureau)	(See notes 2 and 3)	30 days after completion of 40 years of commissioned service or on the first day of the first month after the fifth anniversary of the date of the officer's appointment, whichever is later, the officer shall be separated by the provisions of 10 USC 14514 (See note 2) (10 USC 14508 (d))
Lieutenant General (Navy Vice Admiral)†‡	The last day of the month in which the officer becomes 64 years of age, the officer shall be separated by the provisions of 10 USC 14515 (See notes 2 and 3) (10 USC 14511)	30 days after completion of 38 years of commissioned service or on the fifth anniversary of the date of the officer's appointment, whichever is later, the officer shall be separated by the provisions of 10 USC 14514 (See notes 2 and 4) (10 USC 14508 (c))
Major General (Navy Rear Admiral - Upper)†‡	The last day of the month in which the officer becomes 64 years of age, the officer shall be separated by the provisions of 10 USC 14515 (See note 2) (10 USC 14511)	30 days after completion of 35 years of commissioned service or on the fifth anniversary of the date of the officer's appointment, whichever is later, the officer shall be separated by the provisions of 10 USC 14514 (See notes 2 and 4) (10 USC 14508(b))
Brigadier General (Navy Rear Admiral - Lower) (Not on Promotion List)†‡	The last day of the month in which the officer becomes 62 years of age, the officer shall be separated by the provisions of 10 USC 14515 (See note 2) (10 USC 14510)	30 days after completion of 30 years of commissioned service or on the fifth anniversary of the date of the officer's appointment, whichever is later, the officer shall be separated by the provisions of 10 USC 14514 (See notes 2 and 4) (10 USC 14508(a))
Colonel (Navy Captain) (Not on Promotion List)	The last day of the month in which the officer becomes 62 years of age, the officer shall be separated by the provisions of 10 USC 14515 (See note 2) (10 USC 14509)	On the first day of the month after the month in which the officer completes 30 years of commissioned service, the officer shall be separated by the provisions of 10 USC 14514 (See notes 2 and 4) (10 USC 14507(b)
Lieutenant Colonel (Navy Commander) (Not on Promotion List)	The last day of the month in which the officer becomes 62 years of age, the officer shall be separated by the provisions of 10 USC 14515 (See note 2) (10 USC 14509)	On the first day of the month after the month in which the officer completes 28 years of commissioned service, the officer shall be separated by the provisions of 10 USC 14514 (See notes 2 and 4) (10 USC 14507(a))
Major (Navy Lieutenant Commander)	The last day of the month in which the officer becomes 62 years of age, the officer shall be separated by the provisions of 10 USC 14515 (See note 2) (10 USC 14509)	Either on the first day of the month after the month in which the officer completes 20 years of commissioned service or NLT the first day of the seventh month after the month in which the President approves the report of the board which considered the officer for the second time, whichever is later, the officer shall be separated by the provisions of 10 USC 14513 (See notes 1 and 2) (10 USC 14506)
Captain (Navy Lieutenant)	The last day of the month in which the officer becomes 62 years of age, the officer shall be separated by the provisions of 10 USC 14515 (See note 2) (10 USC 14509)	NLT the first day of the seventh month after the month in which the President approves the report of the board which considered the officer for the second time, the officer shall be separated by the provisions of 10 USC 14513 (See notes 1 and 2) (10 USC 14505)
First Lieutenant (Navy Lieutenant Junior Grade)	The last day of the month in which the officer becomes 62 years of age, the officer shall be separated by the provisions of 10 USC 14515 (See note 2) (10 USC 14509)	NLT the first day of the seventh month after the month in which the President approves the report of the board which considered the officer for the second time (10 USC 14504)
Reserve chaplains and officers in medical specialties	The following officers may be retained in an active Reserve status until the date on which the officer becomes 68 years of age: Army: Chaplain, Medical Corps, Dental Corps, Veterinary Corps, allied health officer, biomedical sciences officer, optometrist, nurse, or the Medical Specialists Corps; Air Force: Chaplain, medical officer, dental officer, Air Force Nurse, Medical Service Corps Officer, biomedical sciences officer; Navy: Chaplain, Medical Corps, Dental Corps, Nurse Corps, and officers designated to perform as a veterinarian, optometrist, podiatrist, allied health officer, or biomedical sciences officer (10 USC 14703)	

NOTES:

†A Reserve officer of the Army or Air Force who is serving as any of the following: Chief of the Army Reserve; Chief of the Air Force Reserve; Director of the Army National Guard; Director of the Air National Guard; an adjutant general, or if a Reserve officer of the Army, commanding general of the troops of a state, shall on the last day of the month in which the officer becomes 66 years of age, be separated in accordance with 10 USC 14512.

‡ The Secretary of the Navy may defer the retirement of an officer above the grade of Colonel (Navy Captain) and retain the officer in an active status until the officer becomes 66 years of age. Not more than 10 officers may be so deferred at one time.

1. Transferred to an inactive status if the Secretary concerned determines that the officer has skills which may be required to meet the mobilization needs of the officer's Service.
2. Transferred to the Retired Reserve, if the officer is qualified and applies for such transfer; or discharged from the officer's Reserve appointment.
3. Mandatory retirement may be deferred: by the President to age 68; and by the Secretary of Defense to age 66.
4. The Secretary of the Army or Air Force may retain a Brigadier General until age 62; Major General until age 64; and a Lieutenant General until age 66.

There are two Reserve retirement systems that parallel the systems for active duty: the Final Basic Pay System and the High-three System. To determine which retirement system a Reserve Component member is under, look at the same criteria that determine the retirement system for the active force—the Uniformed Service Initial Entry Calendar Date (DIEUS). That is the date an individual first became a member of a uniformed service. There is one Reserve retirement system for members with a DIEUS before September 8, 1980 and one Reserve retirement system for those individuals with a DIEUS of September 8, 1980 and later.

When looking at which retirement system applies to a specific individual it is important to note that a DIEUS date will never change. Even a break in service will not affect a DIEUS date. The date an individual first became a member of a uniformed service is the sole determining factor in determining which retirement system is used when computing retired pay.

A member who retires under either system receives longevity credit for those years while a member of the Retired Reserve awaiting pay at age 60 (or a lower age if eligible). However, this does not apply to a former member who is entitled to retired pay under either the Final Basic Pay System or the High-three System. A former member is defined as an individual who elected discharge rather than transfer to the Retired Reserve anytime after receiving notification of eligibility to receive Reserve retired pay at age 60 (or a lower age if eligible). In the case of a former member, regardless of the system under which the individual will receive Reserve retired pay, longevity credit ceases on the date the former member was discharged.

Final Basic Pay System (Reserve). The first system, the Final Basic Pay System, is used for members with a DIEUS date before September 8, 1980. Reserve retired pay is computed using the following method:

1. Dividing all points credited in all years by 360 to convert the points into years of service, including fractions.
2. Using the pay table in effect on the date that the member or former member reaches 60 years of age (or a lower age if eligible), taking the monthly basic pay rate for the member's highest grade satisfactorily held and length of service at the time the member reaches age 60 (or a lower age if eligible); or in the case of a former member, the length of service as determined when the former member was discharged.
3. Multiplying that rate by 2.5 percent of the years of service computed in (1) above.

High-three System (Reserve). The second system, the High-three Retirement System, is used for members with a DIEUS date of September 8, 1980 or later. Within this second system, there are two ways to determine Reserve retired pay. The first computation is for members who, at any point after receiving notification of eligibility to receive retired pay at age 60 (or a lower age if eligible), transfer to the Retired Reserve while awaiting age 60 (or a lower age if eligible) (Retired Reserve Computation). The second computation is for members who choose to separate from the Reserve Components after receiving notification of eligibility to receive retired pay at age 60 (or a lower age if eligible) (Former Member Computation).

Retired Reserve Computation. The retired pay for a member who elects transfer to the Retired Reserve is computed using the average monthly basic pay rate to which the member would have been entitled at the highest grade satisfactorily held, had the member been on active duty during the entire period of the member's high 36 months. In most cases, this would be the average basic pay rate in effect for the member at ages 57, 58, and 59. The formula to compute the retired pay is as follows:

1. Divide all points credited in all years by 360 to convert the points into years of service, including fractions.
2. Take the average of the last 36 months of monthly basic pay for the member's grade and length of service at the time the member becomes entitled to receive retired pay—age 60 (or a lower age if eligible) (remember to include longevity increases while a member of the Retired Reserve).
3. Multiply that rate by 2.5 percent of the years of service computed in (1) above.

Former Member Computation. An individual with a DIEUS date of September 8, 1980 or later, who elected discharge rather than transfer to the Retired Reserve, will have retired pay computed under the High-three System using the average of the monthly basic pay in effect for his or her 36 months of service immediately preceding the date of discharge, or the highest 36 months of basic pay to which entitled based on a higher grade satisfactorily held while a member of the uniformed services. The formula to compute the retired pay is as follows:

1. Divide all points credited in all years by 360 to convert the points into years of service, including fractions.
2. Take the average of the monthly basic pay in effect for the 36 months immediately preceding the date of discharge, or the highest 36 months of basic pay to which entitled while a member of the uniformed services.
3. Multiply that rate by 2.5 percent of the years of service computed in (1) above.

Members receiving Reserve retired pay, regardless of the system under which retired pay is received, are entitled to annual Cost of Living Adjustments as determined by the increase in the Consumer Price Index. This is the same system that is used for retirees from the active forces under the Final Basic Pay and High-three systems.

The choice to elect discharge from the Reserve Components is irrevocable. That choice could cost several thousands of dollars in lost retired pay. Therefore, members with a DIEUS date prior to September 8, 1980 must consider the loss of longevity increases by electing discharge (if applicable). In addition, members with a DIEUS date of September 8, 1980 or later must be certain to elect the choice of discharge only after carefully considering the loss of both longevity increases and the increased multiplier (due to increases in active duty basic pay while awaiting age 60, or lower age, if eligible) used in determining Reserve retired pay for members who elect transfer to the Retired Reserve.

The following illustration compares two 45-year-old Reservists. They both have the same rank and years of service for pay and receive notice of eligibility for retired pay at age 60. Both have 4,000 retirement points. One elects discharge and the other elects transfer to the Retired Reserve. The member who elected discharge is completely separated from the military and cannot be recalled to active duty, even in the event of a full mobilization.

The member that is transferred to the Retired Reserve becomes a "Gray Area" retiree. The time spent in the Retired Reserve is counted for longevity pay purposes. The following are examples, using the January 2009 pay scale, of how choosing one option over the other affects retired pay:

Example 1: *E-7 with 22 years of service for pay at time of discharge or transfer to retired Reserve:*

Discharged from Reserves.

Based on the pay scale in effect on January 1, 2009, monthly retired pay at age 60 for an E-7 over 22 years of service is $1,073.61

Transferred to Retired Reserve.

Based on the pay scale in effect on January 1, 2024, monthly retired pay at age 60 for an E-7 over 26 years of service for pay is $1,896.81

Example 2: *W-3 with 20 years of service for pay at time of discharge or transfer to retired Reserve:*

Discharged from Reserves.

Based on the pay scale in effect on January 1, 2009, monthly retired pay at age 60 for a W-3 over 20 years of service for pay is $1,379.67

Transferred to Retired Reserve.

Based on the pay scale in effect on January 1, 2024, monthly retired pay at age 60 for a W-3 over 26 years of service for pay is $2,458.94

Example 3: *O-5 with 22 years of service for pay at time of discharge or transfer to retired Reserve:*

Discharged from Reserves.

Based on the pay scale in effect on January 1, 2009, monthly retired pay at age 60 for an O-5 over 22 years of service for pay is $2,062.58

Transferred to Retired Reserve.

Based on the pay scale in effect on January 1, 2024, monthly retired pay at age 60 for an O-5 over 22 years of service for pay is $3,326.54

Note: The previous illustrations assume: both members have a DIEUS of September 8, 1980 or later; reached age 45 on January 1, 2009; discharge or transfer to the retired Reserve both occurred on January 1, 2009; and basic pay increased 3.2 percent in each year from 2010 through 2024.

Retirement Eligibility - Special Provisions

Completion of 18 or more, but less than 20 years of service. The provisions of 10 USC, Sections 12646 and 1176 provide that a Reserve commissioned officer or Reserve enlisted member who has completed 18 or more, but less than 20 years of creditable Reserve service toward eligibility for Reserve retired pay may not be discharged or transferred from an active Reserve status (and thereby lose their eligibility for retirement due to having less than 20 years of creditable service) until the earlier of the following dates:

1. The date on which he or she is entitled to be credited with 20 years of creditable Reserve service, or
2. The *third* anniversary of the date on which he or she would otherwise be discharged or transferred from active Reserve status (if he or she has at least 18, but less than 19 years of service), or
3. The *second* anniversary of the date on which he or she would otherwise be discharged from active service (if he or she has at least 19, but less than 20 years of service).

This provision of law protecting a member's retirement eligibility does not apply to members discharged from active status because of disability or age.

Members with physical disabilities not incurred in the line of duty. Title 10, United States Code, Section 12731b provides that the Secretary concerned may treat a member, with more than 15, but less than 20 years of creditable service towards a Reserve retirement, as having completed the requirements for a Reserve retirement. This applies in the case of a member who is unfit for continued service solely due to a physical disability. The physical disability may not be the result of willful neglect, or during a period of unauthorized absence.

COMPUTATION OF RESERVE RETIREMENT PAY

Method of Computation

Retirement pay under the Reserve system is computed by totaling all points credited in all years, whether or not 50 points were earned for a creditable year for retirement eligibility, dividing by 360 (one year for computation purposes) and then multiplying by 2.5 percent to determine the benefit multiplier. For members electing transfer to the Retired Reserve, the multiplier is applied to the basic pay rate in effect for the grade and longevity held when retired pay starts. The longevity includes all years in military service, whether active duty, Reserve, or Retired Reserve status, but not while a former member waiting age 60 (or lower age if eligible), and is not limited to the years of creditable service for retirement purposes. In the case of a member electing discharge from the Reserves instead of transfer to the Retired Reserve, where the member has a DIEUS of September 8, 1980 or later, the multiplier is applied to the high three years of pay in effect for the grade(s) and longevity accrued at the time of discharge.

RESERVE RETIRED PAY FORMULA

If your DIEUS is:	MONTHLY RETIRED PAY BASED ON:	Annually increased by COLA equal to:
Before September 8, 1980	(Years of Satisfactory service) X (2.5%) X (Basic Pay in Effect When Retired Pay Starts) = Monthly Retired Pay Members who separate before becoming entitled to receive retired pay at age 60 (or lower age, if eligible), have longevity computed at time of separation. Members who transfer to the Retired Reserve gain longevity while in the Retired Reserve.	Annual Change in Consumer Price Index (CPI)
On or after September 8, 1980	(Years of Satisfactory service) X (2.5%) X (Average of Highest 36 Months of Basic Pay) = Monthly Retired Pay Members who separate before becoming entitled to receive retired pay at age 60 (or lower age, if eligible), have high 36 computed at time of separation. Members who transfer to the Retired Reserve until age 60 have high 36 computed at age 60.	Annual Change in Consumer Price Index (CPI)

For example, a retiring Reserve E-7 who entered the service before September 8, 1980 has a total of 2,800 retirement points after 22 years of service. Dividing 2,800 by 360 equals 7.777 rounded to 7.78 years of service for Reserve retired pay computation purposes. Then, 7.78 is multiplied by 2.5 percent (.025) to arrive at a figure of .1945 rounded to .195, or 19.5 percent. Monthly basic pay (as of January 1, 2009) of an active duty E-7 with over 22 years of service is $4,142.10. Multiply $4,142.10 by 19.5 percent, or 0.195, and the result is the approximate amount of retired pay the Reservist will receive each month upon reaching age 60—$807.71 in 2009.

For a Reserve member, E-7 with 2,800 retirement points, who entered the service on or after September 8, 1980, the base pay multiplier is computed using the average of the high 36 months of active duty basic pay. Assuming no promotions within the last three years and a retirement date of January 1, 2009 the multiplier is $3,865.38. The result is the amount of monthly retired pay upon reaching age 60—$753.75 in 2009.

The Reserve Retired Pay Formula in this section provides a means for determining the amount of retired pay based on grade, years of service, accumulated retirement points and the current value for each point.

NOTICE OF ELIGIBILITY

Ordinarily members will receive an official notice of eligibility within one year after completion of the required number of qualifying years of service, including information on the Reserve Component Survivor Benefit Plan and the necessary option election forms. Members should be sure that they have sufficient qualifying service before discontinuing active participation to ensure that they will be eligible for retired pay.

APPLICATION FOR RETIRED PAY

In order to actually start receiving retired pay, the member must submit an application for retired pay to the applicable center. Normally an application package is sent to all eligible Reservists approximately four to 10 months prior to becoming eligible to receive retired pay. If, however, this package is not received, it is incumbent upon the Reservist to request the forms and submit them as soon as possible prior to their eligibility date. The following provides information for each of the services.

Army: Upon reaching age 59, retirement-eligible members complete DD Form 108, *"Application for Retired Pay Benefits."* If in an active status, submit through channels, otherwise mail the completed form to: Commander, Human Resources Command – STL, ATTN: AHRC-PAP-T, 1 Reserve Way, St. Louis, MO 63132-5200. Army Reserve and National Guard members may call toll-free 1-800-318-5298 for more information or visit www.hrc.army.mil.

Navy: A retired pay packet is forwarded approximately 10 months in advance to assist members in submitting an application for retired pay. If the packet is not received eight months prior to becoming eligible to begin receiving retired pay, the member should contact the Navy Personnel Command (PERS 912), 5720 Integrity Dr., Millington, TN 38055 or by calling 1-866-827-5672 or at www.npc.navy.mil.

Air Force: Members should apply for retirement pay no sooner than 120 days prior to becoming eligible to begin receiving retired pay. Approximately four months prior to becoming entitled to begin receiving retired pay, HQ ARPC automatically forwards all Air Force Reserve personnel, Air National Guard members, and those in the Retired Reserve the forms necessary to apply for retired pay. Return the completed forms to: HQ ARPC/DPPR, 6760 E. Irvington Pl., Ste. 4000, Denver, CO 80280-4000. Air Force Reserve and Air National Guard members may call 1-800-525-0102. More information is available at: http://arpc.afrc.af.mil.

Marine Corps: Four to six months prior to reaching eligibility to begin receiving retired pay, the member should contact the Marine Corps Separation and Retirement branch at 1-800-336-4649 to verify the member's address. Four months prior to the member's 60th birthday, the Marine Corps will send the member a DD Form 108, *"Application for Retired Pay Benefits."* Return the completed form to: Commandant, Marine Corps, Headquarters Marine Corps, 3280 Russell Rd. (MMSR-5), Quantico, VA 22134-5103. Additional information is available at www.usmc.mil.

Coast Guard: Approximately six months prior to becoming eligible to begin receiving retired pay, the CG Personnel Service Center will mail a Coast Guard Retired Pay Account Worksheet and Survivor Benefit Plan Election Form (CG-PSC 4700), and USCG Reserve Retirement Point Statement (CG4175A) to the member. Members can also access the form and guide on the HRSIC website at www.uscg.mil/hq/psc. The forms should be completed and returned to Commanding Officer/RAS, U.S. Coast Guard Personnel Service Center, 444 SE Quincy St., Topeka, KS 66683-3591 as soon as possible. For more information, call the Coast Guard Personnel Service Center at 1-800-772-8724.

OTHER PAY INFORMATION

Retired pay normally begins on the retiree's 60th birthday. However, in some cases, the age may be lower than age 60. Even if application is made after the initial eligibility date, pay is retroactive. There is a six-year statute of limitations. If the retired pay application is filed more than six years after the initial eligibility date, one day's retired pay is deducted for each day's delay.

Retired pay is subject to federal income taxation. The Defense Finance and Accounting Service will withhold a certain amount each month based on the amount of pay and exemptions claimed on IRS Form W-4. Each January retirees will receive a 1099-R showing how much was withheld and the amount of retired pay that is taxable.

MISCELLANEOUS

After a person has been granted retired pay or been notified of his or her eligibility to receive retired pay at age 60 (or earlier age, if eligible), his or her eligibility may not be denied or revoked because of any error, misinformation, or administrative determination of years of service performed, unless it resulted directly from fraud or misrepresentation by the individual concerned. However, the amount of retired pay may be adjusted to correct any such error, miscalculation, misinformation, or administrative determination, and when such a correction is made, the Reserve retiree is eligible for retired pay computed on the basis of the corrected number of creditable years of service.

Reserve retirees receive annual cost-of-living increases in their retired pay computed on an identical basis as active duty retirees.

Retired Reserve - Grade in Which Retired

Upon retirement, a member shall be placed on the retired list in the highest grade in which he or she has satisfactorily served, as determined by the Secretary of the military department concerned, or in the highest grade for which he or she is eligible according to law.

BENEFITS AND ENTITLEMENTS FOR RETIRED MEMBERS OF THE RESERVES

Reserve Component members are eligible for many federal benefits and entitlements. These federal benefits can usually be divided into four major areas: benefits for inactive duty for training (drill); active duty and active duty for training; Retired Reserve while in the gray area; and Retired receiving retired pay.

Benefits available to members while in a drilling status include: full-time Servicemember's Group Life Insurance (SGLI); unlimited exchange and commissary privileges; limited medical care for injuries incurred during, or while traveling to and from drill; availability of TRICARE Reserve Select health care coverage; access to military clothing sales stores, military theaters, and transient billets if available; Space-A travel (within CONUS, Alaska, Hawaii, Guam, the Virgin Islands, American Samoa and Puerto Rico); and survivor benefits.

Members serving on active duty or active duty for training usually receive the same benefits as active duty servicemembers.

A Reserve Component member, who completes at least 20 years of service creditable toward a Reserve Retirement and is transferred to the Retired Reserve, loses some benefits. This includes eligibility for promotion, the opportunity to earn retirement points; drill pay and allowances; and SGLI eligibility.

Benefits retained by members of the Retired Reserve while in the gray area include: the use of all MWR facilities; unlimited commissary and exchange privileges; access to officer and NCO clubs (local rules permitting); Space-A travel within and between CONUS, Alaska, Hawaii, Guam, the Virgin Islands, American Samoa and Puerto Rico; the option to enroll in Veterans' Group Life Insurance; retention of commissions (officers only); and the opportunity to wear the military uniform at appropriate celebrations and military activities. Members of the Retired Reserve carry a Retired Reserve (red) ID card while in the gray area.

Upon reaching age 60 (or a lower age if eligible), Reservists receive the Retired (blue) ID card and begin receiving monthly retired pay, although they must apply for it. However, regardless of the age when a member is eligible to begin receiving retired pay, eligibility for TRICARE healthcare benefits for themselves and their eligible family members begins at age 60. Additional retirement benefits granted upon reaching age 60 (or lower age if eligible), include worldwide Space-A travel, including eligible family members.

Reservists receiving a pension or other compensation from the Veterans Administration may waive their Reserve retired pay in the amount of the VA pension or compensation, which will keep the compensation or pension (which is exempt from federal income tax) in effect. They may also be eligible for Special Compensation Programs which provide for receipt of both military retirement pay and VA disability compensation.

INCOME TAX INFORMATION FOR RETIRED MILITARY PERSONNEL

Military retirees may be entitled to certain retirement income tax advantages. These are: disability retirement pay exclusion—in part or whole; application of disability income exclusion to a portion of military retirement pay; retirement income tax credits; and Department of Veterans Affairs compensation or pension.

Military retired pay is not exempt from federal income taxes. Disability retirement pay that is computed on the basis of the percentage of disability may be excludable from gross income, but disability retirement pay that is computed based on years of service is excludable only to the extent allowed under the percentage of disability method.

VA Disability Compensation or Pension. VA disability compensation or pension payments are tax exempt. Disability compensation from the VA may include an additional amount payable on behalf of dependents for veterans with a disability rating of 30 percent or higher. Military retirees who do not receive their disability benefits from the VA, may exclude an amount equivalent to the VA benefits to which they are entitled. The percentage of degree of disability awarded by the VA does not exempt the same percentage of military retired pay from taxes as a military disability rating. A military disability rating of 50 percent would exclude 50 percent of the military retired pay. A 50 percent VA disability rating would provide an amount established annually for VA compensation.

Public Law 110-245, June 17, 2008, The "*Heroes Earnings Assistance and Relief Tax Act of 2008*" added a special rule to the IRS Tax Code for military retirees whose retired pay is computed under Section 1406 or 1407 of 10 USC and who receives a delayed disability compensation rating from the VA.

A member who claimed as income, and paid taxes on an amount of military retired pay that was later subject to offset due to a delayed VA disability compensation rating, may file an amended tax return to claim a credit or refund on the overpayment in taxes without being subject to the standard three year limitation on amending a tax return.

The special rule allows a member to amend any tax return that is not more than five taxable years prior to the VA disability rating determination. Such amended returns must be filed within one year of the VA disability rating determination.

SPECIAL ITEMS OF INTEREST

Qualified Military Benefits. Section 134 of the Internal Revenue Code excludes from taxation many military benefits (such as housing and subsistence allowances) received by members for performing active or Reserve duty. Some retirees who teach Junior ROTC for local school districts have their salary based on the full military pay and allowances they received while performing active duty. Because the retiree receives the salary from the local school district for teaching and not from the Federal Government, no portion of the salary constitutes a qualified military benefit. Consequently, the retiree/teacher may not rely on Section 134 to exclude any portion of the retiree's teaching salary from taxation.

Tax Forgiveness. The tax law provides for the forgiveness of federal income tax for military members who die as a result of wounds or injuries incurred in a combat zone.

Credit for the Elderly or the Permanently and Totally Disabled. A tax credit for the elderly or the permanently and totally disabled applies to citizens or residents who are: (1) 65 years of age before the close of the tax year; or (2) under age 65, are retired on disability, were permanently and totally disabled when they retired, and on January 1 of the applicable tax year had not reached mandatory retirement age. The IRS defines permanent and total disability as, "being unable to engage in any substantial gainful activity because of your physical or mental condition." A physician must certify that the condition has lasted or can be expected to last continuously for 12 months or more, or that the condition can be expected to result in death. Professional counsel from a legal assistance office or tax expert is recommended for those who feel they may qualify. For individuals 65 or older, the credit is at most 15 percent of an initial amount provided by the Internal Revenue laws. The initial amount is reduced by the taxpayer's excess gross income and by the taxpayer's nontaxable pensions and benefits. The initial amount varies with filing status, as follows: Single/$5,000; Married (joint return) one spouse qualified/$5,000; Married (joint return) both spouses qualified/$7,500; Married (separate return)/$3,750. Information for figuring the tax credit is available in IRS Publication 524, *"Credit for the Elderly or the Disabled."*

Moving Expenses. Unreimbursed moving expenses from your last duty station to your retirement residence may be deductible if the expenses are more than your moving allowance, even if you use the standard deduction. (See IRS Publication 521 for more information.)

SALE OF HOME

If you sell your home, review IRS Publication 523, "*Selling Your Home*," and instructions to Form 2119, "*Sale of Your Home*."

Sales After May 6, 1997 (rules for military members)

The Taxpayer Relief Act of 1997 as amended by the Military Family Tax Relief Act of 2003, allows taxpayers to exclude up to $250,000 of gain ($500,000 if married filing a joint return in most cases) realized on the sale or exchange of a main home after May 6, 1997. The exclusion is allowed each time a taxpayer sells or exchanges a main home, but generally no more frequently than once every two years. To be eligible, the property must have been owned and used as the taxpayer's main home for a combined period of at least two years out of the five years prior to the sale or exchange. A military member may suspend, for up to 10 years, the time transferred away from home on official extended duty for purposes of applying the five-year portion of the two out of five-year requirement. The law also allows a smaller exclusion for taxpayers who fail to satisfy the two of five-year requirement, but must move for reasons related to work or health.

Sales Before May 7, 1997

All special provisions for military members who sold their homes prior to May 7, 1997 have expired.

INCOME TAX WITHHOLDING

All nondisability retired pay is subject to withholding of federal income taxes. The amount withheld is dependent upon the taxpayer's amount of wages and the number of exemptions claimed.

Generally, federal income taxes will be withheld on annuities such as the SBP and RSFPP. Annuitants may elect not to have any tax withheld by filing the appropriate forms with DFAS.

Over-withholding. To avoid having too much income tax withheld, if you have unusually large itemized deductions, alimony deductions, or tax credits, you may claim additional withholding allowances. To change the number of allowances claimed, complete IRS Form W-4 and submit it to DFAS or the servicing finance center.

Under-withholding. If your spouse is employed or you work at more than one job, you can avoid having insufficient taxes withheld by completing form W-4 claiming fewer exemptions or asking for additional withholding. These actions may preclude the necessity for submitting a declaration of estimated taxes and making quarterly payments. An agreement for additional withholding is considered as having been ratably withheld over the entire year.

Estimated Taxes. You must make estimated tax payments for 2009 if you expect to owe at least $1,000 in tax for 2009 after subtracting your withholding and credits, and you expect your withholding and credits to be less than the smaller of 90 percent of the current year's tax, or 100 percent of the preceding year's tax. (See IRS Publication 505.)

Social Security Tax (FICA). Retired pay is not subject to withholding for Social Security tax, and retired members receive no Social Security credit based on their retired pay.

TAX COMPUTATION
(For 2009 taxable income)

The amount of wages referred to below represents monthly gross taxable wages less $304.17 per exemption for 2009. To compute your monthly income tax withholding, multiply the number of exemptions claimed by $304.17. Subtract this amount from your monthly gross wages (basic pay and other includible items). Then use the result to calculate your taxes from the tax tables.

Monthly Payroll Period Table

SINGLE Person (including head of household):

If the amount of wages (after subtracting withholding allowances) is:		The amount of income tax to withhold is:	
Not over $221		 0	
Over	But not over		of excess over
$221	$867	10%	$221
$867	$2,950	$64.60 plus 15%	$867
$2,950	$7,025	$377.05 plus 25%	$2,950
$7,025	$14,467	$1,395.80 plus 28%	$7,025
$14,467	$31,250	$3,479.56 plus 33%	$14,467
$31,250		$9,017.95 plus 35%	$31,250

MARRIED Person:

If the amount of wages (after subtracting withholding allowances) is:		The amount of income tax to withhold is:	
Not over $667		 0	
Over	But not over		of excess over
$667	$1,996	10%	$667
$1,996	$6,304	$132.90 plus 15%	$1,996
$6,304	$12,067	$779.10 plus 25%	$6,304
$12,067	$18,050	$2,219.85 plus 28%	$12,067
$18,050	$31,725	$3,895.09 plus 33%	$18,050
$31,725		$8,407.84 plus 35%	$31,725

Example: Married officer O-4 with over 14 years of service claiming three exemptions, with taxable pay of $6,534.30 per month.

a. $6,534.30 less $912.51 (3 x $304.17) $5,621.79
b. Tax = $132.90 plus 15% of $3,625.79 ($5,621.79 - $1,996)
c. Total Federal tax withheld monthly $676.77

GROSS INCOME TO BE REPORTED

Retired Pay to the extent it is not: (1) reduced for premiums under the Survivor Benefit Plan; (2) excluded because of disability (usually excludable only if combat related or the member entered the service prior to 1975) resulting from active service; or (3) waived in favor of disability compensation or pension from the VA.

Retainer Pay of enlisted members transferred to the Fleet Reserve.

Compensation for Employment in nonappropriated fund activities.

Cash Payment for accrued leave.

RSFPP, and SBP Survivor Annuities, taxable as employees' annuities.

DEDUCTIONS FROM GROSS INCOME

Moving Expenses. The cost of moving paid by a member as a result of retirement from active duty may be deductible. Moving expenses furnished by the military are not deductible. Unreimbursed moving expenses incurred by the member are deductible. Deductible moving expenses are limited to the cost of: (1) transportation of household goods and personal effects; and (2) travel (including lodging, but not meals) to the new residence. Moving expenses must be reported on Form 3903.

An individual who retires overseas and returns to the U.S., or a survivor (spouse or dependent) of any decedent who worked outside the U.S. at the time of death, is also eligible to deduct moving expenses, if within six months of the decedent's death the survivor moves to the U.S. from a foreign residence that had been shared with the decedent.

Roth IRA. The Roth IRA is available as an alternative to the traditional and nondeductible IRAs. Distributions from a Roth IRA are not required to start at age 70½, as is required with a traditional IRA. This is the difference that makes the Roth IRA particularly attractive. A taxpayer who lives beyond age 70½ and who does not need the funds may continue to make contributions and defer taxes until death. Amounts in the fund continue to be tax deferred until distributed to beneficiaries as required under the distribution rules applicable to beneficiaries.

Under the Roth IRA, a taxpayer whose AGI does not exceed certain limits can make a nondeductible (after-tax) contribution in 2009 of up to $5,000 per year ($6,000 if age 50 or older). Contributions may be made to nondeductible, traditional and Roth IRAs, but a combined annual limit of $5,000 ($6,000 if age 50 or older) applies. Subject to the limits, contributions cannot be for more than 100 percent of compensation earned. Generally, a contribution must be made during the tax year or by the next April 15 if the post-tax-year contribution is designated for the previous year. Proper identification of the tax year for which a contribution is made is necessary to start the counting of a five-year waiting period for withdrawals. For married taxpayers, the full contribution is allowed as long as the couple's AGI does not exceed $166,000. The maximum contribution in 2009 is phased out for couples with AGI between $166,000 and $176,000, and is unavailable for couples with an AGI above $176,000. The full contribution is allowed for single taxpayers with AGI below $105,000, is phased out between $105,000 and $120,000, and is unavailable for single taxpayers with AGI above $120,000. There is no limitation based on active participation in a qualified plan.

If certain conditions are satisfied, contributions may be withdrawn from a Roth IRA at any time, tax and penalty free. Compare this to withdrawals from a nondeductible IRA, where withdrawals are considered to consist in part of previously taxed contributions and in part taxable earnings. A withdrawal of earnings from a Roth IRA will be tax and penalty free only if taken at least five years after the tax year for which the first contribution was made (or five years after the tax year in which a rollover was made), and on or after age 59½, at or after death (to a beneficiary or estate), on account of disability, or for a first time home purchase expense. If these conditions are not satisfied, the earnings on amounts withdrawn will be taxable, and a penalty will apply to the amount includible in income.

Certain taxpayers have the option of converting a traditional IRA to a Roth IRA (by notifying an existing IRA trustee, or by making a direct trustee-to-trustee transfer). To qualify, a taxpayer who is single or married and filing jointly must have AGI of less than $100,000, excluding the income from the rollover. Taxpayers who are married filing separately may not make rollovers to a Roth IRA. Earnings from the traditional or nondeductible IRA and contributions to the extent previously deducted will be taxed at the time of the rollover. The full amount of the taxes will be due in the year of any rollover that takes place. The 10 percent penalty will not apply to amounts rolled over. See IRS Publication 590, "*Individual Retirement Arrangements (IRA)*" for specific details.

Alimony. Payments qualifying as alimony made by a taxpayer to a former spouse are deductible as an adjustment to gross income regardless of whether or not the taxpayer itemizes deductions. Property settlement payments are not deductible. No portion of any payment for support of a minor child is deductible. **If you deduct alimony you paid, you must show the recipient's Social Security number on your return.** Also, for divorce or separation instruments executed after 1984, special rules apply in determining whether or not the payments qualify as alimony. Consult an IRS official or a tax consultant for details.

EXCLUSIONS FROM GROSS INCOME

Disability Retired and Disability Severance Pay. This pay is received for temporary or permanent disability retirement and is received on the basis of a percentage of disability. If computed on years of service further limitations apply. **Note:** The exclusion is in effect only for those military members who were receiving disability retirement pay on or before September 24, 1975; those persons who were members of the Uniformed Services (or Reserve Component) on September 24, 1975; those members who receive military disability retirement by reason of combat related injury; or those who would be entitled to receive disability compensation from the Veterans Administration.

In-Kind Military Benefits. Such as: legal assistance, space-available travel on Government aircraft; medical/dental care; commissary/exchange discounts.

Combat Zone. For retirees recalled to serve in the Persian Gulf or other combat areas, all enlisted pay, and officer pay up to the highest enlisted pay ($7,143.30 per month in 2009) is excluded.

Death Gratuity Pay. Death gratuity payments to the survivor of a servicemember are excludable from the beneficiary's income.

Survivor Benefit Plan. Reductions in retired pay for SBP premiums are not included in gross income of such retired pay. However, SBP payments to the survivor are considered as income. The SBP annuities are subject to federal income tax withholding. Annuitants may elect not to have the tax withheld or make adjustments by filing necessary forms with the Defense Finance and Accounting Service.

Other Exclusions. Other exclusions include such items as: Government or commercial life insurance amounts; insurance dividends and annuities; injury compensation; state veteran bonuses; and others. Check IRS instructions for more information.

STATE INCOME TAXES

The following compilation of state tax information represents the best information available at time of publication. Some states were considering new tax legislation, so care should be exercised and questions should be resolved by checking with the state tax authority. This data should be considered as a general guide to state income tax requirements. The possible requirement for filing a declaration of estimated taxes has not been included. For specific information regarding your state, contact the legal assistance office or the state tax office for your state. For your convenience, a listing of the state tax authorities for the states having some form of income tax may be found on the following pages.

States with no personal income taxes.

Alaska
Florida
Guam
Nevada
New Hampshire*
South Dakota
Tennessee**
Texas
Washington
Wyoming

*Imposes tax on interest or dividend income.
**Imposes tax on certain dividend and interest income.

States that exempt all military disability retired pay.

Arizona***
California
Colorado***
Connecticut***
Delaware***
District of Columbia***
Georgia***
Idaho***
Indiana
Iowa***
Kentucky***
Maine***
Maryland***
Minnesota***
Montana*
Nebraska***
New Jersey**
New Mexico***
North Carolina
North Dakota***
Oklahoma***
Oregon
Puerto Rico
Rhode Island***
South Carolina
Utah***
Vermont***
Virginia***
West Virginia***

*If AGI is less than $15,000—up to $5,200 can be exempted.
**There is a total and permanent disability exclusion.
*** To the extent exempt for federal tax purposes.

States that exempt all military retired pay.

Alabama
Hawaii
Illinois
Kansas
Kentucky*
Louisiana
Massachusetts
Michigan
Mississippi
New York*
North Carolina*
Ohio
Oregon*
Pennsylvania*
Wisconsin

*If certain qualifications are met.

States that exempt all active duty military pay.

Arizona
Illinois
Michigan
Montana
New Mexico
Ohio*
Oregon*
Pennsylvania**
Vermont*

*Exempts all military active duty pay earned outside the state.
**For residents who earn military pay outside of state.

STATE INCOME TAX PROVISIONS FOR RETIRED MILITARY PERSONNEL#

	Due Date for Filing and Making Payments on the Return	Special Provisions Applicable to Armed Services Retired Personnel
ALABAMA Alabama Dept. of Revenue Income Tax Division P.O. Box 327410 Montgomery, AL 36132-7410 www.revenue.alabama.gov	Return and payment due April 15th.	All military retirement income is exempt from taxation, including survivor benefits. Civil service retirement and private defined benefit pension plans exempt. Call 334-242-1000 to order forms.
ALASKA Dept. of Revenue, Tax Division P.O. Box 110420 Juneau, AK 99811-0420 www.revenue.state.ak.us	No State Income Tax.	All military retired pay is exempt. No income tax on RSFPP/SBP benefits.
ARIZONA Dept. of Revenue 1600 W. Monroe St. Phoenix, AZ 85007-2650 www.azdor.gov	Return and payment due April 15th.	Allows deductions of up to $2,500 military retired pay. Disability retired pay is exempt. RSFPP/SBP contributions are excluded from federal income which flows through to the AZ return. AZ does not provide an additional subtraction from AZ gross income.
ARKANSAS Dept. of Finance and Administration P.O. Box 3628 Little Rock, AR 72203-3628 www.arkansas.gov/dfa	Return and payment due April 15th.	First $6,000 nondisability retired pay, and military survivors benefits are exempt.
CALIFORNIA California Franchise Tax Board P.O. Box 1468 Sacramento, CA 95812-1468 www.ftb.ca.gov	Return and payment due April 15th.	Retired pay received by nonresidents not subject to CA tax. All disability retired pay is exempt.
COLORADO Dept. of Revenue 1375 Sherman St. Denver, CO 80261-0005 www.taxcolorado.com	Return and payment due April 15th.	Disability retired pay exempt if exempt federally. Military retirement pay may exclude up to $20,000 if age 55 or older; up to $24,000 if 65 or older.
CONNECTICUT Dept. of Revenue Services 25 Sigourney St. Hartford, CT 06106-5032 www.ct.gov/drs	Return and payment due April 15th. Estimated taxes due April 15, June 16, Sept. 15, and Jan. 15.	Military disability retired pay is exempt from tax to the same extent it is exempt from federal income tax. Retired pay included in federal AGI is subject to CT income tax. RSFPP/SBP benefits are treated same as federal. Social Security benefits fully exempt if federal AGI is less than $50,000 (single or married filing separately) or $60,000 (married filing jointly or head of house). Effective 1/1/08, 50% of military retired pay is exempt from state tax.
DELAWARE Division of Revenue Carvel State Office Bldg. 820 N. French St. Wilmington, DE 19801 www.state.de.us/revenue	Return and payment due April 30th.	First $2,000 nondisability retired pay is exempt if under 60, $12,500 for those 60 or older. $4,000 on low income joint return. All disability retired pay exempt, if excludable on federal tax returns.
DISTRICT OF COLUMBIA Office of Tax & Revenue 941 N. Capitol St., N.E., 1st. Fl. Washington, DC 20002 www.dc.gov	Return and payment due April 15th.	All disability retired pay is exempt. RSFPP/SBP benefits not taxable.
FLORIDA Dept. of Revenue 5050 W. Tennessee St. Tallahassee, FL 32399-0100 http://dor.myflorida.com	No State Personal Income Tax.	No tax on military retired pay or RSFPP/SBP annuities.
GEORGIA Dept. of Revenue Income Tax Division Rm. 15322 1800 Century Ctr. Blvd., N.E. Atlanta, GA 30345-3205 www.dor.ga.gov	Return and payment due April 15th.	VA disability retired pay is exempt for federal income tax purposes. First $35,000 of retirement income is exempt if age 62 or older or permanently and totally disabled. RSFPP/SBP benefits are taxable if they are taxable for federal income tax purposes.

STATE INCOME TAX PROVISIONS FOR RETIRED MILITARY PERSONNEL#

	Due Date for Filing and Making Payments on the Return	Special Provisions Applicable to Armed Services Retired Personnel
GUAM Government of Guam Dept. of Revenue & Taxation P.O. Box 23607 GMF, Guam 96921 www.guamtax.com	No State Income Tax.	Residents of Guam must file in Guam. RSFPP/SBP benefits not taxed.
HAWAII Dept. of Taxation P.O. Box 259 Honolulu, HI 96809 www.state.hi.us/tax	Return and payment due April 21th.	All military retired pay is exempt. RSFPP/SBP benefits are not subject to taxation.
IDAHO Idaho State Tax Commission 800 Park Blvd. Plaza IV P.O. Box 36 Boise, ID 83722-0036 www.tax.idaho.gov	Same as federal.	Disability retired pay exempt if exempt from federal taxes. Full year retirement residents retired benefits paid by the U.S. to a retired member of the U.S. military or the unremarried widow of such member if the recipient is age 65, or age 62 and disabled. The maximum retired benefits which may be deducted for 2008 are: (a) Age 65 filing joint w/spouse, or age 62, disabled and filing jointly, $39,330 (b) Age 65 and single, or age 62, disabled & single, $26,220 (c) Unremarried widow of a retired servicemember age 65, or age 62 and disabled $26,220.
ILLINOIS Dept. of Revenue 101 W. Jefferson St. Springfield, IL 62794-9001 http://tax.illinois.gov/	Same as federal.	Military retired pay is exempt. RSFPP/SBP benefits are not taxable. POW/MIA benefits are not taxed.
INDIANA Dept. of Revenue P.O. Box 7207 100 N. Senate Ave. Indianapolis, IN 46204-2253 www.in.gov/dor	Same as federal.	All disability retired pay is exempt. Up to $5,000 military retired pay may be excluded if age 60. Special rules may apply for permanently disabled retirees. See Tax Information Bulletin #70 for details.
IOWA Iowa Dept. of Revenue Hoover State Office Bldg. 1305 E. Walnut Des Moines, IA 50319 www.state.ia.us/tax	Return and payment due April 30th.	Retired and disability pensions are taxable to same extent as on federal level. Limited eligibility for disability income exclusion if you are totally disabled. Partial exclusion on pension income of up to $12,000 for married filers and up to $6,000 for all other filing statuses is given to qualifying individuals.
KANSAS Director of Taxation Kansas Dept. of Revenue Docking State Office Bldg. 915 S.W. Harrison St. Topeka, KS 66699-1588 www.ksrevenue.org	Return and payment due April 15th.	All disability retired pay is exempt if exempt from federal taxes. Military retired pay is exempt from tax for all taxable years commencing after December 31, 1991.
KENTUCKY Dept. of Revenue 501 High St. Frankfort, KY 40620 http://revenue.ky.gov	Return and payment due April 15th.	Disability retired pay is exempt if exempt for federal tax purposes. Military retired pay is exempt if retired before January 1, 1998. Others must include in gross income that portion attributable to service after December 31, 1997. This portion is eligible for 100% pension exclusion of up to $41,110. Federal guidelines apply to SBP benefits.
LOUISIANA Dept. of Revenue Personal Income Tax Division P.O. Box 201 Baton Rouge, LA 70821-0201 www.rev.state.la.us	Return and payment due May 15th.	All military retired pay is exempt from LA income tax. Use Schedule E to exempt this income. RSFPP/SBP benefits are exempt.
MAINE Revenue Services Income/Estate Tax Division 24 State House Station Augusta, ME 04333-0024 www.maine.gov/revenue	Return and payment due April 15th.	All disability retired pay is exempt, if exempt from federal taxes; RSFPP/ SBP benefits are taxable. Up to $6,000 in eligible pensions may be deductible under certain conditions.

STATE INCOME TAX PROVISIONS FOR RETIRED MILITARY PERSONNEL#

	Due Date for Filing and Making Payments on the Return	Special Provisions Applicable to Armed Services Retired Personnel
MARYLAND Revenue Administration Center Comptroller of Maryland Annapolis, MD 21411 www.marylandtaxes.com	Return and payment due April 15th.	Military retirees can subtract up to $5,000 of military retirement income received by a qualifying individual during the tax year. All disability retired pay is exempt, if exempt for federal tax purposes. In addition to the $5,000 subtraction, up to $24,000 of taxable pension or annuity pay may be exempt for retirees age 65 or older (any age if totally disabled). RSFPP/SBP benefits are taxable. Follow directions for the pension exclusion.
MASSACHUSETTS Dept. of Revenue 100 Cambridge St. Boston, MA 02204 www.dor.state.ma.us	Return and payment due April 15th.	Military retired pay and disability retired pay are exempt. RSFPP/SBP benefits are not taxable.
MICHIGAN Michigan Dept. of Treasury Lansing, MI 48922 www.michigan.gov/treasury	Return and payment due April 15th.	All military retired pay is exempt. RSFPP/ SBP benefits are exempt to the retiree or surviving spouse of the retiree, but not to a surviving child. For the 2008 tax year, private pensions are exempt up to $43,440 single or $86,880 if jointly filed. The maximum deduction for a private pension must be reduced by the amount of pension income received from a public retirement system, including military retired pay. Plus $2,200 additional exemption if over 65, disabled, or having unemployed compensation that amounts to 50% of AGI. Qualified disabled veterans may claim an additional exemption of $250.
MINNESOTA Dept. of Revenue Minnesota Income Tax Division Mail Station 7701 St. Paul, MN 55146-7701 www.taxes.state.mn.us	Return and payment due April 15th.	Same as federal. Income tax credit of up to $750 for past military service. To qualify you must have served at least 20 years or 100% service related disability.
MISSISSIPPI State Tax Commission P.O. Box 1033 Jackson, MS 39215-1033 www.mstc.state.ms.us	Return and payment due April 15th.	Retired pay is exempt. Disability retired pay is exempt.
MISSOURI Dept. of Revenue Personal Tax P.O. Box 2200 Jefferson City, MO 65105-2200 http://dor.mo.gov/tax/personal	Return and payment due April 15th.	Effective for 2008, married couples with MO AGI less than $100,000 and individuals with less than $85,000 may deduct the greater of $6,000 or 35% of their public retirement benefits, to the extent the amounts are included in their federal AGI. Persons with more than those amounts may qualify for a partial exemption. RSFPP/SBP benefits are taxable if taxable by federal.
MONTANA Dept. of Revenue P.O. Box 5805 Helena, MT 59604-5805 www.mt.gov/revenue	Return and payment due April 15th. If serving in a combat zone or contingency area, contact department about filing due dates.	Adjusted federal AGI up to $30,000 provides for a uniform exclusion of up to $3,600 from tax for all types of pension income. The exclusion is reduced by $2 for every $1 of federal AGI received by the taxpayer in excess of $30,000, with the exclusion becoming zero for federal adjusted gross incomes over $31,800.
NEBRASKA Dept. of Revenue P.O. Box 94818 Lincoln, NE 68509-4818 www.revenue.ne.gov	Return and payment due April 15th.	All disability retired pay is exempt to the extent for federal tax purposes. RSFPP/ SBP benefits are taxable. Exemptions from federal or state income taxation provided by federal law will apply.
NEVADA www.tax.state.nv.us	No State Income Tax.	No tax on retired military pay or RSFPP/SBP annuities.
NEW HAMPSHIRE Dept. of Revenue Administration 45 Chenell Dr. P.O. Box 2072 Concord, NH 03302-2072 www.nh.gov/revenue	No State Income Tax.	No tax on retired military pay or RSFPP/SBP benefits.

STATE INCOME TAX PROVISIONS FOR RETIRED MILITARY PERSONNEL#

	Due Date for Filing and Making Payments on the Return	Special Provisions Applicable to Armed Services Retired Personnel
NEW JERSEY Division of Taxation 50 Barrack St. P.O. Box 269 Trenton, NJ 08695-0269 www.state.nj.us/treasury/taxation	Return and payment due April 15th.	For all military pensions and military survivor benefits payments received from DFAS are excluded from the NJ gross income regardless of the recipients age. Disability retired pay exempt until the taxpayer reaches age 65. At that time, the disability pension is treated as a normal pension and may be eligible for pension exclusion.
NEW MEXICO Taxation & Revenue Dept. Tax Information/Policy Office 1100 St. Francis Dr. P.O. Box 630 Santa Fe, NM 87504-0630 www.tax.state.nm.us	Return and payment due April 15th. Electronic filing and electronic payment are due by April 30, 2008.	Disability retired pay is exempt to the extent of federal tax purposes.
NEW YORK Dept. of Taxation & Finance Taxpayer Contact Center W.A. Harriman Campus Albany, NY 12227 www.nystax.gov	Return and payment due April 15th.	Military pensions received in taxable years on or after January 1, 1989 are exempt from NY state taxation.
NORTH CAROLINA Dept. of Revenue P.O. Box 25000 Raleigh, NC 27640-0640 www.dornc.com	Return and payment due April 15th.	Military disability pay is not taxable. Maximum of $4,000 retired pay if less than five years creditable service as of August 12, 1989. If five or more years of service as of August 12, 1989, total exclusion of retiree benefits that are included in federal taxable income.
NORTH DAKOTA Tax Commissioner 600 E. Boulevard Ave. Dept. 127 Bismarck, ND 58505-0599 www.nd.gov/tax	Return and payment due April 15th.	On Form ND-2, up to $5,000 exclusion for retired pay if age 50 or older. Reduce exclusion by Social Security benefits received. No exclusion on Form ND-1. Advised to compare Form ND-1 and Form ND-2 before choosing.
OHIO Dept. of Taxation P.O. Box 182382 Columbus, OH 43218-2382 http://tax.ohio.gov	Return and payment due April 15th.	Beginning 1/1/08, military retired pay and SBP benefits are exempt from state income taxes. Federal civil service retirees who received federal service credit for military service can deduct only that portion attributable to military service. Also, a credit of $50 per return allowed for 65 and older.
OKLAHOMA Oklahoma Tax Commission 2501 N. Lincoln Blvd. Oklahoma City, OK 73194 www.oktax.state.ok.us	Return and payment due April 15th.	Disability retired pay and RSFPP/ SBP benefits are exempt to the extent for federal tax purposes. Individual may exclude 75% of retirement benefits or $10,000, whichever is greater, but not to exceed amount included in federal AGI.
OREGON Oregon Dept. of Revenue Revenue Bldg. 955 Center St., N.E. Salem, OR 97310-2555 www.oregon.gov/dor	Same as federal.	All military retirement and disability pay for service performed before October 1, 1991 is tax exempt. If age 62 or older and low income, there is a possible 9% credit on retirement income taxable by OR.
PENNSYLVANIA Dept. of Revenue Bureau of Individual Taxes Harrisburg, PA 17128-0600 www.revenue.state.pa.us	Same as federal.	If meeting the conditions for retirement, military retired pay and military disability retired pay are exempt. RSFPP/SBP benefits are exempt from taxation.
PUERTO RICO Dept. of the Treasury P.O. Box 9024140 San Juan, PR 00902-4140 www.hacienda.gobierno.pr	Return due April 15th.	All disability pay is exempt. If under 60, $11,000 exemption on retired pay. Over 60, $15,000 exemption. $1,500 deduction for honorable discharged veterans.

STATE INCOME TAX PROVISIONS FOR RETIRED MILITARY PERSONNEL#

	Due Date for Filing and Making Payments on the Return	Special Provisions Applicable to Armed Services Retired Personnel
RHODE ISLAND Division of Taxation One Capitol Hill Providence, RI 02908-5800 www.tax.state.ri.us	Same as federal.	Same as under Federal Internal Revenue Code.
SOUTH CAROLINA Dept. of Revenue 301 Gervais St. P.O. Box 125 Columbia, SC 29214 www.sctax.org	Return and payment due April 15th. Electronically filed state income tax returns are due by May 1, 2009.	Beginning with first year of retirement, a resident may claim a $3,000 deduction from retired income until age 65 when deduction increases to $10,000 from retired income. All residents at age 65 are allowed a deduction of up to $15,000 from any source of income. The $15,000 deduction must be offset by the $10,000 retired deduction.
SOUTH DAKOTA www.state.sd.us/drr2/revenue.html	No State Income Tax.	No tax on military retired pay or RSFPP/SBP benefits exempt.
TENNESSEE Dept. of Revenue Andrew Jackson State Office Bldg. 500 Deaderick St. Nashville, TN 37242-1099 www.Tennessee.gov/revenue	Return and payment due April 15th.	No tax on military retired pay; RSFPP/SBP benefits exempt; individuals 65 or over with total annual income of $16,200 or less, or $27,000 for joint filers, are exempt.
TEXAS www.window.state.tx.us	No State Income Tax.	No tax on military retired pay or RSFPP/SBP annuities.
UTAH State Tax Commission 210 North 1950 West Salt Lake City, UT 84134 www.tax.utah.gov	Return and payment due April 15th.	Disability retired pay exempt to the extent of federal exemption. Taxpayers age 65 or over may deduct up to $7,500 of retirement income ($15,000 for married filing jointly). Taxpayers under age 65 may deduct up to $4,800. RSFPP/SBP benefits are not taxable.
VERMONT Dept. of Taxes 133 State St. Montpelier, VT 05633-1401 www.state.vt.us/tax	Return and payment due April 15th.	Disability retired pay is exempt to the extent of federal exemption. RSFPP/SBP benefits are taxable.
VIRGINIA Virginia Dept. of Taxation P.O. Box 1115 Richmond, VA 23218-1115 www.tax.virginia.gov	Return and payment due May 1st. Fiscal year filers: 15th day of the 4th month after the close of your taxable year.	No special deduction of retired pay. Taxpayers age 65 and older may qualify to claim an age deduction of up to $12,000 subject to filing status and income.
WASHINGTON www.dor.wa.gov	No State Income Tax.	No tax on military retired pay or RSFPP/SBP annuities.
WEST VIRGINIA Dept. of Tax and Revenue Taxpayer Services Division P.O. Box 3784 Charleston, WV 25337-3784 www.wvtax.gov/	Same as federal.	All disability retired pay is exempt to the extent of federal exemption. First $20,000 of military retirement, plus additional $2,000 for a maximum of $22,000. Combined exclusions may not exceed total value of military pension. If 65 or older, an additional $8,000 decreasing modification is allowed, but is reduced dollar for dollar by the amount taken for the additional military retirement.
WISCONSIN Dept. of Revenue Individual Income Tax Assistance P.O. Box 8949 Madison, WI 53708-8949 www.revenue.wi.gov	Return and payment due April 15th.	WI does not tax retirement payments received from: The U.S. military retirement system including payments from the RSFPP; and from the U.S. Government that relate to service with the Coast Guard, the commissioned corps of NOAA, or the commissioned corps of the USPHS.
WYOMING http://revenue.state.wy.us	No State Income Tax.	No tax on military retired pay or RSFPP/SBP annuities.

NOTE: The material contained in these tax tables represent the best information available at time of publication. This presentation should be considered as a general guide to state income tax requirements. For specific information regarding your state, contact the legal assistance office or your state tax office.

PART II

GENERAL INFORMATION

RETIRED ACTIVITIES PROGRAMS

Each of the military services has a Retired Activities Office to provide information and assistance to retired members regarding their rights, privileges, benefits and other information of interest and value such as changes in health care benefits, or legislative actions. The primary means for maintaining contact with retirees is a periodic newsletter, including one published for retired members of the United States Coast Guard. These Retired Activities Offices or their field offices receive and answer thousands of calls and letters from retirees and their families. Although the Retired Activities Offices can generally answer most inquiries, those dealing specifically with retired pay or allotments are referred to the appropriate section of the finance and accounting service center that maintains records for that member.

In addition to the Retired Activities Offices, the services have established either a Retiree Council or Advisory Committee (the Marine Corps participates in the Navy's Advisory Committee on Retired Personnel) composed of officer and enlisted retirees who meet at least annually to review retiree matters, discuss problems and make recommendations to the military services. These councils provide an opportunity to record the concerns of a large and influential group of former servicemembers and their families. The activities of these councils and the actions taken by the services regarding their recommendations are widely reported by the retired activities' newsletters and military publications plus military and retired associations' magazines and periodicals.

The military services encourage continued involvement from retirees with the active forces. All military services conduct periodic individual and group retiree seminars and briefings covering a broad range of subjects. The Army, Navy, Air Force and Marine Corps host annual Retiree Activity Days and provide briefings and counseling to retirees.

Retirees can be of assistance to recruiters of their respective services. Using information provided by local recruiters, they can contact and address individuals or groups of young people and, by relating their knowledge and experiences, encourage enlistment and military careers for these potential members. In addition to assisting recruiters, retirees can continue to serve the active forces, and their civilian communities as well, by making themselves available as speakers to civic, professional, and business groups or by writing articles for local media which can help explain the military position on subjects of current interest and promote public support for the Armed Forces. Support for such activities can be obtained from the information office of local military installations.

Retirees should not overlook the availability of the many retired organizations and associations. These associations have been most influential in obtaining and preserving many benefits for servicemembers and retirees and provide opportunities for social gatherings for individuals with common interests as well. (See listing of ASSOCIATIONS AND ORGANIZATIONS.)

RETIRED ACTIVITIES OFFICES

Service	Retired Activities Office Address	Publication	Retiree Council Committee	TOLL FREE and/or Commercial Phone Numbers
ARMY	Army Retirement Services Attn: DAPE-HRP-RSO 200 Stovall St., Rm. 5N37 Alexandria, VA 22332-0470 www.armyg1.army.mil/rso/mission.asp	Army Echoes	Chief of Staff Army Retiree Council	(703) 325-9158 (800) 336-4909
NAVY	Retired Activities Section 5720 Integrity Dr. Millington, TN 38055-6200 mill_retiredactivities@navy.mil www.npc.navy.mil/Command Support/RetiredActivities	Shift Colors	Secretary of the Navy's Committee on Retired Personnel	(866) 827-5672 (901) 874-4308
AIR FORCE	HQ AFPC/DPSIAR 550 C St. W., Ste. 8 Randolph AFB, TX 78150-4713 afpc.retiree@randolph.af.mil www.retirees.af.mil	AFTERBURNER USAF News for Retired Personnel	Air Force Retiree Council	(800) 531-7502 (210) 565-4663
USMC	Retired Activities Section MMSR-6, HQ USMC Manpower & Reserve Affairs 3280 Russell Rd. Quantico, VA 22134 www.manpower.usmc.mil	Semper Fidelis Memorandum for Retired Marines	Secretary of the Navy's Retiree Council	(800) 336-4649 (703) 784-9310
USCG	Commanding Officer (RAS) U.S. Coast Guard Personnel Service Center 444 S.E. Quincy St. Topeka, KS 66683-3591 www.uscg.mil/hr/psc/ras/	Coast Guard Retiree Newsletter	USCG Retiree Council	(800) 772-8724 (785) 339-3415
In addition to the Retired Activities Offices listed above, the Army, Navy and Air Force maintain Retired Activities Offices at major installations throughout the country.				

IDENTIFICATION CARDS

Identification cards are issued to retirees and eligible family members as a means of identification and as authorization for various benefits and privileges.

The Armed Forces Identification Card, DD Form 2 (Ret), is issued to retired members who are entitled to retired pay.

Dependent ID Cards. DD Form 1172, *"Application for Uniformed Services Identification and Privilege Card,"* is used to apply for issue or reissue of ID cards for dependents (DD Form 1173). DD Form 1172 is automated and generated through DEERS. Information must be verified from the sponsor's personnel records or from appropriate documentation as applicable, such as marriage certificates, birth certificates, certification of student status, adoption papers, or medical statements. Servicemembers and their eligible dependents, age 10 and older, may obtain ID cards at any ID card issuing facility, regardless of their branch of service.

Dependent ID cards generally expire four years from the date of issue or expiration of the sponsor's obligated service. However, a permanent, non-expiring dependent ID card is available for all eligible retiree dependents who are either age 75 and over, or who are permanently disabled.

ID cards must be surrendered when they are replaced or expired or upon demand by an authorized individual, or when the cardholder becomes ineligible.

Eligible dependents are generally defined as:

1. Lawful spouse.

2. Unremarried surviving spouse. (A remarried surviving spouse whose second marriage ends in divorce or death may receive commissary, theater, and exchange benefits, but not medical benefits.)

3. Unmarried children (including adopted or stepchildren) who are:

a. Under 21 years of age.

b. Over 21 but incapable of self-support. (Substantiating documentation is required.)

c. Over 21 but under 23 who are attending an approved learning institution as a full time student (documentation is required).

4. Parent or parent-in-law who is dependent on the servicemember for over one-half of his/her support. For medical care privileges, the parent or parent-in-law must also reside in a dwelling place provided or maintained by the member. Consult an ID card issuing facility for clarification or determination of eligibility.

5. Wards (as defined in appropriate instructions).

6. Unremarried former spouse (one whose final decree of divorce is on or after February 1, 1983, and who was married to a military sponsor for a minimum of 20 years during which time the military member served 20 years of creditable service for retirement purposes).

A comprehensive list of eligible dependents is provided in Department of Defense Instruction 1000.13, *"Identification (ID) Cards for Members of the Uniformed Services, Their Dependents, and Other Eligible Individuals."*

CORRECTION OF MILITARY RECORDS

The Secretary of each military department, acting through a Board of Correction of Military Records, has the authority to amend any military record when necessary to correct an error or remove an injustice.

Some retirees encounter situations where they feel victimized by errors or injustices related to their military service. A correction board may consider applications for correction of a military record, including review of discharges issued by courts martial.

Generally, the retiree, veteran, survivor, or legal representative must file a request for correction within three years after discovery of the alleged error or injustice. However, the board may excuse failure to file within the prescribed time if it finds it would be in the interest of justice to do so. The applicant is responsible to show why the filing of the application was delayed and why it would be in the interest of justice for the board to consider the application in spite of the delay.

In order to justify any correction, the petitioner must show to the satisfaction of the board that the alleged entry or omission in the records was in error or unjust. Application must include any evidence available, such as signed statements of witnesses or a brief of arguments supporting the requested correction. Application is made on DD Form 149, "*Application for Correction of Military Record under the provisions of Title 10, U.S. Code, Section 1552,*" available at military personnel offices or on the Internet at: www.dtic.mil/whs/directives/infomgt/forms/eforms/ dd0149.pdf. The form lists the addresses for each military department to submit the application. A number of military and veterans associations will provide guidance and counseling services without regard to membership or cost. (See ASSOCIATIONS and ORGANIZATIONS.)

RETIREE STATUS AND CATEGORIES

Retirees are an important part of the military and are entitled to the same respect and courtesy shown to active duty members. Their status is similar in many ways to active duty members. They are subject to recall in the event of a national emergency and, therefore, subject to service in a combat zone.

The DoD categorizes all military retirees by age, length of time after retirement, and whether an individual has a disability precluding recall in a mobilization. *Category I* military retirees are those within their first five years of retirement, under age 60 and not disabled. *Category II* are those who have been retired more than five years, under age 60 and not disabled; *Category III* includes all others, including those who are disabled. All Category I and II retirees are considered mobilization assets and are counted as part of the total DoD mobilization base. Category III retirees with selected skills, primarily medical personnel, are considered as mobilization assets on a case-by-case basis.

At any time, the Secretary of a military department may order regular or Reserve 20-year active duty retirees, regardless of the retired list to which assigned, and members of the Fleet Reserve and the Fleet Marine Corps Reserve to active duty without their consent in the interest of national defense. In time of war or national emergency declared by Congress, the Secretary concerned may recall any other retired member of a Reserve Component, with the approval of the Secretary of Defense, when it is determined there are insufficient qualified Reservists in an active status or in the Inactive National Guard.

In this regard, retirees provide an experienced and tested wartime resource in numbers and skills too valuable to ignore. By their faithful service, retirees have earned and are entitled to certain rights, benefits, and privileges. Under specific conditions, retirees are subject to the Uniform Code of Military Justice, standards of conduct, conflict of interest, and other regulations and laws. Retirees receive a monthly payment from the services as a deferred annuity in recognition of their active service and for their vital role in the nation's military readiness.

WEARING THE UNIFORM

Wearing the uniform is a privilege granted to retirees in recognition of faithful service to their country. The uniform may be worn for ceremonies or at official functions when the dignity of the occasion and good taste would dictate the propriety of the uniform. Whenever the uniform is worn, it must be done in such a manner as to reflect credit upon the individual and the service from which retired. Appropriate occasions for wearing the uniform include memorial services, military weddings, military funerals, military balls, patriotic or military parades, and ceremonies in which military units are taking part, meetings or functions of associations where the membership is largely or entirely composed of honorably discharged or retired military personnel, and any other occasion where, in the judgment of the retired member, wearing the uniform is appropriate. On these occasions, the uniform may be worn while traveling to and from the ceremony, providing such travel is completed on the day of the ceremony.

The uniform worn by a retiree may be the uniform prescribed on the date of the member's retirement or any uniform currently authorized for active duty personnel; however, uniform items may not be mixed.

Ribbons, medals and/or other decorations, which were authorized and awarded to the member, may be worn on uniforms in accordance with the uniform directives of the appropriate service. No foreign decoration may be worn on the uniform unless at least one U.S. military decoration or service medal is worn. Any questions regarding the wearing of the uniform should be directed to the nearest installation of the member's service. (See UNIFORM RESTRICTIONS.)

MEDALS ON CIVILIAN DRESS

Miniature medals may be worn with black or white tie in the same way as prescribed for mess dress uniforms. Miniature replicas of ribbons in the form of lapel buttons may be worn on the left lapel of civilian clothes.

MILITARY AWARDS AND DECORATIONS

Send requests for issue or replacement of military service medals, decorations, and awards to the specific branch of the military in which the member served. However, for Army and Air Force personnel, the National Personnel Records Center will verify the awards to which a member is entitled and forward the request with the verification to the appropriate service department for issuance of the medals.

Use Standard Form 180, *"Request Pertaining to Military Records,"* to request medals and awards. An electronic version of the form is available at: www.archives.gov/research/order/standard-form-180.pdf. Provide as much information as possible and send to the address listed on the form.

USE OF MILITARY TITLES

Retired personnel may use their military titles subject to restrictions and the exercise of good judgment. Although military titles may be used in conjunction with commercial enterprises, such use must never cast discredit on the military services or the DoD. The use of military titles is prohibited in connection with commercial enterprises when such use, with or without the intent to mislead, presents the appearance of sponsorship, sanction, endorsement, or approval by the military services or the DoD.

Retired personnel must not use their military titles in connection with public appearances outside the United States unless the appropriate overseas commander authorizes such use.

When military titles are used, the member must show that he or she is in a retired status. Social and business cards must reflect the retired status of the individual.

Contact the Retired Activities Office for your branch of service for the latest guidelines on using military titles.

Some examples of correct titles on business calling cards, or for personal address are as follows:

Army: The Army points out that retirees should indicate Regular, AUS, or Reserve retirement as follows:

USA Retired for Regular Army Retired Personnel

AUS Retired for those on the Army of the United States Retired Lists

USAR Retired for those retired from an Army Reserve Component (USAR or ARNG)

Navy: The U.S. Navy differentiates, for signature block or business calling card placement, as follows:

Full grade and name are centered on the card, U.S. Navy and Retired in lower right corner for Commander and higher.

Full name is centered on the card with other data including rank placed in lower right corner for Lieutenant Commander and lower.

Air Force: Colonel John B. Doe, USAF, Retired (or the abbreviation Ret)

Signature: John B. Doe, Colonel, USAF, Retired.

Marine Corps: The U.S. Marine Corps requires the designation USMC (Ret.) or USMCR (Ret.) as appropriate when the military title is used.

Coast Guard: The Coast Guard requires the designation USCG (Retired) or USCGR (Retired) as appropriate when the military title is used.

HOW TO LOCATE A RETIREE

The following procedures should be followed in order to locate a retired military member. The Privacy Act prohibits the release of personal addresses, however there is an easy way to locate a former servicemember. Simply write a letter to the retiree. Place it in a stamped envelope (only one per envelope) with the retiree's name on it. Indicate your return address.

In a short note provide what information you have about the person you are trying to locate. Include: complete name, rank, Social Security number, birth date (approximate age) and any other identifying information that might help the locator identify the person. State your reason for wanting to contact this person.

Place the postage-paid, sealed envelope and your note in another envelope and address it to the appropriate service (see list of addresses following). A search fee may be required for each name. There is generally no fee charged if the request is from a parent or immediate family member. You will be notified only if the retiree cannot be located. Each branch, excluding the Army, provides this service.

NAVY: Navy Worldwide Locator, Navy Personnel Command, PERS 312, 5720 Integrity Dr., Millington, TN 38055-3120, (901) 874-3388, ($3.50 fee per name.) Make check or money order payable to U.S. Treasurer.

AIR FORCE: Headquarters, AFPC/DPDXIDL, 550 C St. W., Ste. 50, Randolph AFB, TX 78150-4752, (210) 565-2660, Recording, $3.50 fee. Make checks payable to DAO-DE RAFB.

MARINE CORPS: HQ Marine Corps, Manpower and Reserve Affairs (MMSR-6), 3280 Russell Rd., Quantico, VA 22134, (703) 784-9310, (no fee).

COAST GUARD: Commanding Officer (RAS), U.S. Coast Guard Personnel Service Center, 444 S.E. Quincy St., Topeka, KS 66683-3591, 1-800-7728724, (785) 339-3770 FAX, (no fee).

ARMY: None.

PRIVILEGES OF MILITARY FACILITIES

Retired military personnel and their eligible family members are authorized the use of many military facilities, subject to availability as determined by the installation commander. Using the facilities, for the most part, is a privilege that may be granted, rather than a right to which retired personnel are automatically entitled. The presentation of proper identification cards (DD Form 2—gray or blue, or DD Form 1173 for family members) is generally required to use the facilities.

The use of Uniformed Services health and medical care facilities is probably one of the most important rights extended to retired military personnel and eligible family members. Complete coverage of this important subject can be found in the section on UNIFORMED SERVICES HEALTH BENEFITS.

While Department of Defense policy is to afford retired military personnel and their eligible family members the opportunity to use post and base facilities, there are restrictions with regard to use of facilities located in certain overseas areas. Due to certain Status of Forces Agreements and bilateral agreements with host countries, many privileges ordinarily available in CONUS are not authorized for military retirees residing or traveling in foreign countries.

Retirees and their eligible family members may use overseas military postal facilities when such use is authorized through agreements with the host country and where the overseas commander determines that such service can be provided. Requests for authorization to use an overseas military post office should be submitted in advance.

GUIDE TO MILITARY INSTALLATION FACILITIES

The following listing is designed as a reference guide to facilities at most of the military installations in all 50 states. Facilities such as exchanges and commissaries are generally available to all military personnel, while temporary living quarters, family campgrounds, recreation areas and/or lodges may be available only on a space-available basis and some only at certain times of the year. Reservations are generally required for recreation areas and family campgrounds. It is advisable to call or write well in advance to determine if the facility you plan to use or visit will be available to you and your eligible family members. The commercial telephone number for information at each location is provided. Possession of a valid identification card is required to use these facilities. The information in this guide is based on the latest information available at time of publication. All facility listings are subject to change.

ABBREVIATIONS

C	— Commissary	G	— Golf Course	T	— Transient Quarters
F	— Family Center	R	— Recreation Area		

STATE INSTALLATION/LOCATION/ZIP	PHONE	FACILITIES				
		F	C	T	R	G
ALABAMA						
Anniston Army Depot 36201	(256) 235-7501	F			R	
Fort Rucker, Ozark 36362	(334) 255-1110	F	C	T	R	G
Maxwell AFB, Gunter Anx, Montgomery 36112	(334) 953-1110	F	C	T	R	G
Mobile CG Aviation Training Ctr. 36608	(251) 441-6401		C		R	
Redstone Arsenal, Huntsville 35898	(256) 876-2151	F	C	T	R	G
ALASKA						
Eielson AFB, Fairbanks 99702	(907) 377-1110	F	C	T	R	
Elmendorf AFB, Anchorage 99506	(907) 552-1110	F	C	T	R	G
Fort Greely 99731	(907) 873-1110	F	C	T	R	
Fort Richardson, Anchorage 99505	(907) 384-1110	F		T	R	G
Fort Wainwright, Fairbanks 99703	(907) 353-1110	F	C	T	R	G
Kodiak CG Spt. Cmd. 99619	(907) 487-5170	F	C	T	R	G
ARIZONA						
Davis-Monthan AFB, Tucson 85707	(520) 228-3900	F	C	T	R	G
Fort Huachuca, Sierra Vista 85613	(520) 538-7111	F	C	T	R	G
Luke AFB, Glendale 85309	(623) 856-1110	F	C	T	R	G
Yuma MCAS 85369	(928) 269-2011	F	C	T	R	
Yuma Proving Ground 85365	(928) 328-2151	F	C	T	R	
ARKANSAS						
Fort Chaffee 72905	(479) 484-2666			T	R	G
Little Rock AFB, Jacksonville 72099	(501) 987-1110	F	C	T	R	G
Pine Bluff Arsenal 71602	(870) 540-3000	F		T	R	G
CALIFORNIA						
Alameda CG Spt. Cmd. 94501	(510) 437-2904	F		T	R	
Barstow MC Logistics Base 92311	(760) 577-6211	F	C	T	R	G
Beale AFB, Marysville 95903	(530) 634-3000	F	C	T	R	G
Camp Pendleton MCB, Oceanside 92055	(760) 725-4111	F	C	T	R	G
China Lake Naval Air Weapons Station 93555	(760) 939-9011	F	C	T	R	G
Coronado NAB 92135	(619) 437-2011	F	C	T	R	G
Edwards AFB, Rosamond 93524	(661) 277-1110	F	C	T	R	G
El Centro NAF 92243	(760) 339-2220	F	C	T	R	
Fort Hunter Liggett, Jolon 93928	(831) 386-3000		C	T	R	
Fort Irwin, Barstow 92310	(760) 380-1111	F	C	T	R	
Lemoore NAS 93246	(559) 998-0100	F	C	T	R	
Los Angeles AFB, El Segundo 90245	(310) 653-1110	F	C	T	R	
March ARB, Riverside 92518	(951) 655-1110	F	C	T	R	
Miramar MCAS, San Diego 92145	(858) 577-1011	F	C	T	R	G
Monterey Naval PG School 93943	(831) 656-2441	F		T	R	
Naval Medical Center, San Diego 92134	(619) 532-6400				R	
North Island NAS, Coronado 92135	(619) 545-1011	F	C	T	R	G
Petaluma CG Training Ctr. 94952	(707) 765-7215	F		T	R	
Point Mugu NAS 93042	(805) 989-1110	F	C	T	R	
Port Hueneme NCBC 93043	(805) 982-4711	F	C	T	R	G
Presidio of Monterey 93944	(831) 242-5000	F	C	T	R	
San Diego MC Recruit Depot 92140	(619) 524-1011	F		T	R	
San Diego NS 92136	(619) 556-1011	F	C	T	R	G
San Diego NSB (Point Loma) 92106	(619) 553-1011	F			R	
Seal Beach NWS 90740	(562) 626-7011				R	G

STATE INSTALLATION/LOCATION/ZIP	PHONE	FACILITIES
Sierra Army Depot, Herlong 96113	(530) 827-2111	F R
Travis AFB, Fairfield 94535	(707) 424-1110	F C T R G
Twentynine Palms MCB 92278	(760) 830-6000	F C T R G
Vandenberg AFB, Lompoc 93437	(805) 606-1110	F C T R G
COLORADO		
Air Force Academy, Colorado Springs 80840	(719) 333-1110	F C T R G
Buckley AFB, Aurora 80011	(720) 847-9011	F C R
Fort Carson, Colorado Springs 80913	(719) 526-5811	F C T R G
Peterson AFB, Colorado Springs 80914	(719) 556-7321	F C T R G
Schriever AFB 80912	(719) 567-1110	F R
CONNECTICUT		
New London NSB, Groton 06349	(860) 694-3011	F C T R G
U.S. Coast Guard Academy, New London 06320	(860) 444-8444	T R
DELAWARE		
Dover AFB, Dover 19902	(302) 677-3000	F C T R G
DISTRICT OF COLUMBIA		
Anacostia Naval Station 20398	(703) 545-6700	F T R
Bolling AFB 20032	(703) 545-6700	F C T R
Fort McNair 20319	(703) 545-6700	T R
Walter Reed Army Med. Ctr. 20307	(202) 782-3501	F T R
Washington Navy Yard 20374	(703) 545-6700	R
FLORIDA		
Corry Sta, Pensacola 32511	(850) 452-6512	C T R
Eglin AFB, Fort Walton Beach 32542	(850) 882-1110	F C T R G
Hurlburt Field, Fort Walton Beach 32544	(850) 884-1110	F C T R G
Jacksonville NAS 32212	(904) 542-2338	F C T R G
Key West NAS 33040	(305) 293-2268	F C T R
MacDill AFB, Tampa 33621	(813) 828-1110	F C T R G
Mayport NS, Jacksonville 32228	(904) 270-5401	F C T R G
Naval Support Activity, Panama City 32407	(850) 234-4011	F T R
Patrick AFB, Cocoa Beach 32928	(321) 494-1110	F C T R G
Pensacola NAS 32508	(850) 452-0111	F C T R G
Tyndall AFB, Panama City 32403	(850) 283-1110	F C T R G
Whiting Field NAS, Milton 32570	(850) 623-7011	F C T R G
GEORGIA		
Albany MC Logistics Base 31704	(229) 639-5000	F C R
Atlanta NAS, Marietta 30060	(678) 655-5000	F T R
Dobbins Air Reserve Base 30069	(678) 655-5000	F T R
Fort Benning, Columbus 31905	(706) 545-2011	F C T R G
Fort Gillem, Forest Park 30297	(404) 464-2446	F C T R G
Fort Gordon, Augusta 30905	(706) 791-0110	F C T R G
Fort McPherson, Atlanta 30330	(404) 464-3113	F C T R G
Fort Stewart, Hinesville 31314	(912) 767-1411	F C T R G
Hunter Army Air Field, Savannah 31409	(912) 767-1411	F C T R G
Kings Bay NSB 31547	(912) 573-2020	F C T R G
Moody AFB, Valdosta 31699	(229) 257-4211	F C T R G
Robins AFB, Warner Robins 31098	(478) 926-1113	F C T R G

STATE INSTALLATION/LOCATION/ZIP	PHONE	FACILITIES
HAWAII		
Barking Sands Pacific Missile Range Facility, Kekaha 96752	(808) 473-0117	T R
Fort Shafter, Honolulu 96858	(808) 449-7110	R
Hale Koa Hotel, Ft. DeRussy 96815	(800) 367-6027	T R
Hickam AFB, Honolulu 96853	(808) 449-7110	F C T R G
Kaneohe Bay MCB, Kailua 96863	(808) 449-7110	F C T R G
Naval Computer and Telecommunication Area Master Station-Pacific, Wahiawa 96786	(808) 653-5385	F R
Pearl Harbor Naval Station 96860	(808) 449-7110	F C T R G
Schofield Barracks, Wahiawa 96857	(808) 449-7110	F C T R G
Tripler Army Med. Ctr., Honolulu 96859	(808) 433-6662	T R
IDAHO		
Mountain Home AFB 83648	(208) 828-2111	F C T R G
ILLINOIS		
Great Lakes Naval Training Ctr. 60088	(847) 688-3500	F C T R G
Rock Island Arsenal 61299	(309) 782-6001	F C R G
Scott AFB, Belleville 62225	(618) 256-1110	F C T R G
INDIANA		
Grissom Air Reserve Station 46971	(765) 688-5211	F T R
Naval Surface Warfare Ctr., Crane 47522	(812) 854-2511	F C T R G
IOWA		
Camp Dodge, Johnston 50131	(515) 252-4582	F T R
KANSAS		
Fort Leavenworth, Leavenworth 66027	(913) 684-7000	F C T R G
Fort Riley, Junction City 66442	(785) 239-3911	F C T R G
McConnell AFB, Wichita 67221	(316) 759-6100	F C T R G
KENTUCKY		
Fort Campbell, Clarksville 42223	(270) 798-2151	F C T R G
Fort Knox, Radcliffe 40121	(502) 624-1181	F C T R G
LOUISIANA		
Barksdale AFB, Bossier City 71110	(318) 456-2252	F C T R G
Fort Polk, Leesville 71459	(337) 531-2911	F C T R G
New Orleans NAS JRB 70143	(504) 678-3254	F T R
New Orleans NSA 70142	(504) 678-2540	F C T R
MAINE		
Brunswick NAS 04011	(207) 921-1110	F C T R G
MARYLAND		
Aberdeen Proving Ground, Aberdeen 21005	(410) 278-5201	F C T R G
Andrews AFB, Camp Springs 20762	(301) 981-1110	F C T R G
Curtis Bay Coast Guard Yard 21226	(410) 789-1600	T R
Fort Detrick, Frederick 21702	(301) 619-8000	F C R
Fort George G. Meade, Odenton 20755	(301) 677-6261	F C T R G
Indian Head Naval Surface Warfare Center 20640	(301) 744-4000	F T R G
National Naval Medical Ctr., Bethesda 20889	(301) 295-4611	F T
Patuxent River NAS 20670	(301) 342-3000	F C T R G
USNA/NS, Annapolis 21402	(410) 293-1000	F C T R G

STATE INSTALLATION/LOCATION/ZIP	PHONE	FACILITIES
MASSACHUSETTS		
Hanscom AFB, Bedford 01731	(781) 377-4441	F C T R G
MA Military Reservation 02542	(508) 968-1000	C T R G
Westover Air Reserve Station, Chicopee 01022	(413) 557-1110	F T R
MICHIGAN		
Selfridge ANGB, Mt Clemens 48045	(586) 239-4011	C T G
MINNESOTA		
Minneapolis-St. Paul IAP 55450	(612) 713-1000	F T R
MISSISSIPPI		
Columbus AFB, Columbus 39710	(662) 434-7322	F C T R G
Gulfport NCBC 39501	(228) 871-2555	F C T R G
Keesler AFB, Biloxi 39534	(228) 377-1110	F C T R G
Meridian NAS 39309	(601) 679-2211	F C T R G
MISSOURI		
Fort Leonard Wood 65473	(573) 596-0131	F C T R G
Whiteman AFB, Knob Noster 65305	(660) 687-1110	F C T R G
MONTANA		
Malmstrom AFB, Great Falls 59402	(406) 731-1110	F C T R
NEBRASKA		
Offutt AFB, Omaha 68113	(402) 294-1110	F C T R G
NEVADA		
Fallon NAS 89496	(775) 426-5161	F C T R
Nellis AFB, Las Vegas 89191	(702) 652-1110	F C T R G
NEW HAMPSHIRE		
Portsmouth Naval Shipyard, Portsmouth 03804	(207) 438-1000	F C T R
NEW JERSEY		
Cape May CG Training Ctr. 08204	(609) 898-6950	F T R
Earle NWS, Colts Neck 07722	(732) 866-2171	F T R
Fort Dix, Wrightstown 08640	(609) 562-1011	F C T R G
Fort Monmouth, Eatontown 07703	(732) 532-9000	F C T R G
McGuire AFB, Wrightstown 08641	(609) 754-1100	F C T R G
Naval Air Engineering Station, Lakehurst 08733	(732) 323-2011	F C T R G
Picatinny Arsenal 07806	(973) 724-4021	F C T R G
NEW MEXICO		
Cannon AFB, Clovis 88103	(575) 784-3311	F C T R G
Holloman AFB, Alamogordo 88330	(575) 572-1110	F C T R G
Kirtland AFB, Albuquerque 87117	(505) 846-0011	F C T R G
White Sands Missile Range, Las Cruces 88002	(575) 678-2121	F C T R G
NEW YORK		
Fort Drum, Watertown 13602	(315) 772-6011	F C T R
Fort Hamilton, Brooklyn 11252	(718) 630-4101	F C R
Naval Support Unit, Saratoga Springs 12866	(518) 886-0200	F C R
Niagara Falls Air Reserve Station 14304	(716) 236-2000	F T R
US Military Academy, West Point 10996	(845) 938-4011	F C T R G
NORTH CAROLINA		
Camp Lejeune, Jacksonville 28542	(910) 451-1113	F C T R G
Cherry Point MCAS 28533	(252) 466-2811	F C T R G
Elizabeth City CG Spt. Ctr. 27909	(252) 335-6000	F T R
Fort Bragg, Fayetteville 28307	(910) 396-0011	F C T R G

STATE INSTALLATION/LOCATION/ZIP	PHONE	FACILITIES
New River MCAS, Jacksonville 28545	(910) 451-1113	F C T R
Pope AFB, Fayetteville 28308	(910) 394-1110	F T R G
Seymour Johnson AFB, Goldsboro 27531	(919) 722-1110	F C T R G
NORTH DAKOTA		
Cavalier Air Force Station 58220	(701) 993-3292	T R
Grand Forks AFB, Grand Forks 58205	(701) 747-3000	F C T R G
Minot AFB, Minot 58705	(701) 723-1110	F C T R G
OHIO		
Wright-Patterson AFB, Dayton 45433	(937) 257-1110	F C T R G
Youngstown Air Reserve Station, Vienna 44473	(330) 609-1000	F T R
OKLAHOMA		
Altus AFB, Altus 73523	(580) 482-8100	F C T R G
Army Ammunition Plant, McAlester 74501	(918) 420-6591	F T R
Fort Sill, Lawton 73503	(580) 442-8111	F C T R G
Tinker AFB, Oklahoma City 73145	(405) 732-7321	F C T R G
Vance AFB, Enid 73705	(580) 213-5000	F C T R
OREGON		
Portland Air National Guard Base 97218	(503) 335-4000	F R
PENNSYLVANIA		
Air Reserve Station, Coraopolis 15108	(412) 474-8511	F T R
Army War College, Carlisle Barracks 17013	(717) 245-3131	F C T R G
Fort Indiantown Gap, Annville 17003	(717) 861-8500	F T R
New Cumberland Defense Distribution Center 17070	(717) 770-6011	F R G
Tobyhanna Army Depot 18466	(570) 895-7000	F C T R
Willow Grove NAS, JRB, Horsham 19090	(215) 443-1000	F T R
RHODE ISLAND		
Newport NETC 02841	(401) 841-3456	F C T R
SOUTH CAROLINA		
Beaufort MCAS 29904	(843) 228-7100	F R
Beaufort Naval Hospital 29902	(843) 228-5600	R
Charleston AFB, Charleston 29404	(843) 963-6000	F C T R G
Fort Jackson, Columbia 29207	(803) 751-7511	F C T R G
NWS, Charleston 29445	(843) 764-7901	F C T R G
Parris Island MC Recruiting Depot 29905	(843) 228-2111	F C T R G
Shaw AFB, Sumter 29152	(803) 895-1110	F C T R G
SOUTH DAKOTA		
Ellsworth AFB, Rapid City 57706	(605) 385-1000	F C T R G
TENNESSEE		
Arnold AFB, Tullahoma 37388	(931) 454-3000	F C T R G
NSA Mid-South, Millington 38054	(901) 874-5111	F C T R G
TEXAS		
Brooks AFB, San Antonio 78235	(210) 536-1110	F T R
Corpus Christi NAS 78419	(361) 961-2811	F C T R G
Dyess AFB, Abilene 79607	(325) 696-0212	F C T R G
Fort Bliss, El Paso 79916	(915) 568-2121	F C T R G
Fort Hood, Killeen 76544	(254) 287-1110	F C T R G
Fort Sam Houston, San Antonio 78234	(210) 221-1211	F C T R G
Fort Worth NAS JRB 76127	(817) 782-7153	F C T R
Goodfellow AFB, San Angelo 76908	(325) 654-3231	F C T R

STATE INSTALLATION/LOCATION/ZIP	PHONE	FACILITIES
Ingleside NS, Corpus Christi 78362	(361) 776-4200	F T R
Kingsville NAS 78363	(361) 516-6146	F C T R
Lackland AFB, San Antonio 78236	(210) 671-1110	F C T R G
Laughlin AFB, Del Rio 78843	(830) 298-3511	F C T R G
Randolph AFB, San Antonio 78150	(210) 652-1110	F C T R G
Red River Army Depot 75507	(903) 334-3143	F R
Sheppard AFB, Wichita Falls 76311	(940) 676-2511	F C T R G
UTAH		
Dugway Proving Ground 84022	(435) 831-3545	F C T R G
Hill AFB, Ogden 84056	(801) 777-1110	F C T R G
Tooele Army Depot 84074	(435) 833-3211	F T R
VIRGINIA		
Henderson Hall, Arlington 22214	(703) 545-6700	F R
Naval Support Activity NW Annex, Chesapeake 23322	(757) 421-8000	F T R
Fort A.P. Hill, Bowling Green 22427	(804) 633-8760	R
Fort Belvoir, Alexandria 22060	(703) 545-6700	F C T R G
Fort Eustis, Newport News 23604	(757) 878-1212	F C T R G
Fort Lee, Petersburg 23801	(804) 765-3000	F C T R G
Fort Monroe, Hampton 23651	(757) 878-1212	F T R
Fort Myer, Arlington 22211	(703) 545-6700	F C T R
Fort Pickett, Blackstone 23824	(434) 292-8621	F T R
Fort Story, Virginia Beach 23459	(757) 422-7305	F T R
Langley AFB, Hampton 23665	(757) 764-9990	F C T R G
Little Creek NAB, Norfolk 23521	(757) 462-7385	F C T R G
Naval Station Norfolk 23511	(757) 444-0000	F C T R G
Naval Support Facility, Dahlgren 22448	(540) 653-8531	F C T R
Norfolk Naval Shipyard, Portsmouth 23709	(757) 396-3000	F C T R
Oceana NAS 23460	(757) 433-2366	F C T R G
Quantico Marine Corps Base 22134	(703) 784-2121	F C T R G
Training Support Ctr. Hampton Roads, VA Bch 23461	(757) 492-6234	F T R
Wallops Island Surface Combat Systems Ctr. 23337	(757) 824-2068	F T R
Yorktown CG Res Training Ctr. 23690	(757) 856-3500	T R
Yorktown NWS 23691	(757) 887-4000	F T R
WASHINGTON		
Bangor NSB, Silverdale 98315	(360) 396-6111	F C T R
Bremerton Naval Hospital 98312	(360) 475-4000	R
Everett NS, Everett 98207	(425) 304-3000	F C T R
Fairchild AFB, Spokane 99011	(509) 247-1212	F C T R
Fort Lewis, Tacoma 98433	(253) 967-1110	F C T R G
McChord AFB, Tacoma 98438	(253) 982-1910	F C T R G
Puget Sound Naval Shipyard, Bremerton 98314	(360) 476-3711	F C T R
Whidbey Island NAS, Oak Harbor 98278	(360) 257-2211	F C T R G
WEST VIRGINIA		
NIOC Sugar Grove 26815	(304) 249-6302	F C R
WISCONSIN		
Fort McCoy, Sparta 54656	(608) 388-2222	F C T R
WYOMING		
Francis E. Warren AFB, Cheyenne 82005	(307) 773-1110	F C T R G

MANAGEMENT AND MOBILIZATION OF RETIRED MILITARY MEMBERS

Definitions

Key Employee. Any Reservist or any military retiree (Regular or Reserve) identified by his or her employer, private or public, as filling a key position.

Military Retirees or Retired Military Members. Regular and Reserve officers and enlisted members who retire from the military services, Reserve officers and enlisted members who are otherwise eligible for retirement but who have not reached age 60 and who have not elected discharge or are not members of the Ready Reserve or Standby Reserve (including members of the inactive Standby Reserve who meet the above criteria), and members of the Fleet Reserve and the Fleet Marine Corps Reserve.

Military Retiree Categories.

Category I. Non-disability military retirees under age 60 who have been retired fewer than five years.

Category II. Non-disability military retirees under age 60 who have been retired five years or more.

Category III. Military retirees, including those retired for disability, other than categories I or II retirees (includes warrant officers and healthcare professionals who retire from active duty after age 60).

Policy

Department of Defense (DoD) policy is that military retirees be ordered to active duty as needed to perform such duties as the Secretary concerned considers necessary in the interests of national defense. DoD Components and the Commandant of the U.S. Coast Guard plan for the use of military retirees in order to meet national security needs.

Retirees ordered to active duty are used according to guidance prescribed by the Secretary concerned as follows: (1) to fill shortages or to augment deployed or deploying units and activities or units in the CONUS, Alaska, and Hawaii supporting deployed units; (2) to release other military members for deployment overseas; (3) to fill Federal civilian workforce shortages in the Department of Defense, the U. S. Coast Guard, or other Government entities; (4) to meet national security needs in organizations outside the Department of Defense with Defense-related missions; and (5) to perform other duties the Secretary concerned considers necessary in the interests of national defense.

Each of the Secretaries concerned is responsible to prepare plans and establish procedures for mobilization of military retirees according to DoD policy. They determine the extent of military retiree mobilization requirements and develop procedures for identifying retirees in categories I and II. In addition, they maintain records for military retiree Categories I and II, including retirees who are key employees, and their availability for mobilization, civilian employment, and physical condition.

Each military retiree is responsible to provide their branch of service with an accurate mailing address and report any changes in civilian employment, military qualifications, availability for service, and physical condition.

TRAVEL AND TRANSPORTATION ALLOWANCES FOR RETIRED PERSONNEL

Retired personnel are entitled to the travel and transportation allowances explained in this section. The Joint Federal Travel Regulations (JFTR) is the basic directive for entitlements covered in this section. For specific details and additional information, contact the Transportation Office at any military installation. The JFTR is available on the PDTATAC website: http://perdiem.hqda.pentagon.mil/perdiem.

TRAVEL AND TRANSPORTATION INCIDENT TO RETIREMENT

Home of Selection

Retired military personnel are authorized travel allowances from their last duty station to a home location of his or her selection provided the member had eight or more years of continuous (with no single break of more than 90 days) active duty immediately preceding retirement. The home of selection does not have to be the same as the home of record. A retired member is entitled to travel and transportation allowances to a home of selection any place in the United States, or to a home of record or place the member entered active duty outside the United States. A member may select a home of selection outside the United States but costs are limited to those payable had the member selected a home at a CONUS location specified by the member.

Time Limitations

Home of selection travel must be completed within one year after termination of active service.

Exceptions to the one-year limitation are:

1. If the member is hospitalized on date of termination of active service, or;

2. If the member is enrolled in a training or educational program designed to qualify for acceptable civilian employment or begins such a program within the one year period following termination of active service.

If hospitalized, the retiree must perform the travel within one year following discharge. If enrolled in training or an educational program, the retiree must complete travel within one year after completion of the program or within two years after termination of active service, whichever is earlier.

In addition, retired members may request an extension of the one year time limit for travel and transportation entitlements for other than medical or educational reasons. Service Secretaries may grant extensions for other deserving cases where an unexpected event beyond the control of the member occurred which prevented the retiree from moving to the home of selection in the one year period and provided that any extensions are given in terms of a limited period of time. Examples of such exceptions are:

1. A situation where a member is prevented from selecting a home within the one year period because of a spouse's terminal illness.

2. A situation where there has been a death in the immediate family near the time of the planned move.

3. Instances where construction of a retirement home has been delayed because of labor strikes.

Requests for extensions should be submitted prior to expiration of the one year period, for specified periods of time, and show that the circumstances were beyond the control of the member. Procedures for requesting extensions and additional information may be obtained from any transportation office.

The dependents of a retiree who dies on or after January 6, 2006, who had not made a selection at the time of death, have three years from the member's date of retirement to make a selection.

Retirement Outside the CONUS

Travel to a home of selection outside the CONUS is authorized, but requires the issuance of travel orders in addition to retirement orders. Requests for such travel must be submitted in advance to the transportation officer of the member's service. It is important that travel arrangements be discussed with the transportation officer. A certificate of nonavailability is required when a U.S. flag carrier is not available or not used.

Personal Travel Allowances

A member traveling alone by privately owned conveyance (POC) is entitled to a monetary allowance of 27 cents per mile plus a flat rate per diem of $109 per authorized travel day (not to exceed one day for each 350 miles). For transoceanic travel, he or she must submit a request to the transportation officer for Government procured transportation.

Dependent Travel

Ordinarily, eligible family members are entitled to travel under the same conditions of time as the sponsor. Eligible family members are entitled to travel to the same home of selection from either the member's last duty station or from the place to which the dependents were last transported at Government expense. Eligible family members may commence travel prior to the effective date of the retirement orders. Travel must be completed within the authorized time period.

Dependent Travel Allowances

For dependent travel, the Government pays the member a flat per diem rate of $81.75 per day for dependents 12 and over and a flat rate of $54.50 for those under 12, per travel day. If the dependents travel in a second car, the member is paid 27 cents per mile, in addition to the 27 cents per mile for the member's car. Government transportation and Government-procured transportation is normally permitted for these dependents for transoceanic travel.

Claims

After completion of travel, the retired member should submit claims for mileage allowances and/or reimbursement of other travel expenses. Travel Voucher (DD Form 1351-2), along with properly endorsed orders, should be submitted to the disbursing office responsible for that person's pay accounts prior to retirement. All travel must be completed before submitting claims for payment or reimbursement.

Dislocation Allowances

Retirees are not authorized payment of a dislocation allowance for a move from the member's last duty station to the home of selection for retirement.

HOUSEHOLD GOODS, AUTOMOBILES & MOBILE HOMES

Shipping and Storage

Retired members are authorized to have their household goods shipped from their last duty station to their home of selection for retirement. Household goods from storage, or from a previously designated place in the United States or any combination of these places, may also be shipped to the home of selection. The amount of household goods allowed to be shipped depends on the grade of the individual at the time of retirement.

The shipment of household goods to a former spouse, subject to the member's overall weight and cost limitations, is permitted.

Nontemporary storage of household goods is also authorized, prior to shipment, for a period of one year from the date of retirement. Household goods, which are not placed in nontemporary storage, may be placed in temporary storage for a period of 90 days pending shipment to the selected home. Extensions of these storage periods may be authorized under certain conditions, such as hospitalization or if the member is recalled to active duty. For details, contact the transportation office at the military installation processing your retirement sufficiently in advance of the planned move to ensure that all shipments and storage arrangements are properly completed. If the member's home of selection is overseas, it is possible to arrange for shipment of household goods to the overseas location. However, due to customs and duty requirements, it is necessary for the member to contact the foreign embassy to obtain proper clearances, in addition to coordinating with the transportation officer.

PCS HHG WEIGHT ALLOWANCES (Pounds)

Pay Grade OFFICERS:	Without Dependents	With Dependents	Pay Grade ENLISTED:	Without Dependents	With Dependents
O-10	18,000	18,000	E-9	13,000	15,000
O-9	18,000	18,000	E-8	12,000	14,000
O-8	18,000	18,000	E-7	11,000	13,000
O-7	18,000	18,000	E-6	8,000	11,000
O-6	18,000	18,000	E-5	7,000	9,000
O-5	16,000	17,500	E-4	7,000	8,000
O-4	14,000	17,000	E-3	5,000	8,000
O-3	13,000	14,500	E-2	5,000	8,000
O-2	12,500	13,500	E-1	5,000	8,000
O-1	10,000	12,000	Cadets & Midshipmen		350
W-5	16,000	17,500			
W-4	14,000	17,000			
W-3	13,000	14,500			
W-2	12,500	13,500			
W-1	10,000	12,000			

Automobiles

Personnel retiring from overseas locations may ship a privately owned vehicle back to the CONUS. The transportation officer overseas will provide the member with required information regarding delivery of the vehicle to the designated port of embarkation. Land transportation of privately owned vehicles at Government expense is not authorized.

Mobile Homes

Retired members who would otherwise be entitled to transportation of household goods at Government expense are entitled to a mobile home allowance in lieu of transportation of baggage and household goods. Certain rules regarding the condition of the mobile home and its use must be met. The transportation officer can provide the retiree with details concerning movement of mobile homes. The same time restrictions for movement of household goods apply to the movement of a mobile home.

CLAIMS FOR DAMAGE OR LOSS OF HOUSEHOLD GOODS

The Department of Defense has a program that provides compensation, at full replacement value, for the loss or damage of household goods during movement. For personal property that is lost, damaged or destroyed while in care of the transportation service provider, you can recover as much as $5,000 for any single shipment, or $4.00 times the weight of your shipment up to $50,000 (whichever is greater), as compensation for such loss. This benefit applies to shipments picked up on or after October 1, 2007 for international shipments, November 1, 2007 for domestic shipments, and March 1, 2008 for non-temporary storage. You must file a claim directly with carrier within nine months of delivery. Report the loss or damage within 75 days of delivery on DD Form 1840 (at delivery), or DD Form 1840R (after delivery). If the carrier does not respond to your claim, you may file a claim through the military claims office. Detailed information about the Full Replacement Value program is available at the Military Surface Deployment and Distribution Command's (SDDC) website at www.sddc.army.mil. You can also ask your local transportation office or military claims office.

RESTRICTIONS APPLICABLE TO RETIRED MILITARY PERSONNEL

Retired members are not expected to engage in personal or professional activities that are incompatible with the standards of conduct expected of them while on active duty. The Joint Ethics Regulation, DoD Publication 5500.7-R prescribes standards of conduct and provides guidance for present and former members of the DoD, including retired members of all of the Armed Forces.

Since it is not possible to cover all the provisions of law which restrict the activities of retired personnel, only some of the more important provisions are discussed in this section. Whenever a retired member has any doubt as to whether he or she may be in violation of a particular law when accepting employment with the Federal Government or with a firm doing business with the Government, the member is strongly urged to obtain advice on the matter before making a commitment for such employment. The ethics counselor assigned to the organization can provide information and assistance.

PRE-RETIREMENT MATTERS

Retiring members should be aware that negotiating for a position following retirement might involve a financial interest that could be a conflict of interest. Negotiating includes sending a resume; deferring a discussion on employment; or leaving an employment discussion "on the table."

If a conflict of interest is created, the following options are available: immediate consultation with your supervisor and/or ethics advisor to determine if there is a conflict of interest; or, a change of duties so that you have no contact with any actions affecting the potential employer; or the possibility of future employment has been rejected by either party. (See CONFLICT OF INTEREST section.) Travel, meals, and reimbursements may be accepted from prospective employers if such expenses are customarily paid to job applicants.

If you work during terminal leave, remember you are still on active duty, and may not represent anyone before the U.S. Government (18 USC 205).

EMPLOYMENT IN THE DoD

Retired members of the uniformed services may accept civilian employment in the Federal Government. Employment by the DoD is prohibited for a period of 180 days following retirement, including employment by nonappropriated fund instrumentalities under the jurisdiction of the Armed Forces. However, there are certain circumstances when the 180 day restriction may be waived as follows: (1) the appointment is authorized by the Secretary concerned, and if appropriate, the Office of Personnel Management; (2) a state of national emergency exists; or (3) the Office of Personnel Management determines that the provisions of 5 USC 5305 apply to the position (5 USC 3326).

NONGOVERNMENT EMPLOYMENT

Generally there is no provision of federal law or regulation that prevents a retiree from being employed by a domestic business or industry that does not engage in business with the Government. There are no restrictions on the amount of compensation a retired member may receive from private employment, nor does the receipt of salary reduce the amount of retired pay received.

CONFLICT OF INTEREST

There are several conflict of interest provisions that apply to retired military personnel. Three of the main provisions are listed here. For complete details of all provisions, visit the following website: www.defenselink.mil/dodgc/defense_ethics/.

1. Retired military officers may not knowingly make a communication or appearance on behalf of any other person, with the intent to influence, before any officer or employee of any Federal agency or court in connection with a particular matter in which the officer or employee personally and substantially participated, which involved a specific party at the time of the participation and representation, and in which the U.S. is a party or has a direct and substantial interest. (18 USC 207(a)(1))

2. For a period of two years after termination of Government service, former military officers may not knowingly make a communication or appearance on behalf of any other person, with the intent to influence, before any officer or employee of any Federal agency or court, in connection with a particular matter which the employee reasonably should have known was actually pending under his or her official responsibility within one-year before the employee left Government service, which involved a specific party at that time, and in which the U.S. is a party or has a direct and substantial interest. (18 USC 207(a)(2))

3. For a period of one-year after leaving a senior position, former senior officials (flag and general officers) may not make any communication or appearance on behalf of any other person, with intent to influence, before any officer or employee of the agency or agencies in which the individual served within one-year prior to leaving the senior position, in connection with any matter on which official action is sought by such individual. (18 USC 207(c))

FOREIGN GOVERNMENT EMPLOYMENT

The primary restriction on foreign government employment of retired personnel is in Article 1, Section 9a, Clause 8 of the Constitution of the United States. This provision prohibits employment of all retired personnel, commissioned, warrant officers or enlisted personnel, regular or Reserve, by a foreign government unless Congressional consent is first granted. Employment by educational or commercial institutions owned, operated or controlled by a foreign government is within the scope of this restriction.

There are three procedures that may be used in obtaining Congressional consent. These are as follows:

1 By private legislation authorizing specific foreign government employment.

2. Authority may be obtained from the President. The President is permitted to detail members of the Uniformed Services to assist in military matters upon the request of certain countries.

3. The 1978 Foreign Relations Authorization Act grants permission by the Congress to all retired members of the Armed Forces to accept employment and compensation from any foreign government or entity as to which the applicable provisions of the Constitution would require Congressional consent subject to the approval of the Secretary of their military service and the Secretary of State. Foreign government employment may subject a member's retired pay to reduction or forfeit by the amount received from the foreign government. Retired members who wish to accept such employment should apply to their Service Secretary through military channels.

In applying for such foreign government employment the following is required:

- A detailed description of the civil duties to be performed for the foreign government as provided by the prospective employer and the extent of foreign control.
- A statement that you will or will not be paid for duties performed and an explanation of such wages, if applicable.
- A statement that you are unaware of any reason why foreign employment would be inadvisable or reflect unfavorably on the United States.
- A statement as to whether foreign citizenship is required or planned.
- A statement that you will not be required to execute an oath of allegiance to the foreign government involved.
- A statement that you understand your retired pay may be withheld equal to the amount received from the foreign government if you accept employment before proper approval.

If you accept employment different from that in your original or previous request, you must reapply in the same manner and under the same conditions as the original employment.

Note: Representing a foreign government or foreign entity may require registration as a foreign agent (22 USC 612). Contact the Registration Unit, Criminal Division, Department of Justice, Washington, DC 20530, (202) 514-1216 for more information.

The above prohibitions apply to corporations owned or operated by foreign governments, but does not apply to independent foreign companies. You may work as an independent consultant to a foreign government, however it is best to consult with competent legal authority to ensure that you are in compliance with any rules or regulations.

RETIREMENT OVERSEAS

Retired military personnel are not generally restricted from traveling or residing overseas. Retired members are subject to the same State Department requirements as all other United States citizens and must possess valid passports. Certain individuals who have had travel restrictions imposed while on active duty because their duties involved special security information, may be restricted from travel to overseas areas until the period of their restriction is completed. Individuals subject to such travel restrictions will be advised during their security debriefing prior to retirement.

For those who retire overseas or accept foreign employment and are in an area where DoDEA schools are located, their dependents of school age are authorized to attend those DoDEA schools on a space available basis.

In addition to the above, there are a number of factors that must be considered regarding choice of an overseas residence when retiring in a foreign country. Among these are:

- Availability of U.S. military support.
- Availability, cost, and quality of housing.
- Availability, cost, and quality of medical care.
- Availability of banking and financial affairs.
- Availability of transportation and communications.
- Relocation costs, taxes, and cost of living.
- Citizenship, visas, and possible loss of retirement pay if citizenship is changed.

Individuals who plan to reside overseas are generally required to notify DFAS and may be required to file a semi-annual Report of Existence or Certificate of Eligibility to ensure continuation of retirement pay.

FOREIGN CITIZENSHIP AND RETIRED PAY

If you become a citizen of a foreign country by naturalization and voluntarily renounce U.S. citizenship, you may lose your retired pay. Receipt of retired pay depends on continued status as a retired member and availability for additional service if required. Renunciation of U.S. citizenship would make further service impossible. However, the U.S. Comptroller General has ruled that while retired pay would be discontinued, the member could continue participating in the Survivor Benefit Plan if the required deductions from retired pay were being made at the time U.S. citizenship was renounced, and the member continues making premium payments to the U.S. Treasury directly.

UNIFORM RESTRICTIONS

Department of Defense policy prohibits members of the Armed Forces (including Reserve Component and retired members) from wearing the uniform under the following circumstances:

1. At any meeting or demonstration that is a function of, or sponsored by an organization, association, movement, group, or combination of persons

determined to be totalitarian, fascist, communist, or subversive, or as having adopted a policy of advocating or approving the commission of acts of force or violence to deny others their rights under the Constitution of the United States, or as seeking to alter the form of Government of the United States by unconstitutional means;

2. During or in connection with furthering political activities, private employment or commercial interests, when an inference of official sponsorship for the activity or interest may be drawn;

3. When participating in activities such as unofficial public speeches, interviews, picket lines, marches, rallies or any public demonstration, which may imply Service sanction of the cause for which the demonstration or activity is conducted unless approved by an authorized approval authority;

4. When wearing of the uniform may tend to bring discredit upon the Armed Forces; and

5. When specifically prohibited by regulations of the Department concerned.

Any questions regarding the wearing of the uniform should be directed to the nearest installation of the member's service.

SECURITY PROGRAM

All retired military personnel have a continuing responsibility for safeguarding classified information of which they have knowledge. In accordance with DoD Directive 5200.1, each member having access to classified material is required to sign a security termination statement which states that the member has no classified material or documents in his or her possession, and that no classified information will be divulged either orally or by any other means to an unauthorized individual or agency seeking to obtain classified information. Retired members who violate the espionage or internal security laws are subject to prosecution, fine, and imprisonment as well as loss of eligibility for retired pay.

SPEECHES AND PUBLICATION OF ARTICLES

There is no requirement that requires retired military personnel to submit copies of articles or speeches to the DoD or applicable branch of service for clearance. Such writings and public statements are, of course, the responsibility of the individual with regard to protection of information dealing with national security. If any doubt exists, the author may submit the material for review to the Chief of Information of the applicable service. Statements that any opinions expressed are those of the author and not the DoD or service branch are useful in clarifying the status of the article or statement.

MILITARY PERSONNEL RECORDS

Retired members, former members, and those not currently serving on active duty or in the Reserve forces should use Standard Form 180 to request records. A copy of the form can be printed from the National Archives and Records Administration website: www.archives.gov/research/order/standard-form-180.pdf.

Members currently serving on active duty or in the Reserve forces should contact their personnel offices for specific guidance and policy on how to obtain copies of Official Military Personnel Records.

The various categories of military service records are described. For each category there is a code number that indicates the address where requests should be sent.

BRANCH	CURRENT STATUS OF SERVICE MEMBER	ADDRESS CODE	
		Personnel Record	Health Record
AIR FORCE	Discharged, deceased, or retired before 5/1/1994	14	14
	Discharged, deceased, or retired 5/1/1994 to 9/30/2004	14	11
	Discharged, deceased, or retired on or after 10/1/2004	1	11
	Active (including National Guard on active duty in the Air Force), TDRL, or general officers retired with pay	1	
	Reserve, retired reserve in nonpay status, current National Guard officers not on active duty in the Air Force, or National Guard released from active duty in the Air Force	2	
	Current National Guard enlisted not on active duty in the Air Force	13	
COAST GUARD	Discharged, deceased, or retired 1/1/1898 - 3/31/1998	14	14
	Discharged, deceased, or retired on or after 4/1/1998	14	11
	Active, reserve, or TDRL	3	
MARINE CORPS	Discharged, deceased, or retired 1/1/1905 - 4/30/1994	14	14
	Discharged, deceased, or retired 5/1/1994 - 12/31/1998	14	11
	Discharged, deceased, or retired on or after 1/1/1999	4	11
	Individual Ready Reserve	5	
	Active, Selected Marine Corps Reserve, or TDRL	4	
ARMY	Discharged, deceased, or retired 11/1/1912 - 10/15/1992 (enlisted) or 7/1/1917 - 10/15/1992 (officer)	14	14
	Discharged, deceased, or retired on or after 10/16/1992	14	11
	Reserve; or active duty records of current National Guard members who performed service in the U.S. Army before 7/1/72	7	
	Active enlisted (including National Guard on active duty in the U.S. Army) or TDRL enlisted	9	
	Active officers (including National Guard on active duty in the U.S. Army) or TDRL officers	8	
	Current National Guard enlisted and officers not on active duty in Army (including records of Army active duty performed after 6/30/1972)	13	
NAVY	Discharged, deceased, or retired 1/1/1886 - 1/30/1994 (enlisted) or 1/1/1903 - 1/30/1994 (officer)	14	14
	Discharged, deceased, or retired 1/31/1994 - 12/31/1994	14	11
	Discharged, deceased, or retired on or after 1/1/1995	10	11
	Active, reserve, or TDRL	10	
PHS	Commissioned Corps - officers only	12	

ADDRESS LIST OF CUSTODIANS (by code numbers shown below)

Code	Address	Code	Address	Code	Address
1	Air Force Personnel Center HQ AFPC/DPSSRP 550 C St. West, Ste. 19 Randolph AFB, TX 78150-4721	6	National Archives & Records Administration Old Military and Civil Records (NWCTB-Military) Textual Services Division 700 Pennsylvania Ave., N.W. Washington, DC 20408-0001	11	Department of Veterans Affairs Records Management Center P.O. Box 5020 St. Louis, MO 63115-5020
2	Air Reserve Personnel Center DPSSA/B 6760 E. Irvington Pl., Ste. 4600 Denver, CO 80280-4600	7	U.S. Army Human Resources Command ATTN: AHRC-PAV-V 1 Reserve Way St. Louis, MO 63132-5200	12	Division of Commissioned Corps Officer Support ATTN: Records Officer 1101 Wooton Pkwy. Plaza Level, Ste. 100 Rockville, MD 20852
3	Commander CGPC-Adm-3 USCG Personnel Command 4200 Wilson Blvd., Ste. 1100 Arlington, VA 22203-1804	8	U.S. Army Human Resources Command ATTN: AHRC-MSR 200 Stovall St. Alexandria, VA 22332-0444	13	The Adjutant General (of the appropriate state, DC, or Puerto Rico)
4	Headquarters U.S. Marine Corps Personnel Management Support Branch (MMSB-10) 2008 Elliot Rd. Quantico, VA 22134-5030	9	Commander USAEREC ATTN: PCRE-F 8899 E. 56th St. Indianapolis, IN 46249-5301	14	National Personnel Records Center Military Personnel Records 9700 Page Ave. St. Louis, MO 63132-5100
5	Marine Corps Mobilization Command 15303 Andrews Rd. Kansas City, MO 64147-1207	10	Navy Personnel Command PERS-312E 5720 Integrity Dr. Millington, TN 38055-3120		

SPACE-AVAILABLE TRAVEL

Retired military personnel, their spouses and eligible family members who travel with them, are eligible for Space-Available (Space-A) travel on DoD owned or controlled aircraft on flights to, from, and between overseas areas. Family members may also travel within CONUS when accompanied by their sponsor.

Retired members may go to a Passenger Service Center in the passenger terminal to register for Space-A flights or may register by mail, fax, or email. AMC Form 140, "*Space-A Travel Request*" can be found on our website: www.militaryalmanac.com/Travel/Space-A-Form.pdf. The request should list the names of all travelers, a statement that required border clearance documents are current and a list of up to five country destinations. Travelers must possess all required documents: identification cards, passports, immunization records, and visas when required, along with leave orders or other travel authorizations. Travel documents must be presented when selected for travel. Travelers remain on the Space-A register for 60 days after registration.

Each terminal maintains a Space-A register (organized alphabetically, by priority and the date and time of registration for travel) that is updated daily. The register is located in the terminal and directly accessible. Travelers may call the terminal to find out where they stand on the travel register. Once registered, the traveler must wait for notification that their travel category and date/time of sign up has been reached. Space-A seats are normally identified as early as 2-3 hours and as late as 30 minutes prior to departure. Space-A travelers are not required to be present for all scheduled departures, however, when selected for a flight, the traveler must be ready to proceed.

In addition to frequent long waiting periods, it should be noted that the Government is not obligated to provide return or continuing transportation to Space-A passengers. Space-Required passengers or cargo may require the removal of Space-A passengers at any point. If removed en route, you may reregister with your original date and time of registration. Passenger agents will assign a new date and time to any country you change or add on your application.

Adequate funds should be available to procure commercial transportation for return flights or lengthy stays awaiting Space-A flights. Successful Space-A journeys require patience and flexibility.

General Space-Available Information

Space-A air travel seats are normally available after all official duty passengers have been accommodated. Passengers are permitted to sign up for a maximum of five destinations with the last destination being "all" to qualify for any opportune missions to a destination not normally served by the departure terminal.

Space-available passengers are removed from the Space-A register when they: are selected for a flight; have not moved by their leave expiration date; or have been on the Space-A register for 60 days.

Space-A travelers may reregister after they are removed from the list, however, they will be given a new date/time of sign up and drop to the bottom of the list in their respective category.

Space-A passengers should be prepared to spend time awaiting travel, and always be ready to travel at personal expense. As long as seats are available on an aircraft, every attempt is made to use them for Space-A passengers. Passengers will be offered air transportation on a first come, first served basis, based upon established space-available categories:

a. **Categories.** There are six categories of space-available travel. Space-available travelers are placed in one of the six categories based on their status (active duty, uniformed services member, DoDEA teacher, etc.) and their situation (emergency leave, ordinary leave, etc.). Once accepted for movement, a space-available passenger may not be "bumped" by another space-available passenger, regardless of category.

b. **Priority of Movement.** The numerical order of space-available categories indicates the precedence of movement between categories; for example, travelers in Category 3 move before travelers in Category 4. The order in which travelers are listed in a particular category does not indicate priority of movement in that category. In each category, transportation is furnished on a first-in, first-out basis.

c. **Changes to Movement Priorities.** Wherever the issue may arise, the local installation commander may change the priority of movement of any space-available traveler for emergency or extreme humanitarian reasons when the facts provided fully support such an exception. The installation commander may delegate the authority to make such changes to no lower than the Chief of the Passenger Service Center or its equivalent. When a movement priority is changed, the passenger shall be moved no higher than the bottom of Category 1 on the space-available list. Where AMC units are tenants, the senior local AMC authority shall advise the installation commander of this authority and offer assistance, as needed.

Categories. Following are the general categories for space-available travel:

Category 1—Emergency leave unfunded travel transportation by the most expeditious routing only for bona fide immediate family emergencies of the following: Uniformed services members with emergency status indicated in leave orders; U.S. citizen civilians stationed overseas and employees of the Uniformed Services/Nonappropriated Fund (NAF) activities; Dependents of a uniformed services member when accompanied by their sponsor; Command sponsored dependents of: U.S. citizen civilian employees of the uniformed service stationed overseas; U.S. citizen civilians paid by the NAF stationed overseas; or American Red Cross full-time paid personnel serving with a DoD Component overseas.

Category 2—Sponsors in an Environmental Morale Leave (EML) status and their dependents, also in an EML status traveling with them. Sponsors include: Uniformed services members; U.S. citizen civilian employees who are eligible for Government-funded transportation to the United States at tour completion; American Red Cross full-time, paid personnel on duty with a DoD Component overseas; USO personnel on duty with the uniformed services; DoDDS teachers during the school year and for employer-approved training during recess periods.

Category 3—Uniformed services members in a leave or pass status; Reserve Component members on active duty, in leave or pass status; Dependents of a member of the uniformed services when accompanied by their sponsor in a leave status; Uniformed services members traveling under permissive TDY orders for house hunting incident to a pending PCS (one dependent may accompany the member); Medal of Honor recipients, and their dependents when accompanied by their sponsor; Foreign cadets and midshipmen attending U.S. Service academies, in a leave status; Foreign exchange servicemembers on permanent duty with the DoD, and their dependents when accompanied by their sponsor.

Category 4—Unaccompanied family members traveling on EML orders; DoDDS teachers or family members (accompanied or unaccompanied) in an EML status year round; Unaccompanied dependents of deployed military members (including National Guard and Reserve personnel) when the deployment order is for at least 120 consecutive days, including Navy personnel assigned to a deployed ship. The previous sentence applies to travel only during the member's deployment.

Category 5—Permissive TDY (non-house hunting); Dependents (children) 18-23 years of age who are college students attending in residence at an overseas branch of an American (U.S.) university located in the same overseas area in which they reside who are command sponsored and stationed overseas with their sponsor (command-sponsored dependents stationed overseas with their sponsor are permitted to travel unaccompanied to and from the nearest overseas military academy testing site to take scheduled entrance examinations for entry into any of the U.S. Service Academies); Command-sponsored dependents of uniformed services members accompanied or unaccompanied who are stationed overseas.

Category 6—Retired uniformed services members, and dependents of retired uniformed service members when accompanied by their sponsor; Command sponsored dependents stationed overseas with their sponsor. Unaccompanied travel to the U.S. for enlisting in one of the armed forces when local enlistment in the overseas area is not authorized. If an applicant for military service is rejected, return travel to the overseas area may be provided under this eligibility; Reserve Component and National Guard members and authorized Reserve Component members entitled to retired pay at age 60 (gray area retirees) traveling in the CONUS and directly between the CONUS and Alaska, Hawaii, Puerto Rico, the U.S. Virgin Islands, Guam, and American Samoa or traveling within Alaska, Hawaii, Puerto Rico or the U.S. Virgin Islands; Newly commissioned ROTC officers who are awaiting call to extended active duty may travel within and between the CONUS, Alaska, Hawaii, and the U.S. territories.

Individuals traveling in a leave or pass status cannot sign up prior to the effective date on the leave order and must remain in a leave or pass status while awaiting travel and during the entire period of travel.

In addition, when traveling in a pass status (without leave orders), a member can sign up for a maximum of three days.

Currently, services permit appropriate civilian attire on DoD owned or controlled aircraft. When civilian clothing is worn, use common sense. Civilian attire must be neat, clean, and suitable for the occasion. Passengers must wear shoes (no open toe shoes on organic aircraft for safety reasons). Attire should not be in conflict with accepted attire in the overseas country of departure, transit, or destination, as defined by the DoD Foreign Clearance Guide. It should also be capable of keeping you warm, especially on military aircraft.

Each passenger may check two pieces of baggage, 70 pounds each, up to 62 linear inches in size. Family members may pool their baggage allowances. Hand-carried baggage must fit under the seat or in the overhead compartment, if available. Baggage weight, however, may be limited due to type of aircraft or other restrictions.

Passengers should ensure all their baggage is properly tagged with the name and address of the owner. Baggage identification tags are available at the Passenger Service Counter. There is no storage of baggage permitted in the terminal.

All Space-A travelers, except those traveling on official business, and flying Space-A from overseas air terminals, may be charged a user's fee by the U.S. Customs Service.

Individuals planning to use Space-A transportation should check with the Space-A Passenger Service Center at the nearest aerial port to obtain specific information regarding eligibility, attire, priorities, baggage, etc. prior to proceeding.

Following is a partial list of military and commercial gateways from which Space-A travelers may attempt to travel.

AMC PASSENGER OPERATIONS

Altus AFB OK
(580) 481-6428
(580) 481-5573 (FAX)
spacea@altus.af.mil

Andersen AFB Guam
(671) 366-5165/5135
(671) 366-3984 (FAX)
spacea.signup@andersen.af.mil

Andrews AFB MD
(301) 981-1854/3604/3526
(301) 981-4241 (FAX)
passenger@andrews.af.mil

Aviano AB Italy
011 39-0434-30-7680
011 39-0434-30-7782 (FAX)
spacea@aviano.af.mil

Baltimore-Washington IAP
(410) 918-6900
Toll free: 1-877-429-4262
(410) 918-6932 (FAX)
bwipax@mcguire.af.mil

NAS Brunswick ME
(207) 921-2682/2692
(207) 921-2152 (FAX)

Charleston AFB SC
(843) 963-3048/3083
(843) 963-3082 (recording)
(843) 963-2396 (FAX)
spacea@charleston.af.mil
http://public.charleston.amc.af.mil

Christchurch New Zealand
011 64-3-358-1455
011 64-3-358-1458 (FAX)
Christchurch.spacea@hickam.af.mil

Davis-Monthan AFB AZ
(520) 228-2322
(520) 228-7229 (FAX)
Dover AFB DE
(302) 677-4088
(302) 677-2953 (FAX)
436APS.spacea@dover.af.mil
Dyess AFB TX
(325) 696-4505/4903
(325) 696-8199 (recording)
(325) 696-6691 (FAX)
Eielson AFB AK
(907) 377-1854/1250
(907) 377-1623 (recording)
(907) 377-2287 (FAX)
Cato.eielson@eielson.af.mil
Elgin AFB FL
(850) 882-2488
(850) 882-3332 (recording)
(850) 882-3267 (FAX)
Elmendorf AFB AK
(907) 552-8588/4616
(907) 552-3996 (FAX)
spacea@elmendorf.af.mil
Fairchild AFB WA
(509) 247-3406
(509) 247-3399 (FAX)
92lrs.Passenger@fairchild.af.mil
NAS Fort Worth TX
(817) 782-6288/6290
(817) 782-6176 (FAX)
Grand Forks AFB ND
(701) 747-7787
(701) 747-3448 (FAX)
gf319spacea@grandforks.af.mil
Grissom AFRB IN
(765) 688-2254/2861
(765) 688-3643 (FAX)
Guantanamo Bay Cuba
011 53-99-6204/6408/6199
011 53-99-6170 (FAX)
n30atoc@usnbgtmo.navy.mil
Hickam AFB HI
(808) 449-6833
(808) 448-1503 (FAX)
735ams.trps@hickam.af.mil
spacea@hickam.af.mil
www2.hickam.af.mil/units735thams/
Incirlik AB Turkey
011 90-322-316-6424/6425
011 90-322-676-8244 (FAX)
spacea@incirlik.af.mil
MCAS Iwakuni Japan
011-81-6117-53-5509
011-81-6117-53-1854 (recording)
011 81-6117-53-3301 (FAX)
Jackson ANG MS
(601) 405-8761
(601) 405-8698 (FAX)
172aw.spacea@msjack.ang.af.mil
NAS Jacksonville FL
(904) 542-3956/8165
(904) 542-3825 (recording)
(904) 542-3257 (FAX)
Kadena AB Japan
011 81-6117-34-0153
011 81-6117-34-2159 (recording)
011 81-6117-34-4221 (FAX)
space-available@kadena.af.mil
Kaneohe Bay MCB HI
(808) 257-0777 (recording)
(808) 257-2150 (FAX)
www.mcbh.usmc.mil
Keesler AFB MS
(228) 377-2120
(228) 337-4538 (recording)
(228) 377-2488 (FAX)
Kunsan AB Korea
011 82-654-470-4666 or 5403
011 82-654-470-7550 (FAX)
spacea@kunsan.af.mil
Lackland AFB TX
(210) 925-8714/15
(210) 925-2732 (FAX)
spacea@lackland.af.mil
Lajes AB Azores
011 351-295-57-3227
011 351-295-57-3960 (recording)
011 351-295-57-5110 (FAX)
spacea@lajes.af.mil
www.lajes.af.mil
Louisville ANG KY
(502) 413-4473
(502) 413-4629 (FAX)
MacDill AFB FL
(813) 828-2440/2485
(813) 828-7844 (FAX)
(813) 828-2310 (recording)
spacea@macdill.af.mil
March JARB CA
(951) 655-2397/3214
(951) 655-2913 (recording)
(951) 655-3887 (FAX)
452amw.apsf@march.af.mil

Maxwell AFB AL
(334) 953-7372/6454
(334) 953-6760 (recording)
(334) 953-6114 (FAX)
spacea@maxwell.af.mil
McChord AFB WA
(253) 982-7259/7260
(253) 982-6815 (FAX)
eagle@mcchord.af.mil
http://public.mcchord.amc.af.mil
McConnell AFB KS
(316) 759-4810
(316) 759-5404 (recording)
(316) 759-3011 (FAX)
space-a@mcconnell.af.mil
McGuire AFB NJ
(609) 754-2864/5023
(609) 754-4621 (FAX)
wripax@mcguire.af.mil
RAF Mildenhall UK
011 44-1638-54-2248
011 44-1638-54-7426 (FAX)
spacea@mildenhall.af.mil
www.mildenhall.af.mil
Misawa AB Japan
011 81-3117-66-2370/2371
011 81-3117-66-4455 (FAX)
spacea@misawa.af.mil
www.misawa.af.mil
Moron AB Spain
011 34-955-84-8151
011 34-955-84-8354 (FAX)
spacea@moron.af.mil
Naples Italy
(Capodichino Airport)
011 39-081-568-5283/5369
011 39-081-568-5499 (FAX)
spacea@nsa.naples.navy.mil
www.nsa.naples.navy.mil/airops/spacea.htm
Nashville TN (118 AW)
(615) 399-6000 (recording)
(615) 399-5706 (FAX)
spacea@tnnash.ang.af.mil
www.tnnash.ang.af.mil
Nellis AFB NV
(702) 652-2562
(702) 652-6099 (recording)
(702) 652-2561 (FAX)
space-a.signup@nellis.af.mil
NAS New Orleans LA
(504) 678-3602/3603
(504) 678 3156 (FAX)
NAS Norfolk VA
(757) 444-4118/4148
(757) 445-7501 (FAX)
spacea@airtermnorva.navy.mil
www.airtermnorva.navy.mil
NAS North Island CA
(619) 545-9567
(619) 545-9532 (FAX)
Nasni_air_terminal_manager@navy.mil
www.nbc.navy.mil
Offutt AFB NE
(402) 294-8510
(402) 294-0877 (FAX)
pax@offutt.af.mil
Patrick AFB FL
(321) 494-5631/32
(321) 494-7991 (FAX)
Peterson AFB CO
(719) 556-4521
(719) 556-4707 (recording)
(719) 556-4979 (FAX)
space-a@peterson.af.mil
Phoenix Sky Harbor IAP AZ
(602) 302-9058
(602) 302-9181 (FAX)
spacea3@azphoe.ang.af.mil
Pope AFB NC
(910) 394-6527/6528
(910) 394-6526 (FAX)
popespacea@pope.af.mil
RAAF Richmond Australia
011 61-2-4587-1661/1652
011 61-2-4587-1663 (FAX)
Ramstein AB, Germany
011 49-6371-46-4299/4442
011 49-6371-47-2364 (FAX)
spacea@ramstein.af.mil
Randolph AFB TX
(210) 652-5287/3725
(210) 652-1854 (recording)
(210) 652-5718 (FAX)
Rickenbacker ANGB OH
(614) 492-3143
(614) 492-3413 (FAX)
www.ohcolu.ang.af.mil

Robins AFB GA
(478) 926-3166
(478) 926-4446 (recording)
(478) 926-5835 (FAX)
78trans.paxterm@robins.af.mil
www.robins.af.mil
NAS Rota Spain
011 34-956-82-2411/3282
011-34-956-82-2171
(flight information)
011 34-956-82-1734 (FAX)
spacea@navsta.rota.navy.mil
www.rota.navy.mil/navsta
Scott AFB IL
(618) 256-2014/3017
(618) 256-1854 (recording)
(618) 256-1946 (FAX)
spacea@scott.af.mil
http://public.scott.amc.af.mil
Seattle-Tacoma IAP WA
(253) 982-0555
Toll-free 877-863-1463
(253) 982-0557 (FAX)
seattle.gateway@mcchord.af.mil
http://public.mcchord.amc.af.mil/library/space-a/index.asp
Sigonella Airport Sicily
011 39-095-86-5576
011 39-095-86-6725 (FAX)
spacea@nassig.sicily.navy.mil
www.sicily.navy.mil
Singapore (RSAF Paya Lebar)
011 65-6280-0624
011 65-6382-3614 (FAX)
amcsingapore.pax@andersen.af.mil
Stewart ANGB NY
(845) 563-2369
(845) 563-2226 (recording)
(845) 563-2398 (FAX)
StewartSpaceA@nystew.ang.af.mil
www-105aw.ang.af.mil
Souda Bay Crete
011 30-28210-21383/21275
011 30-28210-21525 (FAX)
m-sb-nsa-passengerservice-gd@eu.navy.mil
Tinker AFB OK
(405) 739-4339/2106
(405) 739-3826 (FAX)
spacea@tinker.af.mil
Travis AFB CA
(707) 424-1854
(707) 424-2048 (FAX)
60aps.websa@amc.af.mil
http://public.travis.amc.af.mil/
Westover JARB MA
(413) 557-3453/2622
(413) 557-2549 (recording)
(413) 557-3147 (FAX)
439aw.spacea@westover.af.mil
NAS Whidbey Island WA
(360) 257-2604/6707
Must sign-up in person
http://naswi.ahf.nmci.navy.mil/space-a/schedule.htm
NAS Willow Grove PA
(215) 443-6217
(215) 443-6216 (recording)
(215) 443-6188 (FAX)
Wright-Patterson AFB OH
(937) 257-7741/8549
Toll free: 866-608-2976 (recording)
(937) 656-1580 (FAX)
spacea@wpafb.af.mil
Yokota AB Japan
011 81-3117-55-5660/1/2
011 81-3117-55-9768 (FAX)
730ams.space.available@yokota.af.mil

PART III

UNIFORMED SERVICES HEALTH BENEFITS

This section is a general guide to uniformed services health care in uniformed services medical treatment facilities, under the Uniformed Services Family Health Plan, and under TRICARE. For more specific details and information, contact the nearest military medical treatment facility, TRICARE representative, or visit the TRICARE website at www.tricare.mil.

UNIFORMED SERVICES HEALTH BENEFITS PROGRAM

The uniformed services are the Army, Navy, Air Force, Marine Corps, Coast Guard, Commissioned Corps of the Public Health Service, and the Commissioned Corps of the National Oceanic & Atmospheric Administration.

The primary source of medical care under the health benefits program is the uniformed services medical treatment facilities. If care is not available at a military facility, beneficiaries may use TRICARE.

In general, the priorities for care in uniformed services medical treatment facilities are:

1. Active duty servicemembers.
2. Active duty family members who are enrolled in TRICARE Prime.
3. Retirees, their family members, and survivors who are enrolled in TRICARE Prime.
4. Family members of active duty servicemembers who are NOT enrolled in TRICARE Prime.
5. Retirees, their family members and survivors who are NOT enrolled in TRICARE Prime.
6. All other eligible persons.

DEERS

Eligibility for military benefits, including TRICARE, begins with proper enrollment in the Defense Enrollment Eligibility Reporting System (DEERS). DEERS is a computerized database that identifies military sponsors, families and others worldwide who are entitled to TRICARE benefits. DEERS registration is required for TRICARE eligibility.

Registration in DEERS. Active duty and retired servicemembers are automatically registered in DEERS, but they must register their family members and ensure they are correctly entered into the database. Errors in the DEERS database can cause problems with TRICARE claims, so it is critical to ensure DEERS information is accurate.

Verifying and Updating DEERS Information. Beneficiaries can verify their DEERS information by contacting their TRICARE regional contractor, local TRICARE service center or the nearest uniformed services personnel office (ID card facility). Sponsors or registered family members may make address changes; however, only the sponsor can add or remove a family member from DEERS, and proper documents are required such as a marriage certificate, divorce decree and/or birth certificate. DEERS information may be updated as follows:

- Visiting the local uniformed services personnel office or contacting the Defense Manpower Data Center Support Office at 1-800-538-9552.

• FAX address, email or phone number changes to DEERS at (831) 655-8317.

• Mail the address change to the Defense Manpower Data Center Support Office, ATTN: COA, 400 Gigling Rd., Seaside, CA 93955-6771.

• Online at: https://www.dmdc.osd.mil/appj/address/index.jsp.

You can find the nearest uniformed services personnel office at: www.dmdc.osd.mil/rsl/owa/home.

DEERS Verification for Unremarried Former Spouses. The Social Security Number (SSN) used to verify TRICARE eligibility in DEERS for unremarried former spouses reflects TRICARE eligibility for these beneficiaries using the unremarried former spouses own SSN and not the former sponsors. Health care information is filed under the unremarried former spouses own SSN and name. These beneficiaries use their own name and SSN to schedule medical appointments and to file TRICARE claims.

The Uniformed Services Identification and Privilege Card, DD Form 1173, held by the unremarried former spouse is valid until it expires. Upon renewal, the unremarried former spouse will be issued a replacement Department of Defense/Uniformed Services Identification and Privilege Card, DD Form 2765. Contact or visit the nearest identification card issuing facility for questions or assistance.

Unremarried former spouses should always keep their DEERS information current. For questions regarding medical records, contact the military treatment facility and medical records department where the DoD medical records are maintained.

Permanent Identification Cards. A permanent, non-expiring United States Uniformed Services Identification card is available for all eligible retiree dependents who are age 75 and older, or who are permanently disabled.

Beneficiaries currently in possession of a valid ID card may obtain the new permanent ID card within 90 days of expiration. The availability of the new permanent ID card does not in any way invalidate a current ID card, nor will it impact any associated benefits and entitlements. Beneficiaries should apply for a new ID card when their existing card is within 90 days of expiring.

It is important for surviving family members to update their personal information in DEERS when the active duty or retired sponsor dies.

CHAMPVA. DEERS also applies to military families covered by CHAMPVA, administered by the Department of Veterans Affairs. CHAMPVA claims are now checked against DEERS. To resolve CHAMPVA eligibility problems, contact the CHAMPVA Center, P.O. Box 469063, Denver, CO 80206-9063, or call 1-800-733-8387.

TRICARE FOR RETIRED MILITARY MEMBERS

TRICARE for retired military members can be separated into two categories: coverage for those retirees who are not eligible for Medicare, and coverage for those who are eligible for Medicare.

Retirees Not Eligible for Medicare

Retirees who are TRICARE eligible, but not Medicare eligible, are automatically covered under TRICARE Standard. This category includes most retirees under age 65 and their eligible family members. They have an option to elect coverage under TRICARE Prime if available in their local area; Prime is discussed later in this section. The retiree and eligible beneficiaries are authorized medical care at any MTF on a space-available basis. However, availability varies and is based on location, facilities and staffing. When the services that are needed cannot be obtained through the local MTF, it may be necessary to seek care from a TRICARE authorized or network civilian provider, and have the costs shared by TRICARE Standard or Extra.

While this option provides beneficiaries with the greatest freedom of choice with respect to providers, it also comes with the greatest potential cost. Outpatient deductibles are $150 per person, and $300 per family per year. The required TRICARE Standard annual deductibles must be met before cost sharing begins. Under TRICARE Standard, the cost share for retirees and other than active duty family members is 25 percent of allowable charges, plus 15 percent of any amount assessed by a nonparticipating provider over and above the TRICARE allowable charge. The option to use TRICARE Extra, discussed later in this section, is also available. The annual cap on out-of-pocket deductibles and allowable cost shares for retirees is $3,000.

TRICARE Standard does not cover all health care, and there are special rules and limits on certain care. Therefore, retirees may want to consider a TRICARE Standard supplement to cover the costs of services not covered by TRICARE Standard. This is especially true if the retiree does not have other health insurance through a spouse or employer. TRICARE Standard pays second after other insurance, and will pay the costs the other insurance does not cover, up to the amount it would have paid had no other health insurance been available. Many military associations and organizations offer supplemental insurance. Premiums vary depending on the type of coverage chosen. (See DoD TRICARE MANAGED CARE PROGRAM section for more details about TRICARE options, including options for beneficiaries residing overseas.)

Retirees Eligible for Medicare

Medicare-eligible military beneficiaries, who are also TRICARE eligible, are eligible for TRICARE For Life (TFL). Under TFL, TRICARE acts as a second payer to Medicare. TRICARE beneficiaries must be entitled to Medicare Part A and enrolled in Medicare Part B to be covered under TFL. Under this category, TRICARE/Medicare eligible beneficiaries are further divided into two categories: beneficiaries age 65 and older and those beneficiaries under age 65.

Beneficiaries age 65 and older. Beneficiaries in this category receive TRICARE for Life benefits where TRICARE pays second after Medicare. See WHAT IS COVERED AND WHO PAYS chart for coverage details.

The law requires that all Medicare-eligible beneficiaries, regardless of age, must be enrolled in Medicare Part B to receive TFL benefits. With enrollment in Medicare Part B, TFL benefits for Medicare eligible beneficiaries are as follows:

1. If the medical care received is a benefit of both Medicare and TRICARE, Medicare will pay the allowable amount for the care. TRICARE will pay the amount that is the Medicare cost share, as well as any Medicare deductible. Most, but not all medical services are a benefit covered under both Medicare and TRICARE.

2. If the medical care received is a benefit of Medicare, but NOT a benefit of TRICARE, Medicare will pay its normal amount, and the beneficiary will be responsible only for the Medicare deductible and cost-share. An example of this type of care is certain types of chiropractic care that is covered by Medicare.

3. If the medical care received is a benefit of TRICARE, but NOT a benefit of Medicare, Medicare pays nothing. TRICARE will pay the amount it pays for the same service received by a retiree under age 65. In this case, the beneficiary must pay the applicable TRICARE cost-share and deductible. An example of this type of coverage is the prescription drug benefit.

Beneficiaries under age 65. Beneficiaries under age 65, who are eligible for Medicare, are eligible for TFL benefits. In addition, those living in a Prime service area, which is typically within 40 miles of a military treatment facility, may enroll in TRICARE Prime. This means that the eligible beneficiary receives treatment at the military treatment facility as a TRICARE Prime beneficiary and, if a referral is needed for civilian care, gets other care from a Medicare eligible provider and TRICARE processes the claim under TRICARE Prime. These under age 65 beneficiaries who are TRICARE/Medicare eligible do not have to pay Prime enrollment fees, however, they do have to pay for Medicare Part B.

TRICARE AND MEDICARE

Becoming Medicare-eligible

Medicare eligibility usually begins on the first day of the month in which the beneficiary turns 65. If the 65th birthday falls on the first day of the month, Medicare Part A eligibility begins on the first day of the preceding month. If the Medicare-eligible beneficiary purchases Medicare Part B, he or she will remain eligible for TRICARE through TRICARE For Life (TFL). With TFL, beneficiaries, age 65 and older, retain their eligibility for TRICARE Extra and Standard, but not Prime. TRICARE acts as a second payer to Medicare for services covered by both Medicare and TRICARE. When a retired sponsor reaches age 65 and becomes eligible for TFL, his or her spouse maintains regular TRICARE eligibility until he or she (spouse) becomes 65 years old.

Beneficiaries under age 65 who become entitled to Medicare Part A because of a disability or end stage renal disease, and who are enrolled in Medicare Part B, maintain their eligibility for TRICARE Prime, Extra or Standard. When they reach age 65, they will no longer be eligible for TRICARE Prime; they do, however, remain eligible for TRICARE Extra and Standard and become eligible for TFL. By law, TRICARE pays after Medicare for these eligible persons. Beneficiaries who become eligible for Medicare due to a disability or end stage renal disease should report that eligibility to the nearest military personnel office, on or after their date of eligibility.

Medicare becomes the primary payer for TRICARE beneficiaries and TRICARE becomes the secondary payer at age 65. Medicare is divided into two parts—Part A HOSPITAL INSURANCE PROTECTION and Part B MEDICAL INSURANCE PROTECTION. Part A is provided without cost, until it is used. Part B is available by paying a monthly premium. The basic premium in 2009 is $96.40 per month. Higher premiums may apply for those with income over $85,000 (single) or $170,000 (married couple). **Note:** to participate in TRICARE FOR LIFE, eligible beneficiaries must be eligible for Part A and must be enrolled in Part B. For information on how to enroll in Medicare Part B, visit the Medicare website at www.medicare.gov.

Under Medicare Part B there is an annual deductible of $135 in 2009. However, yearly mammograms for women age 65 and over and certain disabled women are covered with no Part B deductible. In addition, pap smears, pelvic and breast examinations are also covered with no Part B deductible.

TRICARE FOR LIFE: WHAT IS COVERED and WHO PAYS

INPATIENT SERVICES (MEDICARE PART A)				
	Medicare Pays[1]		TRICARE Pays[2]	What You Pay[3]
Inpatient Hospitalization (medical, surgical, and hospital-based psychiatric care) *A new benefit period [9] must begin before Medicare will cover additional days.*	Days 1-60	100% (after $1,068 deductible[4])	$1,068 deductible[4]	Nothing for services paid by Medicare and TRICARE
	Days 61-90	All but $267/day[4]	$267/day[4]	Nothing for services paid by Medicare and TRICARE
	Days 91-150[5]	All but $534/day[4]	$534/day[4]	Nothing for services paid by Medicare and TRICARE
	Days 151+	Not Covered	Allowed amount minus patient's copayment/ cost share	$250/day or 25% of the institutional charges (whichever is less) plus 20% of separately billed professional charges if care is delivered in a TRICARE network hospital $535/day[7] or 25% of the institutional services (whichever is less) plus 25% of allowable professional charges if care is delivered in a non-network hospital
Inpatient Mental Health (Psychiatric Facility)[8]: Inpatient mental healthcare requires preauthorization. Care in excess of 30 days requires a waiver for secondary TRICARE coverage. If authorized, TRICARE pays cost share or deductible. *A new benefit period [9] must begin before Medicare will cover additional days.*	Days 1-60	100% (after $1,068 deductible[4])	$1,068 deductible[4]	Nothing for services paid by Medicare and TRICARE
	Days 61 - 90	All but $267/day[4]	$267/day[4]	Nothing for services paid by Medicare and TRICARE
	Days 91-150	All but $534/day[4]	$534/day[4]	Nothing for services paid by Medicare and TRICARE
	Days 151+[10]	Not Covered	80% if network hospital[6] 75% if non-network hospital	20% of institutional charges plus 20% of professional charges for services received in a network hospital[6] For services in a non-network hospital, contact your HBA for reimbursement details.
Skilled Nursing Facility: *A beneficiary must be admitted to an inpatient hospital during a benefit period [9] for at least three days prior to receiving Medicare authorization to receive this benefit.*	Days 1-20	100%	Remaining Beneficiary Liability (if any)	Nothing for services paid by Medicare and TRICARE
	Days 21-100	All but $133.50/day[4]	$133.50/day[4]	Nothing for services paid by Medicare and TRICARE
	Days 101+	Not Covered	80% if network skilled nursing facility[6] 75% if non-network skilled nursing facility	$250/day or 25% of the institutional charges (whichever is less) plus 20% of separately billed professional charges if care delivered in a TRICARE network facility 25% of TRICARE allowable charges plus 25% of separately billed professional charges if care delivered in a non-network facility
Hospice Care	95%		Remaining Beneficiary Liability 5%	Nothing for services paid by Medicare and TRICARE

TRICARE FOR LIFE: WHAT IS COVERED and WHO PAYS

OUTPATIENT SERVICES (MEDICARE PART B)			
	Medicare Pays[1]	TRICARE Pays[2]	What You Pay[3]
Doctors Visits (Outside MTF)	80%	20%	Nothing for services paid by Medicare and TRICARE
Emergency Room Visit	80%	20%	Nothing for services paid by Medicare and TRICARE
Mental Health Visit	50%	50%	Nothing for services paid by Medicare and TRICARE
Laboratory Services	100%	Remaining Beneficiary Liability (if any)	Nothing for services paid by Medicare and TRICARE
Radiology (X-Rays)	80%	20%	Nothing for services paid by Medicare and TRICARE
Home Health Care	100% for approved services	Remaining Beneficiary Liability (if any)	Nothing for services paid by .Medicare and TRICARE
Durable Medical Equipment	80%	20%	Nothing for services paid by Medicare and TRICARE
Outpatient Hospital Services	80%	20%	Nothing for services paid by Medicare and TRICARE
Blood	Nothing for the first three pints 80% for additional pints	100% of the cost of the first three pints of blood 20% for additional pints	Nothing for services paid by Medicare and TRICARE
Chiropractic Services	80%	Not Covered	20% Medicare cost-share
HEALTHCARE OUTSIDE OF THE UNITED STATES AND ITS TERRITORIES - OUTSIDE MTF			
Inpatient Services	Not covered OCONUS[11]	75%	25% of TRICARE allowable charges
Outpatient Services	Not covered OCONUS[11]	75%	25% of TRICARE allowable charges after the TRICARE annual deductible has been met

[1]All percentages paid by Medicare are for the Medicare approved amounts for services received from Medicare providers who accept Medicare assignment.

[2]TRICARE will pay the difference between Medicare's paid amount and Medicare's limiting charge (up to 115 percent of the allowable amount) for non-participating provider claims.

[3]TRICARE has a $3,000 per fiscal year (October 1- September 30) catastrophic cap (your maximum out of pocket expense).

[4]Medicare amount that will change every calendar year.

[5]Lifetime Reserve days (91-150) are 60 additional days that Medicare will pay for, minus $534/day (in 2009) deductible, when you are in a hospital for more than 90 consecutive days. These 60 reserve days can be used only once.

[6] A network hospital is one that has a contractual agreement with TRICARE.

[7] DRG per diem rate that will change every fiscal year.

[8]190 days in a lifetime are available within a psychiatric facility.

[9]A benefit period begins when a beneficiary is admitted to a hospital or skilled nursing facility and continues until the beneficiary has been out the facility for at least 60 consecutive days.

[10] Medicare ceases to pay after day 150, unless a new benefit period begins. TRICARE will pay 75% or 80% and the beneficiary pays up to 25% depending on whether a network or non-network facility is used.

[11] The original Medicare plan does not cover health care when you travel outside the United States and its territories, except for some emergency situations in Mexico and Canada.

Medicare Part A, hospital insurance, begins automatically upon reaching age 65 if the retiree is unemployed. If you plan to continue to work beyond age 65, you will have to file an application for hospital insurance.

If beneficiaries are not eligible for Medicare Part A when they turn 65, a Social Security Administration "Notice of Disallowance" must be submitted to the uniformed services responsible for issuance of identification cards so that a new ID card showing TRICARE eligibility can be issued. Please visit www.medicare.gov, or call the Social Security Administration at 1-800-772-1213 for more information about Medicare Parts A and B.

Hospital Insurance Benefits

Hospital insurance helps pay for inpatient hospital care, inpatient care in a skilled nursing facility, home health care and hospice care.

Hospital Insurance Deductibles. The amounts a patient will pay for the Medicare hospital insurance deductible for 2009 are:

A total of $1,068 for the first 60 days of inpatient care in a benefit period;

$267 a day for the 61st through the 90th day of inpatient care;

$534 a day for the 91st through 150th day;

$133.50 for the 21st through the 100th day of care in a skilled nursing facility.

TRICARE For Life and Other Health Insurance (OHI)

Beneficiaries with Other Health Insurance (OHI), such as a Medicare supplement or employer-sponsored medical coverage, may use TFL. However, by law, TFL will pay a claim only after all OHIs have paid.

TFL works differently with other health insurance coverage depending on whether a service is payable by Medicare. Most of the time, after Medicare processes the claim, Medicare automatically forwards it to the beneficiary's OHI. After the OHI processes the claim, the beneficiary needs to file a paper claim with TFL. TFL may reimburse any remaining beneficiary liability for TRICARE covered services.

If a beneficiary receives a service that is not covered by Medicare, such as a prescription medication, TFL follows the TRICARE OHI rules to determine what portion to pay. The beneficiary may need to file a claim with TFL after the other coverage pays its portion. TRICARE pays either the amount it would have paid as the primary plan or any remaining beneficiary liability after all OHI has paid, whichever is less.

TRICARE for Life Claims

The TRICARE for Life contract is a single, nationwide contract for claims processing, customer service and administrative services for individuals who are eligible for both TRICARE and Medicare, regardless of whether they are over or under age 65. The TRICARE Management Activity awarded this contract to Wisconsin Physicians Service (WPS) of Madison, Wisconsin. Beneficiaries may call WPS TRICARE For Life (TFL) at 1-866-773-0404 for assistance.

CHAMPVA

Families of veterans who have a 100 percent, permanent disability, or of veterans who died from a service-connected disability, may be covered by CHAMPVA as long as they are not eligible for TRICARE. Eligible former spouses who, when they remarried, lost their TRICARE eligibility and whose marriage ended in divorce or death may also be eligible for CHAMPVA. The Department of Veterans Affairs administers CHAMPVA. Veterans may contact the Department of Veterans Affairs toll-free, 1-800-827-1000, for information. Details on possible CHAMPVA eligibility for family members are available from the Veterans Affairs Health Administration Center. Call toll-free 1-800-733-8387. (See VETERANS BENEFITS section.)

MEDICAL CARE FOR FAMILY MEMBERS IN MILITARY TREATMENT FACILITIES

Medical care is authorized in military treatment facilities for family members of active duty, retired, and deceased personnel, certain former spouses, and certain members, and their dependents, who qualify for transitional health care benefits, when the commander of the facility or his designee determines that space and facilities are available, and the capabilities of the professional staff are sufficient for that purpose.

When the person in charge of the military treatment facility does not have sufficient space, facilities, or professional staff to provide nonemergency care to all eligible persons, nonemergency care is furnished based on established priority. Determinations made by the commander of the medical facility, or a designee, as to the availability of space and facilities and the capabilities of the professional staff shall be conclusive. A dependent will not be denied care on the basis of service affiliation or assignment of the sponsor, or on the basis of location of the residence of the dependents.

UNIFORMED SERVICES FAMILY HEALTH PLAN

The Uniformed Services Family Health Plan (USFHP) is a TRICARE Prime option available to family members of active duty military, retirees and their eligible family members through not-for-profit health care systems in six areas of the country. Eligible beneficiaries of all ages who live in the six areas may enroll, including persons age 65 and over. Members enrolled in USFHP must receive all services through the USFHP facility in their area. These enrollees receive all the benefits offered by the TRICARE Prime program, plus premium advantages and features at no extra cost. More information is available at the USFHP website: www.usfamilyhealthplan.org.

PROVIDER	CONTACT INFORMATION
Service Areas	
Massachusetts and Rhode Island and portions of Northern Connecticut	
Brighton Marine Health Center	P.O. Box 9195, Watertown, MA 02471 1-800-818-8589
Southeast Texas and Southwest Louisiana	
CHRISTUS Health	P.O. Box 924708, Houston, TX 77292 1-800-678-7347
Maryland, Washington DC and parts of Pennsylvania, Virginia and West Virginia	
Johns Hopkins Medical Services Corporation	6704 Curtis Ct., Glen Burnie, MD 21060 1-800-808-7347
Maine, New Hampshire, Vermont and Northeastern New York	
Martin's Point Health Care	P.O. Box 9746, Portland, ME 04104 1-888-241-4556
Puget Sound Area of Washington	
Pacific Medical Centers (PacMed Clinics)	1200 12th Ave., S., Seattle, WA 98144 1-888-958-7347
New York, New Jersey, South Eastern Pennsylvania and Western Connecticut	
Saint Vincent Catholic Medical Centers of New York (SVCMC)	450 W. 33rd St., 12th Fl., New York, NY 10001 1-800-241-4848

Enrollment Eligibility. The following DEERS-eligible beneficiaries who live in USFHP service areas listed in the previous table are eligible to enroll in the USFHP:

- All active duty family members
- Uniformed Services retirees and their eligible family members-including those age 65 and over
- Qualifying activated Reservist family members
- Eligible family members of deceased active duty or retired uniformed services members-including those age 65 and over
- Medal of Honor recipients and their families.

Summary of Benefits. No or low cost health care with predictable costs; pharmacy services and prescription drug coverage, including convenient mail-order options; coordinated programs of health care managed by a primary care physician; large network of primary care and specialized physicians; skilled nursing care; home health care; outpatient care; hospitalization; office visits to all specialists as referred by the primary care physician; annual eye exams; maternity and infant care; preventive care, including well-child care; inpatient and outpatient mental health; diagnostic tests and services; medically necessary ambulance services; medical emergency services worldwide; physical/occupational and speech therapies; durable medical equipment; onsite patient advocates; and ongoing quality management. This is not an all-inclusive list. Check with the Uniformed Services Family Health Plan in your area for complete details of benefit coverage and exclusions.

Enrolling in the USFHP.

- Beneficiaries must complete an application (available online) and submit it to the nearest USFHP program. Beneficiaries must live in that program's service area to enroll.
- Beneficiaries, including retirees, may enroll at any time throughout the year.
- With enrollment in the USFHP, beneficiaries make a one-year commitment and agree to get care only through the plan unless they move out of the area or their eligibility status changes.

For more information about the USFHP, call the toll-free number of the program in the area you are interested, or visit the USFHP website: www.usfamilyhealthplan.org.

DoD TRICARE MANAGED CARE PROGRAM

TRICARE is the Department of Defense managed health care program for active duty military, active duty service families, retirees and their families, and other beneficiaries. Within TRICARE, there are generally three options:

- TRICARE Prime
- TRICARE Extra
- TRICARE Standard

TRICARE PRIME

TRICARE Prime is a managed care option similar to a civilian Health Maintenance Organization (HMO). This option requires enrollment. Active duty servicemembers are required to enroll in TRICARE Prime. Active duty family members, retirees and their family members are encouraged, but not required, to enroll in Prime. However, to receive the TRICARE Prime benefit, they must reside where TRICARE Prime is offered. Ask your local TRICARE

Service Center (TSC) about the TRICARE Prime availability in your area. If you are stationed in a remote area, TRICARE Prime Remote (TPR) and TRICARE Prime Remote for Active Duty Family Members (TPRADFM) may be the option available to you and your family members. This option also requires enrollment.

TRICARE Prime offers less out-of-pocket costs than any other TRICARE option. Active duty members and their families do not pay enrollment fees, annual deductibles or copayments for care in the TRICARE network. Retirees and their family members pay an annual enrollment fee of $230 for an individual or $460 for a family, and minimal copayments apply for care in the TRICARE network. Although TRICARE Prime offers a "point-of-service" option for care received outside of the TRICARE Prime network, receiving care from a nonparticipating provider is not encouraged.

TRICARE Prime enrollees receive most of their care from military providers or from civilian providers who belong to the TRICARE Prime network. Enrollees are assigned a Primary Care Manager (PCM) who manages their care and provides referrals for specialty care. All referrals for specialty care must be arranged by the PCM to avoid point-of-service charges.

Point-of-Service Option

The TRICARE Prime Point-of-Service (POS) option allows TRICARE Prime enrollees to receive nonemergency, TRICARE-covered services from any TRICARE-authorized provider without a referral from their primary care manager or authorization from their regional contractor. The POS option does not apply to active duty service members, newborns in their first 60 days, emergency care or for those who have other health insurance.

Cost-shares and Deductibles. The following chart explains the costs incurred by TRICARE Prime enrollees if they use the point-of-service option.

Charges	Individual	Family
Deductible per fiscal year (10-1 to 9-30) for outpatient care only	$300	$600
Cost-share for outpatient care	50 percent of TRICARE allowable charge, after annual deductible is met	
Cost-share for inpatient care	50 percent of TRICARE allowable charge	
Any additional charges by non-network providers	Beneficiary's responsibility; up to 15 percent above the allowable charge is permitted by law	

The point-of-service charges do not apply to care received under TRICARE Extra or TRICARE Standard.

Explanation of Benefits. The explanation of benefits is a statement received by the beneficiary and the provider after services are provided. The explanation of benefits documents the results of the claims process. Any point-of-service charges are detailed on the explanation of benefits, and the amount applied to the annual deductible is shown.

For more information about the TRICARE Prime point-of-service option or to ask questions about your explanation of benefits, contact your local health benefits adviser, beneficiary counseling and assistance coordinator, or the TRICARE Service Center.

TRICARE EXTRA

With TRICARE Extra, there is no enrollment or annual fee. However, enrollees do have to satisfy an annual deductible for outpatient care. The deductible and cost sharing work the same way for TRICARE Extra and TRICARE Standard.

In the TRICARE Extra program, when care is received from a TRICARE Extra network provider, patients receive a discount on cost sharing, and do not have to file their own claims. Beneficiaries use TRICARE Extra on a case-by-case basis just by using the network providers. TRICARE Extra is not available overseas or for active duty servicemembers.

The annual outpatient deductibles for TRICARE Extra are the same as for TRICARE Standard. For the families of active duty in grade E-4 and below, the deductible is $50 for one person or $100 for a family per fiscal year. For all others, the deductible is $150 for one person or $300 for a family.

After the annual outpatient deductible has been satisfied, the cost share under TRICARE Extra for an active duty family member is 15 percent of the fee for which the TRICARE Extra network provider has contracted to provide the medical service or supply. All other eligible beneficiaries will pay a 20 percent cost share of the contracted fee. The regional TRICARE Service Center can provide a list of TRICARE Extra providers within each region.

There are a few exceptions to the TRICARE Extra cost sharing percentages. The ambulatory surgery cost is $25 for active duty families and 20 percent of the contracted fee for all others. To learn about the exceptions, contact the TRICARE Service Center in your area for details.

TRICARE COVERED SERVICES

Outpatient Services	TRICARE Standard	TRICARE Extra	TRICARE Prime
Ambulance Services **Durable Medical Equipment** **Emergency Room Services** **Enhanced Clinical Preventive Services [1]** **Laboratory and X-ray Services** **Medical Supplies** **Medication Management** **Mental Health (Behavioral Health)** **Occupational Therapy** **Office Visits** **Physical Therapy** **Prosthetic Devices** **Radiology/Pathology/Cardio Studies (ancillary service)** **Urgent Care**	20%	15%	No copay
Ambulatory Surgery (Same Day)	$25*	$25*	No copay
Home Health Care **Hospice Care** **Outpatient Maternity Office Visits**	No cost	No cost	No copay

[1] Enhanced Clinical Preventive Services include: cancer screening, clinical preventive exams (physicals), cardiovascular screening, eye exams, hearing screening, immunizations, infectious disease screening, mammograms, pap smears, and well child care.

* The copay is for facility fees, there is no separate copay for professional fees.

Inpatient Services	TRICARE Prime	TRICARE Extra	TRICARE Standard
Hospitalization for Medical/Surgical care **Inpatient Maternity** **Skilled Nursing Facility Care**	No copay	$15.65/day ($25 min. charge)**	$15.65/day ($25 min. charge)**
Hospitalization for Mental Illness **Substance Use Treatment** **Partial Hospitalization Program - Mental Health** **Residential Treatment Center**	No copay	$20/day ($25 min. charge)**	$20/day ($25 min. charge)**

** Inpatient care at a military treatment facility has a copay of $15.65 per day.

Waiver of Cost-Share Copayments

Beginning in FY 2009, all copayments for preventative care services are waived even if the beneficiary has not met the annual TRICARE deductible during the year in which the preventative care services are received. This applies to: colorectal screening; breast screening; cervical screening; prostate screening; annual physical exam; vaccinations; and any other services that the Secretary of Defense prescribes.

This waiver does not apply to Medicare eligible beneficiaries. However, the Secretary of Defense has discretionary authority to reimburse Medicare eligible beneficiaries for preventative care services obtained during FY 2009.

TRICARE STANDARD

TRICARE Standard is the basic TRICARE health care program, offering comprehensive health care coverage. TRICARE Standard shares most of the costs of care from civilian hospitals and doctors when care is not obtained at a uniformed services hospital or clinic. But there are certain important points that beneficiaries should know about TRICARE Standard before using it.

- The lowest cost medical care is available from military treatment facilities.
- Benefits and costs vary for different categories of eligible beneficiaries.
- While a beneficiary may be eligible to use a military treatment facility, access may be restricted based on the facility's capacity to see patients.
- TRICARE Standard is not free. Patients must pay part of the cost of medical care, as well as everything TRICARE Standard does not cover.
- TRICARE Standard does not cover all health care. There are special rules or limits on certain care, and some care is not covered at all.
- TRICARE Standard pays for only medically necessary care and services that are provided at an "appropriate level of care." Claims for services that do not meet this definition will be denied.
- Providers of care must be certified by the regional contractor as an authorized provider of care under TRICARE Standard for the Government to share the cost of care provided. Being "authorized" or "certified" is not the same as being a "participating" or "nonparticipating" provider of care under TRICARE Standard.
- The patient or the provider must file claims before TRICARE Standard can pay its share of the bills. It is important to fill out the claim form correctly and to include any necessary paperwork.

Preauthorization Requirements. TRICARE Standard beneficiaries are accustomed to self-referring—that is, they schedule their own appointments with primary care providers and specialists. But for anything other than routine care, or for a non-routine procedure (such as an MRI), Standard beneficiaries may need to seek an authorization from the TRICARE regional contractor in their area. The following is a list of the more commonly asked for services that require preauthorization. Please note this list is not all inclusive and beneficiaries should contact their regional contractors to ask whether or not preauthorization is needed.

- Adjunctive dental services
- Home health services
- Hospice care
- Nonemergency inpatient admissions for substance-use disorders or behavioral health
- Outpatient behavioral health care beyond the eighth visit
- Transplants (all solid organ and stem cell)
- TRICARE Extended Care Health Option services

If beneficiaries have Other Health Insurance (OHI), they are not required to obtain pre authorizations for TRICARE covered services, except for adjunctive dental care, the Extended Care Health Option, stem cell and organ transplants, and behavioral health care services.

TRICARE REGIONAL CONTRACTORS

TRICARE North Regional Contractor	TRICARE South Regional Contractor	TRICARE West Regional Contractor
HealthNet Federal Services 1-877-TRICARE	Humana Military Healthcare Services 1-800-444-5445	TriWest Healthcare Alliance 1-888-874-9378
www.healthnetfederalservices.com	www.humana-military.com	www.triwest.com

TRICARE OVERSEAS PROGRAM

The TRICARE Overseas Program is a managed health care program for beneficiaries Outside the Continental United States (OCONUS). The TRICARE Overseas Program blends many of the features of the Continental United States (CONUS) TRICARE program while considering the significant cultural differences unique to foreign countries and their health care practices. The program consists of three regions: Europe, Pacific, and Latin America/Canada. TRICARE Overseas is a dual-option benefit consisting of TRICARE Overseas Prime and TRICARE Overseas Standard.

TRICARE Overseas Prime. TRICARE Prime enrollees select or are assigned a primary care manager (PCM). Your PCM provides all of your primary care or refers you to another provider if he or she cannot provide the care. You must have a referral and authorization from your PCM for routine, specialty, or inpatient care provided by anyone other than your PCM. You will usually receive most of your PCM care, preventive services, and specialty care services from a Military Treatment Facility (MTF).

Accessing Host Nation Specialty Care. When necessary, your PCM will refer you to qualified host nation providers for specialty care. In most cases, these providers will speak English. If not, bilingual patient liaisons are available at each MTF to interact with local providers and assist you.

Many host nation providers expect payment at the time services are provided, often in local currency instead of U.S. dollars. If this is the case, you may then submit a claim to the overseas claims processor, along with a copy of an itemized bill, for appropriate reimbursement from TRICARE.

If referred outside the MTF, host nation providers may assess a fee for copies of medical records, which must be paid by the beneficiary, as the service is not reimbursable by TRICARE.

TRICARE Overseas Standard. The TRICARE Overseas Standard option is similar to the CONUS TRICARE Standard option.

TRICARE PHARMACY PROGRAM

TRICARE provides a world-class pharmacy benefit to all eligible Uniformed Services members. Eligible beneficiaries may fill prescription medications at Military Treatment Facility (MTF) pharmacies; through the TRICARE Mail Order Pharmacy (TMOP); at TRICARE retail network pharmacies (TRRx); and at non network pharmacies. Beneficiaries need a written prescription and a valid Uniformed Services identification card to have a prescription filled.

Pharmacy Benefit Program Copayment Structure

Active duty service members do not pay copayments for prescriptions. However, if they receive medications through an overseas pharmacy or an out-of-network pharmacy, they may need to pay out-of-pocket for the total cost of the medication and then file a claim for reimbursement for the full amount.

Other eligible beneficiaries pay the pharmacy copayment based on whether the prescription medication is classified as a formulary generic (Tier 1), formulary brand name (Tier 2), or non-formulary (Tier 3) drug. The copayment also depends on where the beneficiary chooses to fill their prescription.

Beneficiaries may fill their prescriptions at the MTF, through the TMOP or at one of the approximately 55,000 TRRx in the nationwide network. Beneficiaries may also have prescriptions filled at non-network pharmacies, but will pay significantly more and must meet a deductible.

TRICARE Pharmacy Copayments In the United States (Including Puerto Rico, Guam, Virgin Islands)

Place of Service	Uniform Formulary		
	Formulary		Non-formulary (Tier 3)
	Generic (Tier 1)	Brand Name (Tier 2)	
Military Treatment Facility (MTF) pharmacy (up to a 90-day supply)	$0	$0	Not Applicable
TRICARE Mail Order Pharmacy (TMOP) (up to a 90-day supply)	$3	$9	$22*
TRICARE Retail Pharmacy Network pharmacy (TRRx) (up to a 30-day supply)	$3	$9	$22*
Non-network retail pharmacy (up to a 30-day supply) **Note:** Beneficiaries using non-network pharmacies may have to pay the total amount of their prescription first and then file a claim to receive partial reimbursement.	**For those who are not enrolled in TRICARE Prime:** $9 or 20 percent of total cost, whichever is greater, after deductible is met (E-1 to E-4: $50/person; $100/family; all others, including retirees, $150/person, $300/family) **TRICARE Prime:** 50 percent cost share after point-of-service deductibles ($300 per person/$600 per family deductible)		**For those who are not enrolled in TRICARE Prime:** $22 or 20 percent of total cost, whichever is greater, after deductible is met (E-1 to E-4: $50/person; $100/family; all others, including retirees, $150/person, $300/family) **TRICARE Prime:** 50 percent cost share after point-of-service deductibles ($300 per person/$600 per family deductible)

Beneficiary Copayments At Overseas Locations

	Active duty servicemembers	Active Duty Family Members	
		Enrolled in Prime	Not enrolled in Prime
Copayment	No copayment	No copayment**	20 percent cost share after deductible of $50/100 for E-1 to E-4 ADFMs or $150/300 for E-5 and above ADFMs

* If medical necessity is established for a non-formulary drug, patients may qualify for the $9 copayment for up to a 30-day supply in the TRRx or a 90-day supply in the TMOP program.

**ADFMs enrolled in TRICARE Overseas Program Prime (at an MTF) who use host nation pharmacies are subject to Prime point-of-service deductibles of $300/600 and 50 percent cost shares.

MTF Pharmacy

Prescriptions may be filled (up to a 90-day supply for most medications) at an MTF pharmacy at no cost to the beneficiary, if the medication is on the MTF formulary. Beneficiaries should contact their local MTF to find out what is on the formulary and for specific details about filling prescriptions at the MTF pharmacy. See the TRICARE Formulary Search Tool at www.tricareformularysearch.org to find out what medications are available at military pharmacies.

TRICARE Mail Order Pharmacy (TMOP)

TMOP is administered by Express Scripts Inc. (ESI), and is available for prescriptions that beneficiaries use on a regular basis. Beneficiaries may save a considerable amount of money by using the TMOP for maintenance medications instead of retail network pharmacies. Beneficiaries may receive up to a 90-day supply for most medications. Prescriptions may be refilled by mail, phone or online. Beneficiaries who have prescription drug coverage from another health insurance plan may not use TMOP unless the medication is not covered under the other plan, or the beneficiary exceeds the dollar limit of coverage under the other plan.

To use TMOP, beneficiaries register at www.express-scripts.com/TRICARE. They should follow the instructions on the ESI website to submit the form. Beneficiaries must then mail their health care provider's written prescription and the appropriate copayment to ESI. The patient's physician may FAX or phone in new prescriptions. Medications are sent directly to the beneficiary within 10 to 14 days.

For more information about TMOP, beneficiaries may visit the ESI website at www.express-scripts.com/TRICARE or contact TMOP member services toll-free 1-866-363-8667. The TMOP website is www.tricare.mil/pharmacy/tmop.cfm.

TRICARE Retail Pharmacy Program (TRRx)

TRRx is also administered by ESI. Beneficiaries in the CONUS and its territories (Guam, Puerto Rico, U.S. Virgin Islands) may use the nationwide network of approximately 55,000 retail pharmacies to fill prescriptions. Search for local pharmacies on the ESI website, or call 1-866-363-8779 for assistance.

Non-Network Pharmacies

A "non-network pharmacy" is a retail pharmacy that is not part of the TRICARE network. To verify that a pharmacy is or is not a part of the TRICARE retail pharmacy network, beneficiaries should check the ESI website, or call 1-866-363-8779. Filling prescriptions at non-network pharmacies is the most expensive option. Beneficiaries may have to pay for the total amount first, and file a claim to receive partial reimbursement.

Medical Necessity

If the doctor believes that it is medically necessary for a patient to receive a non-formulary (Tier 3) medication instead of any of the formulary alternatives that are on the Uniform Formulary, the medication may be provided at the formulary brand name (Tier 2) copayment if the medical necessity can be substantiated. To establish medical necessity, complete the form for establishing medical necessity. The form, including instructions for completing and submitting, is available on the TRICARE website at www.tricare.mil/pharmacy/medical-nonformulary.cfm.

Filing Claims

Beneficiaries with other health insurance, filing a pharmacy claim with TRICARE, should mail the claim to: Express Scripts Inc., P.O. Box 66518, St. Louis, MO, 63166-6518. Claims must be filed within one year of the date of service. The TRICARE claim form (DD 2642) is available at www.dtic.mil/whs/directives/infomgt/forms/eforms/dd2642.pdf. Beneficiaries may call 1-866-363-8779 for questions about filing a pharmacy claim.

Using Other Health Insurance (OHI)

When using insurance other than TRICARE, the OHI is the first payer. Beneficiaries may then be eligible for full or partial reimbursement from TRICARE for out-of-pocket costs, including copayments. Beneficiaries who have OHI should use a retail pharmacy under their private insurer that is also in the TRICARE retail network to avoid paying the TRICARE non-network deductible. Beneficiaries who have prescription drug coverage from OHI may not use TMOP unless the medication is not covered under the other plan, or unless the beneficiary exceeds the dollar limit of coverage under the other plan. When beneficiaries have OHI, the rules of that insurer apply. Beneficiaries should call ESI at 1-866-363-8779 for specific instructions about filing pharmacy claims if they have other health insurance.

TRICARE RETIREE DENTAL PROGRAM

Delta Dental of California administers the TRICARE Retiree Dental Program (TRDP). The TRDP is separate from the TRICARE Dental Program (TDP), which is available only to active duty family members and National Guard and Reserve members and their family members.

TRDP enrollment is open to Reserve Component retirees and their family members, including gray area retirees who are entitled to retired pay but not yet receiving it. In addition, a retiree who elects to enroll within four months after retirement from active duty or transfer to Retired Reserve is eligible for a waiver of the 12-month waiting period for the full scope of Enhanced Program benefits.

Coverage is available in the 50 States, the District of Columbia, Puerto Rico, Guam, the U.S. Virgin Islands, American Samoa, the Commonwealth of the Northern Mariana Islands and Canada. All premiums are paid by the enrollee and vary depending on where the enrollee lives.

There are more than 100,000 participating locations in the Delta Dental Select and Delta Dental PPO networks. If the enrollee visits a Delta Dental Select or Delta Dental PPO dentist, all claims are submitted for the enrollee, and the enrollee is responsible only for the enrollee's percentage of the TRDP-allowed amount.

Enrollees may use any licensed dentist within the designated service area for treatment. However, for an out-of network dentist, Delta Dental pays the same percentage of covered services but cannot guarantee the dentist's fees. Out-of-network dentist's bill for their normal fees, which may be higher than the program allowed amount for the service. The enrollee is responsible for paying the copayment plus any difference between the program allowed amount and the dentist's billed charge.

For more information about the TRDP call toll-free 1-888-838-8737, or visit Delta Dental online at www.trdp.org.

TRICARE RETIREE DENTAL PROGRAM BENEFITS

Benefits available during the first 12 months of enrollment:	**Delta Dental Pays***
Diagnostic services (such as exams)	100%
Preventive services (such as cleanings)	100%
Basic Restorative services (such as fillings)	80%
Endodontics (such as root canals)	60%
Periodontics (such as gum treatment)	60%
Oral Surgery (such as extractions)	60%
Emergency (such as treatment for minor pain)	80%
Dental Accident Coverage	100%
Additional services available after 12 months of continuous enrollment:	
Cast Crowns, Onlays & Bridges	50%
Partial/Full Dentures	50%
Dental Implant Services	50%
Orthodontics	50%
Deductibles & Maximums	
Annual Deductible (per person, $150 cap per family, per benefit year)	$50
Annual Maximum (per person, per benefit year)	$1,200
Orthodontic Maximum (per person, per lifetime)	$1,500
Dental Accident Maximum (per person, per benefit year)	$1,000
Benefit Year: October 1 - September 30	

*The percentage paid by Delta is based on the allowed amount for each procedure. Your out-of-pocket costs may be higher if care is received from a non-participating provider.
Covered benefits are subject to certain limitations.

CONTINUED HEALTH CARE BENEFIT PROGRAM

The Continued Health Care Benefit Program (CHCBP) provides benefits similar to TRICARE Standard for a specific period of time (18-36 months) to former servicemembers and their family members, some unremarried former military spouses, and emancipated children who enroll and pay quarterly premiums. CHCBP is not part of TRICARE.

Individuals who lose TRICARE eligibility or other coverage under the Military Health System are eligible for temporary health care coverage in CHCBP. Enrollment in CHCBP must occur within 60 days after separation from active duty or loss of eligibility for military health care. The premiums for this coverage are $933 per quarter for individuals and $1,996 per quarter for families.

The Department of Defense has contracted with Humana Military Healthcare Services, Inc. to administer the CHCBP. Contact CHCBP for information on eligibility for enrolling in the program, to request a copy of the CHCBP enrollment application, to obtain information regarding the health care benefits that are available to CHCBP enrollees, and to obtain information regarding the premiums and out-of-pocket costs once you are enrolled. Write to: Humana Military Healthcare Services, Inc., Attn. CHCBP, P.O. Box 740072, Louisville, KY 40201, or call 1-800-444-5445, option 4. Detailed information regarding the CHCBP and an enrollment application is also available on the website: www.humana-military.com. For more information about the CHCBP prior to separation, contact your health benefits advisor or military personnel and transition office.

SPECIAL CIRCUMSTANCES FOR TRICARE ELIGIBILITY

Check with your local ID card issuing facility or unit personnel office about eligibility requirements for the following beneficiary categories:

• Certain family members of active duty servicemembers who were discharged as a result of a court-martial conviction or separated for spouse or child abuse.

• Certain abused spouses, former spouses and dependent children of servicemembers who were eligible for retirement, but whose retirement benefits were revoked as a result of abuse of the spouse or child.

• Spouses and children of representatives of North Atlantic Treaty Organization (NATO) and Partners for Peace (PFP) nations that are signatories to the respective Status of Forces Agreements (SOFAs) with the United States, while stationed in or passing through the U.S. on official business. These family members are eligible for outpatient benefits only.

Dependent Parents and Parents-in-law

Dependent parents and parents-in-law are eligible for care in Military Treatment Facilities (MTFs) and can enroll in TRICARE Plus on a space/resource available basis. TRICARE Plus is a local MTF-based primary care enrollment program that may provide TRICARE-eligible beneficiaries, who are not enrolled in Prime, an opportunity to enroll with their local MTF for primary care services. Dependent parents and parents-in-law are not eligible for TRICARE Prime, Standard, Extra or TRICARE For Life.

TRICARE ELIGIBILITY

Beneficiary Category	Description
Active duty and retired servicemembers	From any of the seven uniformed services Army, Navy, Air Force, Marine Corps, Coast Guard, National Oceanic & Atmospheric Administration, or the United States Public Health Service.
Spouses and unmarried children (including stepchildren) of active duty or retired servicemembers Note: Stepchildren lose eligibility after a divorce unless adopted by the sponsor.	* Remain eligible even if parents divorce or remarry. * Eligibility ends at age 21 unless the child is a full-time student (validation of student status required) then eligibility ends at age 23 or when the full-time student status ends. * Eligibility may extend past age 21 if the child is incapable of self-support because of a mental or physical incapacity and the condition existed prior to age 21, or if the condition occurred between the ages of 21 and 23 while the child was a full-time student. * Illegitimate children of current or former servicemembers or their spouses may be eligible under certain conditions. * Children placed in the custody of a servicemember or former member, either by a court or by a recognized adoption agency, in anticipation of legal adoption by the member.
Reserve Component members on active duty for more than 30 days—under Federal orders	From any of the Reserve Components: Army, Navy, Air Force, Marine Corps, Coast Guard, or the United States Public Health Service.
Spouses and unmarried children of Reserve Component servicemembers	* Covered while Reserve Component sponsor is on active duty for more than 30 consecutive days. * Covered if Reserve sponsor was injured or dies during, or on the way to or from, active duty training for a period of 30 days or less.
Retired Reserve Component servicemembers and their family members	When the retired Reserve Component servicemember is eligible for retirement pay (usually at age 60), the member and his/her eligible family members become TRICARE eligible.

Beneficiary Category	Description
Widows or widowers and unmarried children of deceased active duty or retired service members	* Are eligible as family members of deceased member if sponsor was serving or was ordered to active duty for more than 30 days at time of death. * Claims will be cost-shared at the active duty family member rate for three years after death of active duty sponsor, and thereafter at the retiree rate. * Widows or widowers remain eligible until they remarry (loss of benefits remains applicable even if remarriage ends in death or divorce) * Children remain eligible until age 21, unless they meet the exceptions above. If the sponsor dies on or after October 7, 2001 while serving on active duty, the dependent child is covered at the active duty family rate until no longer eligible for TRICARE.
Medal of Honor recipients and their family members	Any servicemember who has been awarded the Medal of Honor, his eligible family members and widows are eligible for medical and dental benefits under TRICARE.
Certain eligible former spouses of active duty or retired servicemembers	* Must not have remarried. (If they remarry, the loss of benefits remains applicable even if remarriage ends in death or divorce) * Eligibility and medical records are listed under former spouses own name and Social Security Number—not their sponsor's. * Must not be covered by an employer-sponsored health plan. * Must not be the former spouse of a North Atlantic Treaty Organization (NATO) or "Partners for Peace" (PFP) nation member. * Must meet the requirements of one of the following three situations:
Situation 1	
* Must have been married to the SAME member or former member for at least 20 years, and at least 20 of those years must have been creditable in determining the member's eligibility for retirement pay. * If the date of the final decree of divorce or annulment was on or after February 1, 1983, the former spouse is eligible for TRICARE coverage of health care that is received after the date of the divorce or annulment. * If the date of the final decree is before February 1, 1983, the former spouse is eligible for TRICARE coverage of health care received on or after January 1, 1985. * Eligibility continues as long as the preceding requirements continue to be met.	
Situation 2	
* Must have been married to the SAME military member or former member for at least 20 years, and at least 15, but less than 20, of those married years must have been creditable in determining the member's eligibility for retirement pay. * If the date of the final decree of divorce or annulment is before April 1, 1985, the former spouse is eligible only for care received on or after January 1, 1985, or the date of the decree, whichever is later. * Eligibility continues as long as the preceding requirements continue to be met. However, if the date of the final divorce decree or annulment is on or after April 1, 1985, but before September 29, 1988, the former spouse was eligible for care received from the date of the decree until December 31, 1988, or two years from the date of the decree, whichever was later.	
Situation 3	
* Must have been married to the SAME military member or former member for at least 20 years, and at least 15, but less than 20, of those married years must have been creditable in determining the member's eligibility for retirement pay. * If the date of the final decree of divorce or annulment is on or after September 29, 1988, the former spouse is eligible only for care received for one year from the date of the decree.	

TRICARE CONTRACTORS

The following is a list of TRICARE regions, what states/areas they include, and the names, claims mailing addresses, telephone numbers and websites of the contractors for each region.

Generally, claims addresses and telephone numbers are for the firms that are sub-contracted to process claims for each regional TRICARE contractor.

If you need to check on the status of a TRICARE claim with your state's claims processor listed, the process will be speeded up considerably if you have the following information at hand when you call: patient's full name; sponsor's full name, rank, branch of service and status (active or retired); sponsor's Social Security number; sponsor's and patient's addresses and telephone numbers; name and address of provider of the care; and the dates of service.

Whether you submit a claim form depends on the type of TRICARE program option and provider you choose to use.

TRICARE Prime and Prime Remote. In-network providers file claims on your behalf. Non-network providers or the provider of emergent/urgent care outside your region may file claims on your behalf or require you to pay out-of-pocket and file your own claim for reimbursement.

TRICARE Extra. When you use network providers, the providers file claims on your behalf.

TRICARE Standard. The provider may file claims on your behalf or require you to pay out-of-pocket and file your own paper claim for reimbursement. Participating providers agree to accept TRICARE payment and any cost share as payment in full. Nonparticipating providers do not agree to accept the TRICARE allowable charge as the total charge for services; they can bill you up to 115 percent of the TRICARE allowable charge. TRICARE will send reimbursement to you and you, in turn, pay the provider.

TRICARE Contractors and Claims Information

TRICARE West—includes Alaska, Arizona, California, Colorado, Hawaii, Idaho, Iowa (except for the Rock Island Arsenal area), Kansas, Minnesota, Missouri (not including St. Louis), Montana, Nebraska, Nevada, New Mexico, North Dakota, Oregon, South Dakota, Texas (El Paso only), Utah, Washington, Wyoming.

Contractor: TriWest Healthcare Alliance
Information: 1-888-TRIWEST, 1-888-874-9378 Option 2
Claims Address: West Region Claims
P.O. Box 77028, Madison, WI 53707-7028
Phone for Claims: 1-888-874-9378
Contractor's Website: www.triwest.com

TRICARE North—includes Connecticut, Delaware, District of Columbia, Illinois, Indiana, Iowa (only the Rock Island Arsenal area), Kentucky, Maine, Maryland, Massachusetts, Michigan, Missouri (only the St. Louis area), New Hampshire, New Jersey, New York, North Carolina, Ohio, Pennsylvania, Rhode Island, Tennessee (only the Ft. Campbell area), Vermont, Virginia, West Virginia, and Wisconsin.

Contractor: Health Net Federal Services, LLC
Information: 1-877-874-2273
Claims Address: Health Net Federal Services, Inc.
c/o PGBA, LLC TRICARE
P.O. Box 870140, Surfside Beach, SC 29587-9740
Phone for Claims: 1-877-874-2273

Contractor's Website: www.healthnetfederalservices.com
Claims Processor's Website: www.mytricare.com

TRICARE South—includes Alabama, Arkansas, Florida, Georgia, Louisiana, Mississippi, Oklahoma, South Carolina, Tennessee (does not include Ft. Campbell, KY), Texas (not including El Paso).
Contractor: Humana Military Healthcare Services, Inc.
Information: 1-800-444-5445
Claims Address: TRICARE South Region
P.O. Box 7031, Camden, SC 29020-7031
Phone for Claims: 1-800-403-3950
Contractor's Website: www.humana-military.com
Claims Processor's Website: www.mytricare.com

Western Pacific/Far East—includes China, Thailand, Korea, Asia, Guam, Japan, New Zealand, India and Western Pacific.
Information: 1-888-777-8343
Claims Address: WPS, P.O. Box 7985, Madison, WI 53707-7985
Phone for Claims: (608) 301-2310

TRICARE Europe—includes Europe, Africa and the Middle East.
Information: 1-888-777-8343
Claims Address: WPS
P.O. Box 8976 (active duty family)
P.O. Box 7968 (active duty members)
Madison, WI 53708-(P.O. Box #)
Phone for Claims: (608) 301-2310
Website: www.europe.tricare.osd.mil/

TRICARE Latin America and Canada—includes Canada, Mexico, Central and South America, Puerto Rico, Bermuda, and West Indies, Virgin Islands and the Caribbean Basin.
Information: 1-888-777-8343
Claims Address: WPS, P.O. Box 7985, Madison, WI 53707-7985
Phone for Claims: (608) 301-2311

OTHER TRICARE CONTRACTORS

TRICARE DENTAL PLAN
(for active duty families, and for Reservists and their families)
Contractor: United Concordia Companies, Inc.
Information: 1-800-866-8499
Claims Address: United Concordia, TDP
Claims Processing
P.O. Box 69411, Harrisburg, PA 17106-9411
Enrollment Applications and Information: 1-888-622-2256
Contractor's Website: www.ucci.com

TRICARE RETIREE DENTAL PLAN
Contractor: Delta Dental
Information: 1-888-838-8737
Claims Address: Delta Dental of California, Federal Services
P.O. Box 537008, Sacramento, CA 95853-7008
Contractor's Website: www.ddpdelta.org

TRICARE RETAIL PHARMACY
Contractor: Express Scripts, Inc.
Information: 1-866-363-8667
Claims Address: Express Scripts, Inc.
P.O. Box 66518, St. Louis, MO 63166-6518
Contractor's Website: www.express-scripts.com

TRICARE FOR LIFE
Claims Processor: WPS/TFL
Information: 1-866-773-0404
Claims Address: WPS/TFL, P.O. Box 7890, Madison, WI 53708-7890
Contractor's Website: www.tricare4u.com

WEBSITES RELATED TO HEALTH as of January 7, 2009

Gateways and Publications

Healthfinder.gov: www.healthfinder.gov
The Federal Government's gateway to consumer health and human services information

The Surgeon General: www.surgeongeneral.gov/
Office of the Surgeon General of the United States

National Institutes of Health: www.nih.gov The National Institutes of Health's 20 institutions and seven centers, and research trials. (An NIH site specifically devoted to cancer trials: www.cancer.gov/clinical_trials)

Health Topics A to Z: www.cdc.gov/az.do
Health Topics from the Centers for Disease Control and Prevention provides a listing of disease and health topics found on the agency's website

Links to State and Local Health Departments: www.cdc.gov
Search on "State or Local Health Departments"

Health Web Links: www.consumer.gov/health.htm
Health consumer information from the Federal Government

Government and Consumer Publications on Health: www.pueblo.gsa.gov/health.htm
General health information from the Federal Citizen Information Center

Healthcare Cost & Utilization Project: www.ahrq.gov/data/hcup
Identify, track, analyze, and compare statistics on hospitals at the national, regional, and state level

USA.GOV for Seniors: www.usa.gov/Topics/Seniors.shtml
Include information on benefits, health, consumer protection, services, employment & volunteer activities, a retirement planner online, taxes, travel & leisure, education & training, and other links.

Food and Nutrition

Food Information Gateway: www.consumer.gov/food.htm
Consumer information on food from the Federal Government

Interactive Healthy Eating Index: www.mypyramid.gov/
Provides a quick assessment of the quality of your diet, including nutrition information targeted to your specific score

FoodSafety.Gov: www.foodsafety.gov
Gateway to Government food safety information

Dietary Guidelines for Americans: www.health.gov/dietaryguidelines
A joint publication of the U.S. Departments of Health & Human Services and Agriculture

Specific Topics or Concerns

Travelers' Health: www.cdc.gov/travel/default.aspx
How to protect yourself from disease when traveling outside the U.S. and alerts about disease outbreaks

U.S. MILITARY MEDICAL CENTERS AND HOSPITALS

The following do not have military medical centers or hospitals: Arkansas, Delaware, Indiana, Iowa, Maine, Massachusetts, Michigan, Minnesota, Montana, New Jersey, New Mexico, North Dakota, Oregon, Pennsylvania, Puerto Rico, Rhode Island, South Dakota, Tennessee, Utah, Vermont, West Virginia, Wisconsin, and Wyoming.

ALABAMA

Fox Army Health Center, Redstone Arsenal	(256) 955-8888

ALASKA

3rd Medical Group Naval Liaison, Elmendorf AFB, Anchorage	(907) 580-6260
Bassett Army Community Hospital, Fort Wainwright, Fairbanks	(907) 361-5172

ARIZONA

56th Medical Group, Luke AFB, Phoenix	(623) 856-2273

CALIFORNIA

David Grant Medical Center, 60th Medical Group, Travis AFB, Fairfield	(707) 423-3000
Naval Hospital, Camp Pendleton, Oceanside	(760) 725-1288
Naval Hospital, Lemoore	(559) 998-4481
Naval Hospital, Twentynine Palms	(760) 830-2190
Naval Medical Center, San Diego	(619) 532-6400
Weed Army Community Hospital, Fort Irwin	(760) 380-3108

COLORADO

10th Medical Group Hospital, USAF Academy, Colorado Springs	(719) 333-5111
Evans Army Hospital, Fort Carson, Colorado Springs	(719) 526-7000

CONNECTICUT

Naval Ambulatory Care Center, Groton	(860) 694-4123

DISTRICT OF COLUMBIA

Walter Reed Army Medical Center	(202) 782-3501

FLORIDA

6th Medical Group Hospital, MacDill AFB, Tampa	(813) 828-2273
96th Medical Group Hospital, Eglin AFB, Valpariso	(850) 883-8242
Naval Hospital, Jacksonville	(904) 542-7300
Naval Hospital, Pensacola	(850) 505-6601

GEORGIA

Dwight David Eisenhower Army Medical Center, Fort Gordon, Augusta	(706) 787-5811
Martin Army Community Hospital, Fort Benning, Columbus	(706) 544-2273
Winn Army Community Hospital, Fort Stewart, Savannah	(912) 767-6633

GUAM

Naval Hospital, Guam	(671) 344-9202

HAWAII

Tripler Army Medical Center, Honolulu	(808) 433-2778

IDAHO

366th Medical Group Hospital, Mountain Home AFB	(208) 828-7600

ILLINOIS

375th Medical Group Hospital, Scott AFB, Belleville	(618) 256-7000

KANSAS

Irwin Army Community Hospital, Fort Riley, Junction City	(785) 239-7000

KENTUCKY

Blanchfield Army Community Hospital, Fort Campbell, Hopkinsville	(270) 798-8400
Ireland Army Community Hospital, Fort Knox	(800) 493-9602

LOUISIANA

Bayne-Jones Army Community Hospital, Fort Polk, Leesville	(337) 531-3118

MARYLAND

Malcolm Grow Medical Center, Andrews AFB, Camp Springs	(240) 857-5911
National Naval Medical Center, Bethesda	(301) 295-4611
Wyman Park Medical Center, John Hopkins, Baltimore	(410) 338-3000

MISSISSIPPI

81st Medical Group Hospital, Keesler AFB, Biloxi	(800) 700-8603

MISSOURI

General Leonard Wood Army Community Hospital, Ft. Leonard Wood	(573) 596-1490

NEBRASKA

Ehrling Bergquist Hospital, 55th Medical Group, Offutt AFB, Omaha	(402) 294-9760

NEVADA

99th Medical Group Hospital, Nellis AFB, Las Vegas	(702) 653-2260

NEW YORK

Keller Army Community Hospital, USMA, West Point	(845) 938-5169

NORTH CAROLINA

Naval Hospital, Camp Lejeune	(910) 450-4300
Womack Army Medical Center, Fort Bragg, Fayetteville	(910) 432-3719

OHIO

88th Medical Group Hospital, Wright-Patterson AFB, Dayton	(937) 257-0837

OKLAHOMA

Reynolds Army Community Hospital, Fort Sill, Lawton	(580) 458-2000

SOUTH CAROLINA

Naval Hospital, Beaufort	(843) 525-5600
Moncrief Army Community Hospital, Fort Jackson, Columbia	(803) 751-2160

TEXAS

82nd Medical Group Hospital, Sheppard AFB, Wichita Falls	(940) 676-1847
Brooke Army Medical Center, Fort Sam Houston, San Antonio	(210) 916-4141
Darnall Army Medical Center, Fort Hood, Killeen	(254) 288-8000
Naval Hospital, Corpus Christi	(361) 961-2688
Wilford Hall Medical Center, 59th Medical Wing, Lackland AFB, San Antonio	(210) 292-7100
William Beaumont Army Medical Center, Fort Bliss, El Paso	(915) 569-2121

VIRGINIA

1st Medical Group Hospital, Langley AFB, Hampton	(757) 764-6833
Dewitt Army Community Hospital, Fort Belvoir	(703) 805-0510
McDonald Army Health Center, Fort Eustis	(757) 314-7500
Naval Medical Center, Portsmouth	(757) 953-5008

WASHINGTON

Madigan Army Medical Center, Tacoma	(253) 968-1110
Naval Hospital, Bremerton	(360) 475-4000
Naval Hospital, Whidbey Island, Oak Harbor	(360) 257-9500

U.S. MILITARY MEDICAL CLINICS

The following do not have medical clinics: Guam, Idaho, Indiana, Iowa, Minnesota, Nebraska, Ohio, Vermont, Wisconsin, and the Virgin Islands.

ALABAMA

42nd Medical Group Clinic, Maxwell AFB	(334) 953-7801
Lyster Army Health Clinic, Fort Rucker	(334) 255-7359
U.S. Coast Guard ATC Clinic, Mobile	(251) 441-6725

ALASKA

354th Medical Group Clinic, Eielson AFB	(907) 377-1847
U.S. Army Troop Medical Clinic, Fort Richardson	(907) 384-0600
USCG Air Station Clinic, Sitka	(907) 966-5430
USCG Air Station, Rockmore-King Medical Clinic, Kodiak	(907) 487-5757
USCG Integrated Support Command Clinic, Juneau	(907) 463-2140
USCG Integrated Support Command Clinic, Ketchikan	(907) 228-0320

ARIZONA

355th Medical Group, Davis-Monthan AFB, Tucson	(520) 228-2778
Branch Medical Clinic, MCAS, Yuma	(928) 269-2772
Raymond W. Bliss Army Health Center, Fort Huachuca	(520) 533-9200
U.S. Army Health Clinic, Yuma Proving Ground	(928) 328-2502

ARKANSAS

314th Medical Group Clinic, Little Rock AFB	(501) 987-8811
U.S. Army Health Clinic, Pine Bluff Arsenal	(870) 540-3409

CALIFORNIA

9th Medical Group Clinic, Beale AFB, Marysville	(530) 634-4831
30th Medical Group Clinic, Vandenberg AFB, Lompoc	(805) 606-2273
61st Medical Group Clinic, Los Angeles AFB	(310) 653-2873
95th Medical Group Clinic, Edwards AFB	(661) 277-7118
Branch Health Clinic, NAWS, China Lake	(760) 939-8000
Branch Medical Annex, NAWS, Seal Beach	(562) 626-7285
Branch Medical Clinic, Bridgeport	(760) 932-1616
Branch Medical Clinic, Brig Medical, Area 24, Camp Pendleton	(760) 725-2811
Branch Medical Clinic, Camp Chappo, Area 22, Camp Pendleton	(760) 725-3784
Branch Medical Clinic, Camp Delmar, Area 21, Camp Pendleton	(760) 725-2141
Branch Medical Clinic, Camp Las Pulgas, Area 43, Camp Pendleton	(760) 725-3268
Branch Medical Clinic, Camp Los Flores, Area 41, Camp Pendleton	(760) 725-2873
Branch Medical Clinic, Camp Margarita, Area 33, Camp Pendleton	(760) 725-8099
Branch Medical Clinic, Camp San Mateo, Area 62, Camp Pendleton	(760) 725-7100
Branch Medical Clinic, Camp San Onofre, Area 52, Camp Pendleton	(760) 725-7200
Branch Medical Clinic, Edson Range, Area 31, Camp Pendleton	(760) 725-7135
Branch Medical Clinic, Mainside, Area 13, Camp Pendleton	(760) 725-6682
Branch Medical Clinic, MCLB Barstow	(760) 577-6491
Branch Medical Clinic, MCRD, San Diego	(619) 524-4036
Branch Medical Clinic, NAB Coronado, San Diego	(619) 545-9473
Branch Medical Clinic, NAF El Centro	(760) 339-2674
Branch Medical Clinic, NAS Miramar, San Diego	(858) 577-4656
Branch Medical Clinic, NAS North Island, San Diego	(619) 545-9473
Branch Medical Clinic, Naval Station, San Diego	(619) 556-8084
Branch Medical Clinic, Point Mugu	(805) 989-8815
Naval Medical Clinic, Port Hueneme	(805) 982-2633
TRICARE Outpatient Clinic, Chula Vista	(619) 744-5355

TRICARE Outpatient Clinic, Clairemont Mesa (619) 645-0155
U.S. Army Health Clinic, Presidio of Monterey (831) 242-5663
U.S. Army Occupational Health Clinic, Sierra Army Depot, Herlong (530) 827-4141
USCG Clinic, McKinleyville (707) 839-6176
USCG Clinic, Petaluma (707) 765-7200
USCG Integrated Support Command Clinic, San Pedro (310) 521-6050
USCG Integrated Support Command Health and Safety Division, Alameda (510) 437-3581

COLORADO

21st Medical Group Clinic, Peterson AFB (719) 556-1955
460th Medical Squadron Clinic, Buckley ANGB (720) 847-9355

CONNECTICUT

Naval Clinic, Groton (860) 694-4123
USCG Academy Health Services Clinic, New London, Groton (860) 444-8402

DELAWARE

436th Medical Group Clinic, Dover AFB (302) 677-2525

DISTRICT OF COLUMBIA

579th Medical Group Clinic, Bolling AFB (202) 767-5532
Branch Medical Clinic, Naval Research Lab (202) 767-3592
Branch Medical Clinic, Navy Yard (202) 433-3407
DiLorenzo Army Health Clinic, Pentagon (703) 692-8800
Naval Branch Medical Clinic, ARLEX Federal Office Bldg., Wing 3 (703) 614-2726
Naval Branch Medical Clinic, NAF, Andrews AFB (240) 857-2850
U.S. Army Health Clinic, Fort McNair (202) 685-3100

FLORIDA

1st Medical Group Clinic, Hurlburt Field (850) 884-7882
45th Medical Group Clinic, Patrick AFB (321) 494-8241
325th Medical Group Clinic, Tyndall AFB (850) 283-2778
Branch Health Clinic, Jacksonville (904) 542-3406
Branch Health Clinic, NAS, Pensacola (850) 452-5662
Branch Health Clinic, NTTC, Corry Station, Pensacola (850) 452-6326
Branch Health Clinic, Whiting Field NAS, Milton (850) 623-7568
Branch Medical Clinic, Key West (305) 293-4605
Branch Medical Clinic, Naval Station, Mayport (904) 270-5303
Branch Medical Clinic, NATTC, Pensacola (850) 452-8970
Branch Medical Clinic, Panama City (850) 234-4177
USCG Air Station Clinic, Opa Locka (305) 953-2265
USCG Clinic, Clearwater (727) 535-1437, Ext. 4
USCG Medical Services Clinic, Miami Beach (305) 535-4350

GEORGIA

23rd Medical Group Clinic, Moody AFB (229) 257-2778
78th Medical Group Clinic, Robins AFB (478) 327-7995
Branch Medical Clinic, Albany (229) 639-5976
Branch Medical Clinic, Athens (706) 354-7321
Branch Medical Clinic, NAS Atlanta, Marietta (678) 655-5311
Lawrence Joel Army Health Clinic, Fort McPherson (404) 464-2778
Naval Medical Clinic, Kings Bay (912) 573-2619
Tuttle Army Health Clinic, Hunter Army Airfield (912) 767-6633

GUAM

36th Medical Group, Andersen AFB, Yigo	(671) 366-2873

HAWAII

15th Medical Group Clinic, Hickam AFB	(808) 448-6000
Battalion Medical Clinic, Camp H.M. Smith	(808) 477-3773
Branch Medical Clinic, Kekaha	(808) 471-6203
Branch Medical Clinic, Makalapa, Pearl Harbor	(808) 473-0247
Branch Medical Clinic, NSY, Pearl Harbor	(808) 474-3136
Naval Health Clinic, Pearl Harbor	(808) 473-0247
U.S. Army Health Clinic, Schofield Barracks	(808) 433-2778
U.S. Navy Health Clinic, MCAS, Kaneohe Bay	(808) 257-3365
USCG Integrated Support Command Clinic, Honolulu	(808) 433-9800

ILLINOIS

Branch Medical Clinic, RTC, Great Lakes	(847) 688-5568
Fisher Branch Health Clinic, NTC, Great Lakes	(847) 688-2092
Naval Health Clinic, Great Lakes	(847) 688-4560
U.S. Army Health Clinic, Rock Island Arsenal	(309) 782-0805

KANSAS

22nd Medical Group Clinic, McConnell AFB	(316) 759-6300
Munson Army Health Clinic, Fort Leavenworth	(913) 684-6000

KENTUCKY

Army Occupational Health Clinic, Blue Grass Army Depot, Richmond	(859) 779-6559

LOUISIANA

2nd Medical Group Clinic, Barksdale AFB	(318) 456-6555
Naval Ambulatory Care Center-East Bank, NAS, New Orleans	(504) 678-8148
Naval Ambulatory Care Center, NAS JRB Belle Chasse, New Orleans	(504) 678-3660
Naval Ambulatory Care Center-West Bank, NAS, New Orleans	(504) 678-2400

MAINE

Branch Health Clinic, NAS, Brunswick	(207) 921-2991

MARYLAND

Branch Medical Clinic, Naval Ordnance Station, Indian Head	(301) 744-4601
Branch Medical Clinic, NAVSURFWARCEN Carderock	(301) 295-2663
Kimbrough Ambulatory Care Center, Fort Meade	(301) 677-8392
Kirk U.S. Army Health Clinic, Aberdeen Proving Ground	(410) 278-5475
Naval Health Clinic, Patuxent River	(301) 342-1506
Naval Medical Clinic, Annapolis	(410) 293-2273
U.S. Army Health Clinic, Fort Detrick	(301) 619-7175
USCG Yard Clinic, Curtis Bay, Baltimore	(410) 636-3144

MASSACHUSETTS

66th Medical Group Clinic, Hanscom AFB	(888) 628-9633 Option 6
Kaehler Memorial Medical Clinic, USCG Air Station, Otis AFB	(508) 968-6582
USCG Support Center Clinic, Boston	(617) 223-3121

MICHIGAN

Army Occupational Health Clinic, Warren	(586) 574-5771
U.S. Army Health Clinic, Selfridge ANGB, Mount Clemens	(586) 239-4011
USCG Clinic, USCG Air Station, Traverse City	(231) 922-8282

MISSISSIPPI

14th Medical Group Clinic, Columbus AFB	(662) 434-2334
Branch Medical Clinic, NAS, Meridian	(601) 679-2633
Naval Branch Health Clinic, NCBC, Gulfport	(228) 871-2810

MISSOURI

509th Medical Group, Whiteman AFB	(660) 687-2188
TRICARE Medical Clinic, Kansas City	(816) 843-3666

MONTANA

341st Medical Group Clinic, Malmstrom AFB	(406) 731-4633

NEVADA

Branch Health Clinic, NAS, Fallon	(877) 843-2735

NEW HAMPSHIRE

Naval Medical Clinic, Portsmouth	(888) 628-9633 Option 4

NEW JERSEY

305th Medical Group Clinic, McGuire AFB	(609) 754-9200
Branch Medical Clinic, Naval Air Warfare Center, Lakehurst	(732) 323-2231
Earle Naval Health Clinic, Colts Neck	(732) 866-2300
Patterson Army Health Clinic, Fort Monmouth	(732) 532-1764
USCG Training Center Clinic, Cape May	(609) 898-6959

NEW MEXICO

27th Medical Group Clinic, Cannon AFB, Clovis	(505) 784-4582
49th Medical Group Clinic, Holloman AFB, Alamogordo	(505) 572-2778
377th Medical Group Clinic, Kirtland AFB, Albuquerque	(505) 846-3200
McAfee, U.S. Army Health Clinic, White Sands Missile Range	(505) 678-1138

NEW YORK

Ainsworth Army Health Clinic, Fort Hamilton	(718) 630-4129
Branch Medical Clinic, Saratoga Springs	(518) 583-5300
Guthrie Ambulatory Care Clinic, Fort Drum	(315) 772-2778
U.S. Army Health Clinic, Watervliet Arsenal	(518) 266-4195

NORTH CAROLINA

4th Medical Group Clinic, Seymour Johnson AFB, Goldsboro	(919) 722-1802
43rd Medical Group Clinic, Pope AFB	(910) 394-4258
Branch Medical Clinic, Camp Geiger, Camp Lejeune	(910) 449-0371
Branch Medical Clinic, Camp Johnson	(910) 450-0836
Branch Medical Clinic, Camp Lejeune	(910) 451-5830
Branch Medical Clinic, Courthouse Bay MCB, Camp Lejeune	(910) 450-7338
Branch Medical Clinic, French Creek, MCB, Camp Lejeune	(910) 451-5125
Branch Medical Clinic, MCAS New River, Jacksonville	(910) 449-6400
Naval Health Clinic, Cherry Point	(252) 466-0266
USCG Clinic, USCG Support Clinic, Elizabeth City	(252) 335-6460

NORTH DAKOTA

5th Medical Group Clinic, Minot AFB, Minot	(701) 723-5633
319th Medical Group Clinic, Grand Forks AFB, Grand Forks	(701) 747-5601

OKLAHOMA

71st Medical Squadron Clinic, Vance AFB	(580) 213-7416
72nd Medical Group Clinic, Tinker AFB, Oklahoma City	(405) 734-2778
97th Medical Group Clinic, Altus AFB, Altus	(580) 481-5235
U.S. Army Occupational Health Clinic, McAlester	(918) 420-6495

OREGON

U.S. Army Health Clinic, Umatilla Army Depot, Hermiston	(541) 564-5215
USCG Air Station Clinic, North Bend	(541) 756-9234
USCG Air Station Clinic, Warrenton	(503) 861-6240

PENNSYLVANIA

Branch Medical Clinic PNBC, Philadelphia Navy Yard	(215) 897-8147
Branch Medical Clinic, Mechanicsburg	(717) 605-2636
Branch Medical Clinic, NAS JRB, Willow Grove	(215) 443-6360
Dunham Army Health Clinic, Carlisle Barracks	(717) 245-3400
U.S. Army Health Clinic, Fort Indiantown Gap, Annville	(717) 861-2091
U.S. Army Health Clinic, New Cumberland	(717) 770-7281
U.S. Army Health Clinic, Tobyhanna	(570) 895-7225
USA Health Clinic, Letterkenny Army Depot, Chambersburg	(717) 267-8416

PUERTO RICO

Rodriguez Army Health Clinic, Fort Buchanan	(787) 707-2043
USCG Clinic, Aquadilla	(787) 890-8477

RHODE ISLAND

Naval Ambulatory Care Center, Newport	(401) 841-3771

SOUTH CAROLINA

20th Medical Group Clinic, Shaw AFB	(803) 895-2273
437th Medical Group Clinic, Charleston AFB	(843) 963-6880
Branch Medical Clinic, MCRD, Parris Island	(843) 228-3793
Naval Health Clinic, North Charleston	(843) 743-7000
NWS Health Clinic, North Charleston	(843) 743-3709
U.S. Navy Branch Medical Clinic, MCAS, Beaufort	(843) 228-7424

SOUTH DAKOTA

28th Medical Group Clinic, Ellsworth AFB, Rapid City	(605) 385-3201

TENNESSEE

Branch Medical Clinic, NAVSUPPACT, Millington	(901) 874-6100

TEXAS

7th Medical Group Clinic, Dyess AFB, Abilene	(325) 696-4677
12th Medical Group Clinic, Randolph AFB	(210) 652-4373
17th Medical Group Clinic, Goodfellow AFB	(325) 654-3149
47th Medical Group Clinic, Laughlin AFB	(830) 298-3578
59th Medical Squadron Clinic, Brooks AFB	(210) 536-1847
Branch Health Clinic, NAS, Kingsville	(361) 516-6160
Branch Medical Clinic, Ingleside	(361) 776-4580
Naval Branch Medical Clinic, Fort Worth	(817) 782-5900
U.S. Army Occupational Health Clinic, Red River Army Depot, Texarkana	(903) 334-2155

UTAH

75th Medical Group Clinic, Hill AFB	(801) 728-2600
U.S. Army Health Clinic, Dugway Proving Ground, Salt Lake City	(435) 831-3311
U.S. Army Health Clinic, Tooele Army Depot	(435) 833-2572

VIRGINIA

Admiral Joel T. Boone Branch Medical Clinic, Norfolk	(866) 645-4584
Andrew Rader Army Health Clinic, Fort Myer	(703) 696-7951
Branch Medical Clinic, Dam Neck, Virginia Beach	(757) 953-6666
Branch Medical Clinic, NAS, Oceana, Virginia Beach	(757) 953-3903
Branch Medical Clinic, Naval Station Sewells Point, Norfolk	(757) 953-9020

Branch Medical Clinic, Naval Weapons Station, Yorktown	(757) 953-8430
Branch Medical Clinic, NAVSECGRUACT, Northwest Chesapeake	(757) 314-8999
Branch Medical Clinic, Norfolk Naval Shipyard, Portsmouth	(757) 953-5000
Branch Medical Clinic, OCS Quantico	(703) 784-2062
Branch Medical Clinic, The Basic School, Quantico	(703) 784-5542
Dilorenzo TRICARE Health Clinic, The Pentagon	(703) 692-8800
Kenner Army Health Clinic, Fort Lee	(804) 734-9000
Naval Medical Clinic, Quantico	(703) 784-1699
NNMC Branch Medical Clinic, Dahlgren	(540) 653-8241
TRICARE Prime Clinic, Chesapeake	(757) 953-6366
TRICARE Prime Clinic, Virginia Beach	(757) 953-6708
Craven Army Health Clinic, Fort Monroe	(757) 314-8023
U.S. Army Health Clinic, Fort Story	(757) 422-7822
USCG Clinic, USCG Reserve Training Center, Yorktown	(757) 856-2230
USCG Integrated Support Command Clinic, Portsmouth	(757) 483-8596

WASHINGTON

62nd Medical Group Clinic, McChord AFB	(800) 404-4506 Option 5
92nd Medical Group Clinic, Fairchild AFB	(509) 247-2361
Branch Health Clinic, Everett	(425) 304-4060
Branch Medical Clinic, Bangor, Silverdale	(360) 315-4391
Naval Branch Clinic, Puget Sound Naval Station	(360) 476-2508
USCG Clinic, USCG Air Station, Port Angeles	(360) 417-5895
USCG Integrated Support Command Clinic, Seattle	(206) 217-6432

WEST VIRGINIA

Naval Branch Medical Clinic, Sugar Grove	(304) 249-6381

WYOMING

90th Medical Group Clinic, Francis E. Warren AFB, Cheyenne	(307) 773-3461

PART IV

VETERANS BENEFITS

This section provides general information concerning the most important federal benefits that Congress has provided specifically for all veterans, their dependents and beneficiaries. More detailed information concerning any benefit may be obtained from the federal agency administering each respective program. Applications for VA benefits may be filed at, or further information may be obtained from your nearest VA office. A regional representative may be contacted toll-free at 1-800-827-1000. The VA also has many other nationwide toll-free telephone services, including: Life Insurance, 1-800-669-8477; the VA Debt Management Center, 1-800-827-0648; Telecommunication Device for the Deaf (TDD), 1-800-829-4833; and CHAMPVA, 1-800-733-8387. Writing directly to the VA Central Office in Washington, DC, may result in significant delays resolving your questions or concerns.

HEALTH CARE

For most veterans, entry into the VA health care system begins with applying for enrollment. Veterans do not have to be enrolled if they: (1) have a service-connected disability of 50 percent or more; (2) want care for a disability the military determined was incurred or aggravated in the line of duty, but which VA has not yet rated, during the 12-month period following discharge; or (3) want care for a service-connected disability only. However, to permit better planning of health resources, these veterans also are urged to enroll.

To apply, veterans must complete VA Form 10-10EZ, *"Application for Health Benefits."* The form may be obtained from any VA health care facility or regional benefits office, or by calling the VA Health Benefits Service Center toll-free at 1-877-222-8387. It is also available on the VA website: www.va.gov/1010ez.htm. Veterans may complete the form in person at a VA health care facility, or mail it to a local VA health care facility for processing. Once enrolled, a veteran is eligible to receive services at VA facilities anywhere in the country. Additional information can be found on the VA website www.va.gov/healtheligibility.

Priority Groups. The following are enrollment priorities only. VA services and treatment available to enrolled veterans generally are not based on priority groups. Enrollment is reviewed each year and veterans are notified in writing of any change in their enrollment status. Call the nearest VA health care facility or the Health Benefits Service Center, 1-877-222-8387, to obtain the latest information.

Priority Group 1: Veterans with service-connected disabilities rated 50 percent or more.

Priority Group 2: Veterans with service-connected disabilities rated 30 or 40 percent.

Priority Group 3: Veterans who are former POWs or were awarded a Purple Heart, veterans with disabilities rated 10 and 20 percent, and veterans awarded special eligibility for disabilities incurred in treatment.

Priority Group 4: Veterans receiving aid and attendance or housebound benefits and veterans determined by VA to be catastrophically disabled, although some may be responsible for copayments.

Priority Group 5: Veterans who are determined to be unable to defray the expenses of needed care.

Priority Group 6: All other eligible veterans not required to make copayments. This includes veterans of the Mexican border period or World War I; veterans seeking care solely for certain conditions associated with exposure to radiation, for any illness associated with combat service in a war after the Gulf War or during a period of hostility after November 11, 1998; and veterans with zero percent service-connected disabilities who are nevertheless compensated, including veterans receiving compensation for inactive tuberculosis.

Priority Group 7: Nonservice-connected veterans and noncompensable zero percent service-connected veterans with income above the VA's national means test threshold and below the VA's geographic means test threshold for the fiscal year ending on September 30 of the previous calendar year, or with income below both the VA national threshold and the VA geographically based threshold, but whose net worth exceeds the VA's ceiling who agree to pay copayments.

Priority Group 8: All other nonservice-connected veterans and zero percent noncompensable service-connected veterans who agree to pay copayments. As of January 17, 2003, this group is closed to new veterans.

However, P.L. 110-329, the *Consolidated Security, Disaster Assistance, and Continuing Appropriations Act, 2009,* provides the VA with additional funding to allow expanded enrollment opportunity for certain Priority Group 8 veterans who may have been previously denied enrollment in VA's health care system because their income exceeded VA's means tests thresholds.

This will allow veterans whose incomes do not exceed the means tests thresholds by more than 10 percent to enroll in VA's health care system. The new provision is expected to take effect in June, 2009 and will be applied retroactively to all enrollment applications received on or after January 1, 2009. This new provision does not open enrollment to all Priority Group 8 veterans.

Special Access to Care

Service Disabled Veterans: Veterans with service-connected disabilities rated 50 percent or more based on one or more disabilities or unemployability and veterans receiving care for a service-connected disability receive priority in the scheduling of appointments for outpatient medical services and admissions for inpatient hospital care.

Combat Veterans: Veterans who served in combat locations during active military service after November 11, 1998, and were discharged or released from active service after January 28, 2003, are eligible for free health care services for conditions potentially related to combat service for five years following separation from active duty. In addition, a combat veteran who was discharged on or before January 28, 2003, and who did not enroll in the VA health care system, has three years of health care services from January 28, 2008 until January 27, 2011.

Billing Insurance Companies

When applying for medical care, all veterans are asked to provide information pertaining to health insurance coverage. VA is authorized to submit claims to insurance carriers for the costs of medical care provided to nonservice-connected veterans and service-connected veterans for nonservice-connected conditions. Veterans are not held responsible for the deductible requirements and copayments established by their insurance carriers. They are also not responsible for portions of an insurance claim not covered. Veterans above certain income levels, however, are responsible for the copayments required by federal law.

FILING FOR VETERANS ADMINISTRATION BENEFITS

Servicemembers separating from active service, who are suffering from an ailment that may be service-connected should file VA Form 21-526, *"Veterans Application for Compensation and/or Pension"* with the VA upon separation from service. Application can also be filed by those already separated by filing an application with the VA Regional Office or on the Internet by using the Veterans ON-line APPlication (VONAPP) website. The address is http://vabenefits.vba.va.gov/vonapp. In some instances, servicemembers may begin the application process during pre-separation processing. To shorten the necessary time for application processing by the VA, a copy of DD Form 214 should be attached as well as copies of pertinent military medical records, including the member's separation physical.

If the VA awards compensation and the member is entitled to retired pay from the service, but not eligible for Concurrent Retirement and Disability Payments or Combat Related Special Compensation (see section on MILITARY RETIREMENT), an application for waiver of retired pay for that amount of compensation awarded by the VA should be submitted in order to receive the VA compensation which is tax exempt.

If there is no basis for a claim to be originated at time of separation, but a condition develops later in life, the member should file a claim at that time and let the VA determine if the condition is service-connected or compensable.

If the veteran has filed or considered filing a claim for service-connected ailments, it is recommended that they leave instructions with their family that at the time of death an autopsy be performed to determine if the cause of death was service-connected. An autopsy may disclose service-connected contributory causes of death that would not be indicated on a normal death certificate. If the death is determined to be service-connected, the surviving family members should file an application for Dependency and Indemnity Compensation or Death Pension. It is extremely important to ensure that the VA makes a full review of the military medical record. In the event the VA rules that the member's death was service-connected, the surviving spouse should question the VA regarding eligibility for a guaranteed home loan and educational benefits for the surviving spouse and any eligible children.

It must also be emphasized that none of the benefits for survivors are paid automatically but each must be applied for individually (i.e., Application for Headstone or Marker, Application for Burial Allowance).

Information concerning VA home loan guaranty and related benefits is provided by the VA, along with information relating to conversion of SGLI to VGLI (Veterans Group Life Insurance) upon separation. If this information is not received, the member should inquire at the nearest VA office.

DISABILITY BENEFITS FOR PRESUMPTIVE SERVICE-CONNECTION

The VA presumes that specific disabilities diagnosed in certain veterans were caused by their military service. A veteran who suffers from one of the conditions listed below, for their period or category of service, may receive VA disability compensation.

Former Prisoners of War. (1) A veteran who was *imprisoned for any length of time* who has a disability at least 10 percent disabling for any of the following: psychosis; any of the anxiety states; dysthymic disorder; organic residuals of frostbite; post-traumatic osteoarthritis; heart disease or hypertensive vascular disease and their complications; stroke and its residuals; osteoporosis if the veteran has PTSD. (2) A veteran who was *imprisoned for at least 30 days*, who has a disability at least 10 percent disabling for any of the following: avitaminosis; beriberi; chronic dysentery; helminthiasis; malnutrition (including optic atrophy); pellagra; any other nutritional deficiency; irritable bowel syndrome; peptic ulcer disease; peripheral neuropathy; cirrhosis of the liver.

Vietnam Veterans (exposed to Agent Orange). A veteran who served in the Republic of Vietnam between January 9, 1962 and May 7, 1975 who has any of the following: chloracne or other acne form disease similar to chloracne*; porphyria cutanea tarda*; soft-tissue sarcoma (other than osteosarcoma, chondrosarcoma, Kaposi's sarcoma or mesothelioma); Hodgkin's disease; multiple myeloma; respiratory cancers (lung, bronchus, larynx, trachea); non-Hodgkin's lymphoma; prostate cancer; acute and subacute peripheral neuropathy*; type 2 diabetes; chronic lymphocytic leukemia.

*Must become manifest to a degree of 10 percent or more within a year after the last date on which the veteran was exposed to an herbicide agent during active military, naval, or air service.

Atomic Veterans (exposed to Ionizing Radiation). Veterans who participated in atmospheric nuclear testing; occupied or was a POW in Hiroshima or Nagasaki; has service before February 1, 1992 at a diffusion plant in Paducah, KY, Portsmouth, OH, or Oak Ridge, TN; or had service before January 1, 1974 at Amchitka Island, AK, and who has any of the following: all forms of leukemia (except for chronic lymphocytic leukemia); cancer of the thyroid, breast, pharynx, esophagus, stomach, small intestine, pancreas, bile ducts, gall bladder, salivary gland, urinary tract (renal pelves, ureter, urinary bladder and urethra), brain, bone, lung, colon, ovary; bronchiolo-alveolar carcinoma; multiple myeloma; lymphomas (other than Hodgkin's disease); primary liver cancer (except if cirrhosis or hepatitis B is indicated).

Gulf War Veterans (undiagnosed illness). Veterans who served in the Southwest Asia Theater of operations during the Persian Gulf War with a condition at least 10 percent disabling by December 31, 2011. Included are medically unexplained chronic multi-symptom illnesses defined by a cluster of signs or symptoms that have existed for six months or more, such as: chronic fatigue syndrome; fibromyalgia; irritable bowel syndrome; any diagnosed or undiagnosed illness that the Secretary of VA determines warrants a presumption of service connection.

Note: Signs or symptoms of an undiagnosed illness include: fatigue, skin symptoms, headaches, muscle pain, joint pain, neurological symptoms, respiratory symptoms, sleep disturbance, GI symptoms, cardiovascular symptoms, weight loss, and menstrual disorders.

In addition, veterans who are within one year of release from active duty and who are diagnosed with chronic diseases (such as arthritis, diabetes, or hypertension) and veterans deployed to the Southwest Asia Theater of operations from August 2, 1990, to July 31, 1991 diagnosed with amyotrophic lateral sclerosis (ALS)/Lou Gehrig's disease are also encouraged to apply for disability compensation.

COMPENSATION FOR SERVICE-CONNECTED DISABILITY

Nature of Benefit. Disability compensation consists of a monthly payment by the VA to a veteran for a service-connected disability. The payments are made for disabilities of 10 percent up to 100 percent. Certain serious disabilities are given special monthly rates. When the extent of disability falls between two such groups, an intermediate rate may be awarded. The current nontaxable minimum compensation is $123 per month and the maximum is currently $2,673 per month. The Federal Government does not count compensation from the VA as taxable income. Social Security payments do not affect VA compensation.

Basis of Eligibility. Disability must result from disease or injury incurred in, or aggravated during active duty or active duty for training. Benefits may also be paid for injuries sustained during inactive duty for training. In all cases, the disability must have occurred in the line of duty and discharge or release from duty must have been under other than dishonorable conditions.

Monthly Rates of Compensation.

Service-Connected Disability	Effective December 1, 2008
1. 10 percent	$ 123
2. 20 percent	243
3. 30 percent	376
4. 40 percent	541
5. 50 percent	770
6. 60 percent	974
7. 70 percent	1,228
8. 80 percent	1,427
9. 90 percent	1,604
10. 100 percent	2,673

Additional Amounts Payable for:

a. Total disability, plus additional service-connected disabilities independently rated at 60 percent or veteran is permanently housebound by reason of service-connected disability **$2,993**

b. Loss, or loss of use, of a creative organ, or one foot, or one hand, or both buttocks, or blindness of one eye, or has suffered a complete loss of speech or bilateral deafness (but in no event to exceed **$3,327** exclusive of additional compensation for dependents) **$96**

c. Loss, or loss of use, of both feet, or one hand and one foot, or blindness of both eyes, or is permanently bedridden or so helpless as to be in need of regular aid and attendance **$3,327**

d. Loss, or loss of use, of both hands, or both legs at a level, or with complications, preventing natural knee action with prostheses in place, or of one arm and one leg at levels, or with complications, preventing natural elbow and knee action with prostheses or suffered blindness in both eyes, resulting in need of regular aid and attendance **$3,671**

e. Loss, or loss of use, of both arms at levels, or with complications, preventing natural elbow action with prostheses in place, has suffered the anatomical loss of both legs so near the hip as to prevent the use of prosthetic appliances, or has suffered the anatomical loss of one arm and one leg so near the shoulder and hip as to prevent the use of prosthetic appliances, or suffered total blindness or anatomical loss of both eyes **$4,176**

f. Suffered disability under conditions which would entitle the veteran to two or more rates in (a) through (d) above, no condition being considered twice, or suffered bilateral deafness (and hearing impairment in either or both ears is service-connected) rated at 60 percent or more, plus disabling and service-connected total blindness, or if the veteran has suffered the anatomic loss of both arms so near the shoulder as to prevent the use of prosthetic appliances **$4,667**

Allowances for Dependents. Veterans whose service-connected disabilities are rated at 30 percent or more are entitled to additional compensation for dependents. The number of dependents and the degree of disability determines the additional amount.

The current rates are listed below and are based upon 100 percent disability. The rates for 30 percent or more are payable at the same ratio that the degree of disability bears to 100 percent.

Dependency	Rate
Spouse and —	
No children	$150
1 child	259
2 children	334
3 children	409
Additional children (each)	75
No spouse and	
1 child	101
2 children	176
3 children	251
Additional children (each)	75
Dependent parents	
1 parent	120
2 parents	240
Child over age 18 attending school (in lieu of benefits shown above)	240

PENSION

Veterans with low incomes who are permanently and totally disabled, or are age 65 and older, may be eligible for monetary support if they have 90 days or more of active military service, at least one day of which was during a period of war. Generally, veterans who entered active duty on or after September 8, 1980, or officers who entered active duty on or after October 16, 1981, may have to meet a longer minimum period of active duty. The discharge from active duty must have been under conditions other than dishonorable. The permanent and total disability must be for reasons other than the veteran's own willful misconduct. Payments are made to qualified veterans to bring their total income, including other retirement or Social Security income, to a minimum level set by Congress. Unreimbursed medical expenses may reduce countable income. Veterans of a period of war who are age 65 or older and meet service and income requirements are also eligible to receive a pension, regardless of current physical condition.

The Aid and Attendance, or Housebound allowance may be payable to a wartime veteran who has in-home care or who lives in a nursing home or assisted-living facility.

Aid and Attendance. A veteran may be eligible for Aid and Attendance when the veteran:

1. requires the aid of another person in order to perform personal functions required in everyday living, such as bathing, feeding, dressing, attending to the wants of nature, adjusting prosthetic devices, or protecting the person from the hazards of his/her daily environment, or

2. is bedridden, in that his/her disability or disabilities requires that he/she remain in bed apart from any prescribed course of convalescence or treatment, or

3. is a patient in a nursing home due to mental or physical incapacity, or

4. is blind, or so nearly blind as to have corrected visual acuity of 5/200 or less in both eyes, or concentric contraction of the visual field to 5 degrees or less.

Housebound. A veteran may be eligible for Housebound benefits when the veteran:

1. has a single permanent disability evaluated as 100-percent disabling and, due to such disability, he/she is permanently and substantially confined to his/her immediate premises, or

2. has a single permanent disability evaluated as 100-percent disabling and, another disability or disabilities, evaluated as 60 percent or more disabling.

The Improved Pension Program provides for the maximum annual rates listed below. The payment is reduced by the amount of the countable income of the veteran and the income of the spouse or dependent children. When a veteran without a spouse or a child is being furnished nursing home or domiciliary care by the VA, the pension is reduced to an amount not to exceed $90 per month after three calendar months of care. The reduction may be delayed if nursing home care is being continued for the primary purpose of providing the veteran with rehabilitation services.

Maximum Annual Pension Rates for Veterans:

Veteran without dependent spouse or child $11,830
Veteran with one dependent (spouse or child) $15,493
Veteran in need of regular aid and attendance w/o dependents $19,736
Veteran in need of regular aid and attendance with one dependent $23,396
Veteran permanently housebound without dependent $14,457
Veteran permanently housebound with one dependent $18,120
Two veterans married to one another $15,493
World War I veteran add to the applicable annual rate $2,686
Increase for each additional dependent child $2,020

GI BILL BENEFITS

The Post 9/11 GI Bill

The Post 9/11 GI Bill is a new education benefit program for individuals who served on active duty on or after September 11, 2001. Post 9/11 GI Bill benefits are payable for training pursued on or after August 1, 2009. No payments can be made under this program for training pursued before that date.

Eligibility. You may be eligible if you served at least 90 aggregate days on active duty after September 10, 2001, and you are still on active duty or were: (1) discharged from active duty; (2) released from active duty and placed on the retired list or temporary disability retired list; (3) released from active duty and transferred to the Fleet Reserve or Fleet Marine Corps Reserve; or (4) released from active duty for service in a Reserve Component. Under all circumstances, your period of service must have been under honorable conditions.

You may also be eligible if you were honorably discharged from active duty for a service-connected disability and served at least 30 continuous days after September 10, 2001.

Coordination with other educational programs. If, on August 1, 2009, you are eligible for the Montgomery GI Bill, Montgomery GI Bill – Selected Reserve, or the Reserve Educational Assistance Program and you qualify for the Post 9/11 GI Bill, you may make an irrevocable election to receive benefits under the Post 9/11 GI Bill. However, once you elect to receive benefits under the Post 9/11 GI Bill, you will no longer be eligible to receive benefits under the program from which you elected the Post 9/11 GI Bill.

Benefits. Based on your length of active duty service after September 10, 2001, you are entitled to a percentage of the following:

• Cost of tuition and fees, not to exceed the most expensive in-state undergraduate tuition at a public institution of higher education (paid to the school);

• Monthly housing allowance* equal to the basic allowance for housing payable to an E-5 with dependents, in the same ZIP code as the school (paid to you);

• Annual books and supplies stipend of up to $1,000 per year (paid to you); and

• A one-time payment of $500 paid to certain individuals relocating from highly rural areas.

*The housing allowance and books and supplies stipend are not payable to individuals on active duty; those pursuing training at half time or less; or to individuals enrolled in distance learning.

Aggregate period of active duty after September 10, 2001		Percentage of Maximum Benefit
30 or more continuous days and discharged due to service connected disability		100
At Least	But Less Than	
36 months	N/A	100
30 months	36 months	90
24 months	30 months	80
18 months	24 months	70
12 months	18 months	60
6 months	12 months	50
90 days	6 months	40

In general, you may receive up to 36 months of benefits under the Post 9/11 GI Bill. You will be eligible for benefits for 15 years from your last period of active duty of at least 90 consecutive days. If you were released for a service-connected disability after at least 30 days of continuous service, you will also be eligible for benefits for 15 years.

Example: Tuition and fees for full time enrollment: $6,700. Highest in-state tuition and fees: $7,000.

Scenario 1: If you separated after three years on active duty and are going to school full-time, you would receive $6,700 for tuition and fees, $1,000 for books and supplies, and the monthly housing allowance.

Scenario 2: If you have an aggregate of 12 months of active duty in a Reserve Component and are going to school full-time, you would receive $4,020 (60 percent of $6,700) for tuition and fees, $600 (60 percent of $1,000) for books and supplies, and 60 percent of the monthly housing allowance.

Programs Covered. You may receive benefits for any approved program offered by a school in the U.S. that is authorized to grant an associate (or higher) degree. To verify the program at the school you are interested in, go to: www.gibill.va.gov/GI_Bill_Info/search_programs.htm. You may also receive benefits for tutorial assistance or up to $2,000 for the reimbursement of one licensing or certification test.

In addition, if you transferred to the Post 9/11 GI Bill from the Montgomery GI Bill – Active Duty, Montgomery GI Bill – Selected Reserve, or the Reserve Education Assistance Program, you may also receive Post 9/11 GI Bill benefits for flight training, apprenticeship or on-the-job training programs, and correspondence courses.

Transferability. If you are a member of the Armed Forces on August 1, 2009, DoD may offer you the opportunity to transfer benefits to your spouse or dependent children. DoD and the military services are developing policy on entitlement to transferability. For more information, go to www.gibill.va.gov.

Montgomery GI Bill (Active Duty)

The Montgomery GI Bill is a program of educational assistance enacted by Congress to attract high quality men and women into the Armed Forces. It provides education and training opportunities to eligible persons. There are four categories of eligibility:

Category 1—(a) you entered active duty for the first time after June 30, 1985; (b) your military pay was reduced by $100 a month for first 12 months; (c) you served continuously for three years, or two years if that is what you first enlisted for, or two years if you entered the Selected Reserve within one year of leaving active duty and served four years in the Selected Reserve.

Category 2—(a) you entered active duty before January 1, 1977; (b) you served at least one day between October 19, 1984 and June 30, 1985, and stayed on active duty through June 30, 1988, (or June 30, 1987, if you entered the Selected Reserve within one year of leaving active duty and served four years in the Selected Reserve; (c) on December 31, 1989, you had entitlement left from the Vietnam Era GI Bill.

Category 3—(a) you are not eligible under Category 1 or 2; (b) you were on active duty on September 30, 1990, **and** were involuntarily separated after February 2, 1991, **or** you were involuntarily separated on or after November 30, 1993, **or** you were voluntarily separated under either the Voluntary Separation Incentive (VSI) or Special Separation Benefit (SSB) program; (d) before separation, you had military pay reduced by $1,200.

Category 4—(a) you were on active duty on October 9, 1996 **and** you had money remaining in a VEAP account on that date **and** you elected the Montgomery GI Bill by October 9, 1997 **or** you entered full-time National Guard duty under title 32, USC, between July 1, 1985 and November 28, 1989 **and** you elected the Montgomery GI Bill during the period October 9, 1996 through July 8, 1997; (b) you had military pay reduced by $100 a month for 12 months or made a $1,200 lump-sum contribution.

Additional Requirements. Your discharge must be "honorable." Discharges "under honorable conditions" and "general" discharges do not establish eligibility for the Montgomery GI Bill (38 USC, Chapter 30).

You are not eligible under Categories 1 or 2 if you graduated from a service academy and received a commission after December 31, 1976. You are not eligible under Categories 1 or 2 if you received a commission after completing an ROTC scholarship program after December 31, 1976. If you are a graduate of an ROTC program, you may still be eligible for the Montgomery GI Bill (38 USC, Chapter 30) under one of the following conditions if: you received a commission after becoming eligible for benefits; you completed ROTC without a full scholarship; or, you received a commission after September 30, 1996. You must have received less than $2,000 during any one year of your ROTC program. The amount increased to $3,400 as of January 2002.

You may be entitled to receive up to 36 months of education benefits under this program after completing the full period of your enlistment. This applies if you were discharged for the convenience of the Government after completing: 20 months of an enlistment of less than three years; or 30 months of an enlistment of three years or more.

Your monthly check may include a "Kicker." Kickers are amounts contributed by DoD to an education fund on behalf of individuals to encourage enlistment or retention in the Armed Forces, usually in specialized areas. VA only pays Kickers to individuals authorized by DoD.

If you first entered active duty after June 30, 1985, and your first service obligation was more than two years, you will receive the following monthly rates.

BASIC MONTHLY RATES (Effective August 1, 2008)			
TYPE OF TRAINING	**FULL-TIME**	**THREE-QUARTER TIME**	**ONE-HALF TIME**
INSTITUTIONAL	$1,321.00	$990.75	$660.50
LESS THAN HALF-TIME	Tuition and fees ONLY, not to exceed the rate of $660.50 for Less Than Half-Time, but more than One-Quarter Time and $330.25 for One-Quarter Time		
COOPERATIVE	$1,321.00 (Full-Time Only)		
CORRESPONDENCE	Entitlement charged at the rate of one month for each $1,321.00 paid		
APPRENTICESHIP ON-THE-JOB TRAINING	First six months $990.75 Second six months $726.55 Remainder of the program $462.35		
FLIGHT	Entitlement charged at the rate of one month for each $1,321.00 paid		

You will also receive the rates in the previous table if you first entered active duty after June 30, 1985, with a service obligation of two years plus an additional four years in the Selected Reserve; you first entered active duty after June 30, 1985, with a service obligation of two years, and you extended your obligation to three years or more; or you are eligible under Category 3.

If you first entered active duty after June 30, 1985, and your first service obligation was two years, you will receive the following monthly rates.

BASIC RATES FOR PERSONS WHOSE INITIAL ACTIVE DUTY OBLIGATION WAS LESS THAN THREE YEARS, AND WHO SERVED LESS THAN THREE YEARS, EXCLUDING 2 X 4 PARTICIPANTS (Effective August 1, 2008)			
TYPE OF TRAINING	**FULL-TIME**	**THREE-QUARTER TIME**	**ONE-HALF TIME**
INSTITUTIONAL	$1,073.00	$804.75	$536.50
LESS THAN HALF-TIME	Tuition and fees ONLY, not to exceed the rate of $536.50 for Less Than Half-Time, but more than One-Quarter Time and $268.25 for One-Quarter Time		
COOPERATIVE	$1,073.00 (Full-Time only)		
CORRESPONDENCE	Entitlement charged at the rate of one month for each $1,073.00 paid		
APPRENTICESHIP ON-THE-JOB TRAINING	First six months $804.75 Second six months $590.15 Remainder of the program $375.55		
FLIGHT	Entitlement charged at the rate of one month for each $1,073.00 paid		

If you are eligible under Category 2, you may receive a higher rate of payment. In addition to your basic Montgomery GI Bill (38 USC, Chapter 30) rate, you will receive an amount equal to one-half of the rate in effect for benefits under 38 USC, Chapter 34. You will receive this rate only as long as you have Chapter 34 entitlement remaining. After that, you will receive the basic Montgomery GI Bill (38 USC, Chapter 30) rate. The following chart shows the rates in effect under 38 USC, Chapter 34.

BASIC INSTITUTIONAL RATES FOR PERSONS WITH REMAINING ENTITLEMENT UNDER CHAPTER 34 OF TITLE 38, USC (Effective August 1, 2008)				
TRAINING TIME	NO DEPENDENTS	ONE DEPENDENT	TWO DEPENDENTS	EACH ADDITIONAL DEPENDENT
FULL	$1,509.00	$1,545.00	$1,576.00	$16.00
THREE-QUARTER	$1,132.25	$1,158.75	$1,182.25	$12.00
ONE-HALF	$754.50	$772.50	$788.00	$8.50
LESS THAN ONE-HALF BUT MORE THAN ONE-QUARTER	Tuition and fees ONLY, not to exceed the rate of $754.50			
ONE-QUARTER	Tuition and fees ONLY, not to exceed the rate of $377.25			
COOPERATIVE	$1,509.00	$1,545.00	$1,576.00	$16.00
CORRESPONDENCE	55 percent of the approved charges			
FLIGHT	60 percent of the approved charges			
APPRENTICESHIP ON-THE-JOB TRAINING				
1st Six Months	$1,093.50	$1,105.88	$1,116.75	$5.25
2nd Six Months	$782.93	$792.28	$799.98	$3.85
3rd Six Months	$486.15	$492.28	$497.00	$2.45
Remainder of Program	$474.25	$480.03	$485.28	$2.45

An individual who is eligible for Montgomery GI Bill (active duty) benefits, may receive an accelerated payment of the basic educational assistance allowance if enrolled in an approved program of education that leads to employment in a high technology industry (as determined by the VA); and is charged tuition and fees for the program of education that, when divided by the number of months (and fractions thereof) in the enrollment period, exceeds 200 percent of the monthly rate of educational assistance. Contact the VA for additional information on this program.

Counseling. Counseling may be available for individuals who are eligible for VA educational assistance, who are on active duty and within 180 days of discharge, or who have been discharged one year or less. The VA will help these individuals understand their educational and vocational strengths and weaknesses, and plan an educational, training or employment goal. The VA may also help individuals plan a job search.

More Information. The VA Education website, www.gibill.va.gov, has a section that answers frequently asked questions. You can also call the VA toll-free from all 50 states, Puerto Rico, and the U.S. Virgin Islands at 1-888-GIBILL-1 (1-888-442-4551). If you do not have access to this number, call 1-800-827-1000, or for the hearing impaired call 1-800-829-4833.

Reserve Educational Assistance Program (REAP)

Education Benefits for Activated Reservists. A Reserve Component member who serves on active duty on or after September 11, 2001 under Title 10, United States Code for a contingency operation (as defined by DoD) and who serves at least 90 consecutive days or more is eligible for Chapter 1607 benefits. National Guard members also are eligible if their active duty is under Section 502(f), Title 32, United States Code and they serve for 90 consecutive days when authorized by the President or Secretary of Defense for a national emergency that is supported by federal funds. Individuals are eligible as soon as they reach the 90 day point whether or not they are currently serving on active duty.

Benefits cannot be paid from more than one program for the same period of educational training. However, a member receiving the Chapter 1606 Kicker benefit may continue to receive that benefit while receiving Chapter 1607 benefits.

The Chapter 1607 benefit provides 36 months of full-time entitlement at the given rate. However, a Chapter 1607 participant may not use more than 48 months of entitlement under any combination of VA Educational programs. For example, if the member already used 20 months of Chapter 1606 benefits, only 28 months of Chapter 1607 benefits are available.

The rate payable under Chapter 1607 is a percentage of the Chapter 30, Title 38, United States Code three year rates based on the number of continuous days served on active duty. The full-time rate, effective August 1, 2008 is $528.40 for 90 days; $792.60 for one year; and $1,056.80 for two years or more of continuous, or three years or more of aggregate service, as described above.

In addition to the above rates, a member who is eligible for Chapter 1607 benefits may contribute a maximum of $600 to the program in order to receive increased benefits. For each $20 a member contributes, the monthly full-time benefit is increased by $5. Contributing the maximum would result in an additional monthly benefit of $150. The additional monthly benefit would begin with the enrollment period beginning after the contribution is made.

A member who is eligible for Chapter 1607 benefits may continue to receive benefits, until exhausted, for up to 10 years after separating from the Selected Reserve. However, a member who separates from the Selected Reserve on or after January 28, 2008, and as of October 14, 2008 had not used any portion of their REAP benefits, must have been separated under honorable conditions in order to use the benefits after separation.

VOCATIONAL REHABILITATION PROGRAM

Eligibility. Veterans and servicemembers are eligible for vocational rehabilitation if three conditions are met:

1. They suffered a service-connected disability or disabilities in active service, and are receiving at least 20 percent compensation or would do so but for receipt of military retirement pay. Veterans with a 10 percent disability may also be found eligible if they have a serious employment handicap.
2. They were discharged or released under other than dishonorable conditions or are hospitalized awaiting separation for a service-connected disability of at least 20 percent.
3. They need vocational rehabilitation to overcome an employment handicap caused at least partly by their service-connected disability.

Benefits. The VA will evaluate a disabled veteran to establish eligibility. An eligible disabled veteran may receive employment assistance, self-employment assistance, training in a rehabilitation facility, and college or other training. Severely disabled veterans may receive assistance to improve their ability to live independently.

Period of Rehabilitation Program. The veteran must complete a rehabilitation program within 12 years from the date the veteran has been notified of a service-connected disability rating. This period may be extended if a medical condition prevented the veteran from training, or if the veteran has a serious employment handicap. Disabled veterans may receive assistance services until they have reached their rehabilitation goal, up to 48 months. VA may provide counseling, job placement and post-employment services for up to 18 additional months.

Rehabilitation Program Costs. VA will pay the costs of tuition and required fees, books, supplies and equipment. VA may also pay for special support, such as tutorial assistance, prosthetic devices, lip reading training and signing for the deaf. VA will help the veteran pay for at least part of the transportation expenses unique to disabled persons during training or employment services. VA can also provide an advance against future benefit payments for veterans who have financial difficulties during training.

PROGRAM FOR UNEMPLOYABLE VETERANS

Veterans awarded 100 percent disability compensation based upon unemployability may request an evaluation and, if found eligible, may participate in a vocational rehabilitation program and receive assistance in securing employment. A veteran with an unemployability rating who secures gainful employment under the special program will continue to receive disability compensation without reduction until the veteran has worked continuously for at least 12 months.

SURVIVORS AND DEPENDENTS EDUCATION

Educational assistance is provided to a spouse who has not remarried and children of: (1) a veteran who died, or is permanently and totally disabled as the result of a disability sustained during a period of qualifying service, to include a servicemember who is hospitalized or receiving outpatient medical care, services, or treatment and who is likely to be discharged or released from such service for such disability; (2) a veteran who died from any cause while rated permanently and totally disabled from service-connected disability; (3) a servicemember listed for more than 90 days as currently missing in action or captured in the line of duty by a hostile force; or (4) a servicemember listed for more than 90 days as currently detained or interned by a foreign government or power.

Benefits may be awarded for pursuit of associate's, bachelor's or graduate degrees at colleges and universities including independent study, cooperative training and study abroad programs. Courses leading to a certificate or diploma from business, technical or vocational schools may also be taken. Benefits may be awarded for apprenticeships, on-the-job training programs and farm cooperative courses. Benefits for correspondence courses under certain conditions are available to spouses only. Secondary school programs may be pursued if the individual is not a high-school graduate. An individual with a deficiency in a subject may receive tutorial assistance benefits if enrolled half-time or more. Deficiency, refresher, and other training may also be available.

Monthly Payments. The rate is $915.00 a month for full-time school attendance, with lesser amounts for part-time training. A person may receive educational assistance for full-time training for up to 45 months or the equivalent in part-time training.

Payments to a spouse generally end 10 years from the date the individual is found eligible, or from the date of the death of the veteran. However, the spouse of a member who dies while serving on active duty may receive 20 years in which to use the benefits. In addition, the spouse of a member who suffers from a 100 percent service-connected disability that was awarded within three years after discharge from service may receive 20 years to use the benefits if the person remains the spouse of the disabled member throughout the entire period. Children generally must be between 18 and 26 to receive education benefits, however, extensions may be granted.

MONTHLY RATES FOR PERSONS TRAINING UNDER THE SURVIVORS AND DEPENDENTS ASSISTANCE PROGRAM (Effective October 1, 2008)			
TYPE OF TRAINING	FULL-TIME	THREE-QUARTER TIME	ONE-HALF TIME
INSTITUTIONAL	$915.00	$686.00	$456.00
LESS THAN HALF-TIME*	Tuition cost, not to exceed the rate of $456.00 for less than Half-Time but more than One-Quarter Time and $227.75 for One-Quarter Time		
FARM COOPERATIVE	$737.00	$553.00	$368.00
CORRESPONDENCE	Entitlement charged at the rate of one month for each $915.00 paid		
APPRENTICESHIP ON-THE-JOB TRAINING*	First six months $666.00 Second six months $499.00 Third six months $329.00 Remainder of the program $166.00		
SPECIAL RESTORATIVE TRAINING		$915.00	
Accelerated Charges - Cost of Tuition and Fees in Excess of		$284.00	
Entitlement Reduced One Day for Each		$30.50 (1/30th of Full-Time rate)	

* Tuition and Fees ONLY. Payment cannot exceed the amount listed above.

Counseling Services. VA may provide counseling services to help an eligible dependent pursue an educational or vocational objective.

Special Benefits. An eligible person with a physical or mental disability that impairs pursuit of an educational program may receive special restorative training to lessen or overcome that impairment. This training may include speech and voice correction, language retraining, lip reading, auditory training, Braille reading and writing, and similar programs. Specialized vocational training is also available to an eligible spouse or child who is handicapped by a physical or mental disability that prevents pursuit of an educational program.

Spina Bifida Assistance. A child with spina bifida can receive vocational training to guide the child, parent, or guardian in choosing a vocational training program. VA also will provide up to 24 months of training to achieve a vocational goal. Call the VA at 1-888-820-1756 for more information about benefits available to dependent children with spina bifida.

WORK-STUDY PROGRAM

The work-study program is available to any student receiving VA educational benefits. Participants must train at the three-quarter or full-time rate. They will be paid an amount equal to the applicable minimum wage multiplied by the number of hours worked during the applicable period. Participants under the supervision of a VA employee may provide outreach services, prepare and process VA paperwork, and work at a VA medical facility or perform other approved activities.

HOME LOAN GUARANTEES

The VA guarantees loans made to servicemembers, veterans, Reservists, and unremarried and certain remarried surviving spouses for the purchase or refinancing of homes, condominiums, and manufactured homes. The VA guarantees part of the total loan, permitting the purchaser to obtain a mortgage with a competitive interest rate, even without a down payment if the lender agrees. The VA requires a down payment for the purchase of a manufactured home. The VA also requires a down payment for a home or condominium if the purchase price exceeds the reasonable value of the property or the loan has a graduated payment feature. With a VA guaranty, the lender is protected against loss up to the amount of the guaranty if the borrower fails to repay the loan. A VA loan guaranty can be used to buy a home; buy a residential condominium; build a home; repair, alter or improve a home; refinance an existing home loan; buy a manufactured home with or without a lot; buy and improve a manufactured home lot; install a solar heating or cooling system or other weatherization improvements; purchase and improve a home simultaneously with energy efficient improvements; refinance an existing VA loan to reduce the interest rate; or refinance a manufactured home loan to acquire a lot.

General Rules for Eligibility. For wartime service, you must have at least 90 days on active duty and been discharged under other than dishonorable conditions. If you served less than 90 days, you may be eligible if discharged for a service-connected disability.

Periods of wartime service include: WWII—September 16, 1940 to July 25, 1947; Korea—June 27, 1950 to January 31, 1955; and Vietnam—August 5, 1964 to May 7, 1975.

For service on active duty during the Gulf War (August 2, 1990 to a date yet to be determined) you must have: (1) completed 24 months of continuous active duty or the full period (at least 90 days) for which you were called or ordered to active duty and discharged under conditions other than dishonorable; (2) completed at least 90 days of active duty and been discharged under the specific authority of 10 USC 1173 (hardship), or 10 USC 1171 (early out), or have been determined to have a compensable service-connected disability; or (3) discharged with less than 90 days of service for a service-connected disability. Individuals may also be eligible if they were released from active duty due to an involuntary reduction in force, certain medical conditions, or, in some instances, for the convenience of the Government.

For peacetime service, you must have served at least 181 days of continuous active duty and been discharged under other than dishonorable conditions. If you served less than 181 days, you may be eligible if discharged for a service-connected disability. Periods of peacetime service include: July 26, 1947 to June 26, 1950; February 1, 1955 to August 4, 1964; May 8, 1975 to September 7, 1980 (enlisted); and May 8, 1975 to October 16, 1981 (officer).

For service after September 7, 1980 (enlisted), or October 16, 1981 (officer), you must have: (1) completed 24 months of continuous active duty or the full period (at least 181 days) for which you were ordered or called to active duty and been discharged under conditions other than dishonorable; (2) completed at least 181 days of active duty and been discharged under the specific authority of 10 USC 1173 (hardship), or 10 USC 1171 (early out), or have been determined to have a compensable service-connected disability; or (3) discharged with less than 181 days of service for a service-connected disability. Individuals may also be eligible if they were released from active duty due to an involuntary reduction in force, certain medical conditions, or, in some instances for the convenience of the Government.

If you are now on active duty (not active duty for training), you are eligible after having served 181 days unless discharged or separated from a previous qualifying period of active duty service.

A member of the Selected Reserve, who is not otherwise eligible and has completed a total of six years in the Selected Reserve (member of an active unit, attended required weekend drills and annual training) and has an honorable discharge; or was placed on the retired list; or was transferred to the Standby Reserve or an element of the Ready Reserve other than the Selected Reserve after service characterized as honorable; or continues to serve in the Selected Reserve, may be eligible. Individuals who completed less than six years may be eligible if discharged for a service-connected disability.

The unremarried spouse of a veteran who died while in service or from a service-connected disability; the spouse of a servicemember missing in action or a prisoner or war; and a surviving spouse who remarries on or after reaching age 57 may also be eligible.

Applicants must have a good credit rating, have an income sufficient to support mortgage payments, and agree to live in the property. To obtain a VA certificate of eligibility, complete VA Form 26-1880, *"Request for Determination of Eligibility and Available Loan Guaranty Entitlement,"* and submit it to the nearest VA regional office. The website www.va.gov has links to Form 26-1880 and other VA forms. To determine service eligibility, contact your VA regional office.

Guaranty Amount. The amount of the VA guaranty available to an eligible veteran is called the entitlement and may be considered the equivalent of a down payment by lenders.

Loans up to $417,000. Up to $104,250 (2009 rate) in entitlement may be available to veterans purchasing or constructing homes and to veterans who obtain an Interest Rate Reduction Refinancing Loan. Lenders will generally lend up to four times the amount of a veterans entitlement without requiring a down payment. This means that qualified veteran's could get a no down payment purchase loan of up to $417,000 (2009 rate). The amount of entitlement varies with the loan amount.

Loans for more than $417,000. If the original principal loan amount is greater than $417,000, VA will guarantee 25 percent of the original principal loan amount, up to the maximum guaranty amount. The maximum guaranty amount varies depending upon the location of the property as follows:

1. For all locations in the U.S. other than Alaska, Guam, Hawaii, and the U.S. Virgin Islands, the maximum guaranty amount is the greater of 25 percent of $417,000 or 125 percent of the area median price for a single-family residence. However, the guaranty may not exceed 175 percent of the Freddie Mac loan limit for a single family residence in the county in which the property securing the loan is located. This translates to a potential maximum loan amount of $1,094,625.

2. In Alaska, Guam, Hawaii, and the U.S. Virgin Islands, the maximum guaranty amount is the greater of 25 percent of $625,500 or 125 percent of the area median price for a single-family residence. However, the guaranty may not exceed 175 percent of the Freddie Mac loan limit for a single family residence in the county in which the property securing the loan is located. This translates to a potential maximum loan amount of $1,641,937.50.

The VA does not establish a maximum loan amount. However, no loan for the acquisition of a home or refinancing an existing mortgage or other liens secured on a dwelling may exceed 100 percent of the appraised value of a home.

Required Occupancy. Veterans must certify that they intend to live in the home they are buying or building with a VA guaranty. A veteran who wishes to refinance or improve a home with a VA guaranty also must certify to being in occupancy at the time of application. A spouse may certify occupancy if the buyer is on active duty. If refinancing a VA guaranteed loan solely to reduce the interest rate, veterans need only certify prior occupancy.

Closing Costs. Payment in cash is required on all home loan closing costs, including title search and recording, hazard insurance premiums, prepaid taxes and a one percent origination fee which may be required by lenders in lieu of certain other costs. In the case of refinancing loans, all such costs may be included in the loan, as long as the total loan does not exceed 100 percent of the reasonable value for the property. Interest rate reduction refinancing loans may include closing costs and a maximum of two discount points.

Funding Fee. The VA charges a funding fee for loans, including refinancing loans. The funding fee is waived for (1) a veteran who is receiving VA compensation for service-connected disabilities (or who, but for the receipt of retirement pay, would be entitled to receive compensation); (2) a veteran who is rated eligible to receive VA compensation as a result of a pre-discharge disability examination; and (3) a surviving spouse of any veteran who died from a service-connected disability. The VA funding fee is based on the loan amount and, at the discretion of the veteran and the lender, may be included in the loan.

FUNDING FEES

<table>
<tr><th colspan="2" rowspan="3">LOAN CATEGORY</th><th colspan="4">FUNDING FEES (Percentage of Loan Amount)</th></tr>
<tr><th colspan="2">Loan closed on or after 10/1/2004 and before10/1/2011</th><th colspan="2">Loan closed on or after 10/1/2011</th></tr>
<tr><th>Active Duty Veteran</th><th>Reservist</th><th>Active Duty Veteran</th><th>Reservist</th></tr>
<tr><td rowspan="4">To purchase or construct a dwelling</td><td>Initial Loan with 0 percent down</td><td>2.15</td><td>2.40</td><td>1.40</td><td>1.65</td></tr>
<tr><td>Subsequent loan with 0 percent down</td><td colspan="2">3.30</td><td colspan="2">2.15 Reduced to 1.25 percent for loan closed on or after 10/1/2013</td></tr>
<tr><td>Initial or subsequent loan with at least five but less than 10 percent down</td><td>1.50</td><td>1.75</td><td>0.75</td><td>1.00</td></tr>
<tr><td>Initial or subsequent loan with at least 10 percent down</td><td>1.25</td><td>1.50</td><td>0.50</td><td>0.75</td></tr>
<tr><td colspan="2">Interest rate reduction refinancing loan</td><td colspan="4">.50</td></tr>
<tr><td colspan="2">Manufactured Home Loan other than interest rate reduction refinancing loan</td><td colspan="4">1.00</td></tr>
<tr><td colspan="2">Assumption of VA guaranteed loan</td><td colspan="4">.50</td></tr>
</table>

Financing, Interest Rates, and Terms. Veterans obtain VA guaranteed loans through the usual lending institutions, including banks, saving and loan associations, building and loan associations, and mortgage loan companies. Veterans may obtain a loan with a fixed interest rate which may be negotiated with the lender. If the lender charges discount points on the loan, the veteran may negotiate with the seller as to who will pay points or if they will be split between buyer and seller. Points paid by the veteran may not be included in the loan, except that a maximum of two points may be included in interest rate reduction refinancing loans. The term of the loan may be for as long as 30 years and 32 days.

The VA normally does not require that a down payment be made. However, they do require a down payment for a manufactured home or lot loan, for a loan with graduated payment features, and to prevent the amount of a loan from exceeding VA's determination of the property's reasonable value. If the sale price exceeds the reasonable value, the veteran must certify that the difference is being paid in cash without supplementary borrowing. A cash down payment of five percent of the purchase price is required for manufactured homes or lot loans.

Release of Liability. When a veteran sells a home financed through a VA guaranty to a purchaser who assumes the loan, the veteran may request release from liability to the Federal Government provided the loan is current, the purchaser has been obligated by contract to purchase the property and assume all of the veteran's liabilities, and VA is satisfied that the purchaser is a good risk. A release of liability does not mean that a veteran's guaranty entitlement is restored. The VA usually restores entitlement only when it is no longer liable. If the new veteran buyer, however, agrees to substitute entitlement for that of the original veteran buyer, entitlement may be restored to the original veteran buyer.

For loans made on or after March 1, 1988, a release from liability determination must be made in every case involving the assumption of a VA guaranteed loan. This will involve a determination of the good credit of the buyer assuming the loan by the holder of the loan or VA. A VA loan for which a commitment was made on or after March 1, 1988, is not assumable without prior approval of VA or its authorized agent. The person who assumes a VA loan for which a commitment was made on or after March 1, 1988, must pay a fee to VA equal to one half of one percent of the balance of the loan being assumed. If a person disposes of the property securing a VA guaranteed loan for which a commitment was made after March 1, 1988, without first notifying the holder of the loan, the holder may demand immediate and full payment of the loan.

Repossessed Houses. The VA sells homes that have been acquired after foreclosure of a VA guaranteed loan. These homes are available to both veterans and nonveterans. Contact local real estate agents for available listings.

Safeguards for Veterans.

1. Homes completed less than a year before purchase with VA financing and inspected during construction by either VA or HUD must meet VA requirements.

2. The VA may suspend from the loan program those who take unfair advantage of veteran borrowers or decline to sell a new home or make a loan to an eligible veteran of good credit because of race, color, religion, sex, disability, family status or national origin.

3. The builder of a new home is required to give the purchasing veteran a one year warranty that the home has been constructed to VA-approved plans and specifications. A similar warranty must be given for new manufactured homes.

4. In cases of new construction completed under VA or HUD inspection, VA may pay or otherwise compensate the veteran borrower for correction of structural defects seriously affecting livability if assistance is requested within four years of a home loan guaranty.

5. The borrower obtaining a VA loan may only be charged the fees and other charges prescribed by VA as allowable.

6. The borrower can pre-pay without penalty the entire loan or any part not less than the amount of one installment or $100.

7. The VA encourages holders to extend forbearance if a borrower becomes temporarily unable to meet the terms of the loan.

SPECIALLY ADAPTED HOUSING FOR DISABLED VETERANS

These special areas of assistance are in addition to GI loan benefits to which the individual may be entitled. Contact the VA for more detailed information.

A disabled veteran or member of the Armed Forces with a service-connected disability may be entitled to a grant from VA for a home specially adapted to their needs or for adaptations. There are three types of grants that are available to assist severely disabled individuals in adapting housing to their special needs. The Specially Adapted Housing Grant pays up to $60,000 and the Special Home Adaptations Grant pays up to $12,000. In addition, a qualified individual may also receive a Home Improvements and Structural Alterations Grant. An individual may apply up to three times for a grant, although the total amount of all grants may not exceed the established limits.

Specially Adapted Housing Grant. VA may approve a grant of not more than 50 percent of the cost of constructing a home or remodeling an existing home with needed adaptations, or toward the unpaid balance of a home already adapted without the assistance of a VA grant, up to a maximum of $60,000. Individuals must be entitled to compensation for permanent and total service-connected disability due to:

1. loss, or loss of use of both lower extremities, such as to preclude locomotion without the aid of braces, crutches, canes, or wheelchair; or
2. disability which includes (a) blindness in both eyes, having only light perception, plus (b) loss, or loss of use, of one lower extremity; or
3. loss, or loss of use of one lower extremity together with (a) residuals of organic disease or injury or (b) the loss or loss of use of one upper extremity, which so affects the functions of balance or propulsion as to preclude locomotion without the aid of braces, crutches, canes or a wheelchair; or
4. loss, or loss of use, of both upper extremities such as to preclude use of the arms at or above the elbows; or
5. a severe burn injury.

Special Home Adaptations Grant. The VA may approve a grant for the actual cost, up to a maximum of $12,000 for adaptations to an individual's residence that are determined by VA to be reasonably necessary. The grant may also be used to assist eligible individuals in acquiring a residence that has already been adapted with special features for the individual's disability. Individuals must be entitled to compensation for permanent and total service-connected disability due to:

1. blindness in both eyes with 5/200 visual acuity or less, or
2. anatomical loss, or loss of use, of both hands.

Temporary Residence Assistance Grants. In addition, until December 31, 2011, an individual with a service-connected disability who meets the criteria for either a Specially Adapted Housing Grant or Special Home Adaptations Grant and is temporarily residing in a home owned by a family member may receive up to $14,000 or $2,000, respectively, to assist with adapting the family member's home to meet their special needs.

Home Improvements and Structural Alterations Grant. A veteran or servicemember with a service-connected disability may receive assistance for any home improvement necessary for the continuation of treatment or for disability access to the home and essential lavatory and sanitary facilities. This grant is available to individuals who have received a medical determination indicating that improvements and structural alterations are necessary or appropriate for the effective and economical treatment of the disability. An individual may receive both this grant and either a Specially Adapted Housing Grant or Special Home Adaptations Grant. The amount of the grant is up to $4,100 for service-connected individuals, and up to $1,200 for nonservice-connected veterans.

Supplemental Financing. Individuals with available loan guaranty entitlement may also obtain a guaranteed loan or a direct loan from VA to supplement the grant to acquire a specially adapted home.

Housing Insurance. Individuals with a specially adapted housing grant may be eligible for Veterans Mortgage Life Insurance.

AUTOMOBILES OR OTHER CONVEYANCES

Veterans qualify for this benefit if they have service-connected loss of one or both hands or feet, or permanent loss of use, or permanent impairment of vision of both eyes. Veterans entitled to compensation for ankylosis (immobility) of one or both knees, or one or both hips, also qualify for adaptive equipment for an automobile. There is a onetime payment by VA of not more than $11,000 toward the purchase of an automobile or other conveyance. VA will pay for adaptive equipment, and for repair, replacement, or reinstallation required because of disability, and for the safe operation of a vehicle purchased with VA assistance. To apply, contact a VA regional office or a VA medical center.

CLOTHING ALLOWANCE

Any veteran who is entitled to receive compensation for a service-connected disability for which he or she uses prosthetic or orthopedic appliances may receive an annual clothing allowance. The annual rate effective December 1, 2008 is $716. The allowance is also available to any veteran whose service-connected skin condition requires prescribed medication that damages the veteran's outer garments. To apply, contact a VA regional office.

PERSIAN GULF WAR, AGENT ORANGE AND IONIZING RADIATION

Registry Programs. VA maintains databases called registries to help analyze health conditions reported by veterans.

Gulf War/Operation Iraqi Freedom Registry is for veterans who served in Southwest Asia during the Gulf War (August 2, 1990 to a date not yet established, including Operation Iraqi Freedom). Veterans who served in the following areas are eligible for the lab test that measures depleted uranium in urine: Iraq; the Neutral Zone (between Iraq and Saudi Arabia); Saudi Arabia; Kuwait; Qatar; The United Arab Emirates; Oman; Gulf of Oman; Gulf of Aden; Waters of the Persian Gulf, Arabian Sea; and Red Sea.

Depleted Uranium Registry provides a depleted uranium evaluation for veterans who are identified by DoD because of possible depleted uranium exposure during military activities in: the 1991 Gulf War; Bosnia; Operation Iraqi Freedom; Operation Enduring Freedom; or for Veterans who contact the VA because they are concerned about potential exposure to depleted uranium.

Agent Orange Registry is for veterans possibly exposed to dioxin or other toxic substances in herbicides used during the Vietnam War (between 1962 and 1975), while serving in Korea (between 1968 and 1969), or as a result of testing, transporting, or spraying herbicides for military purposes.

Ionizing Radiation Registry is for veterans possibly exposed to atomic radiation during the following activities: participation in tests involving the atmospheric detonation of a nuclear device; occupation of Hiroshima or Nagasaki from August 6, 1945 through July 1, 1946; internment as a prisoner of war in Japan during World War II; serving in official military duties at the Department of Energy gaseous diffusion plants at Paducah, KY; Portsmouth, OH; or the K-25 area at Oak Ridge, TN, for at least 250 days before February 1, 1992, or in Longshot, Milrow or Cannikin underground nuclear tests at Amchitka Island, AK, before January 1, 1974; or treatment with nasopharyngeal (NP) radium during active military service.

Veterans eligible for participation in any VA registry program may receive free, comprehensive registry medical examinations, including laboratory and other diagnostic tests deemed necessary by an examining clinician. Eligible veterans do not have to be enrolled in VA health care to participate in registry examinations. To participate, contact the nearest VA health care facility. More information is available at www.va.gov/environagents.

DOMICILIARY CARE

Domiciliary care provides rehabilitative and long-term, health-maintenance care for veterans who require minimal medical care but who do not need the skilled nursing services provided in nursing homes. VA may provide domiciliary care to veterans whose annual income does not exceed the maximum annual rate of VA pension or to veterans the Secretary of Veterans Affairs determines have no adequate means of support.

NURSING HOME CARE

The Veterans Health Administration provides nursing home services to veterans through three national programs: VA-owned and operated nursing homes, state veterans' homes owned and operated by the states, and the community nursing home program. Each program has admission and eligibility criteria specific to the program.

VA Nursing Homes. VA-owned and operated nursing homes typically admit patients requiring short-term care, in need of placement for a service-connected disability, or those who have a 70 percent or greater service-connected disability. All other admissions are based on available resources.

State Veterans' Home Program. The state veterans' home program is a cooperative venture between the states and VA whereby the states petition VA for matching construction grants and once granted, VA pays a portion of the per diem. State veterans' homes establish eligibility criteria for long-term or short-term nursing home care. Specialized services offered are dependent upon the capability of the home to render them.

Community Nursing Home Program. VA maintains contracts with community nursing homes though every VA medical center. The program is designed to meet the nursing home needs of veterans who require long-term nursing home care in their own community, close to their families.

Eligibility. The general admission criteria for nursing home placement requires that a resident must be medically stable, i.e. not acutely ill, have sufficient functional deficits to require inpatient nursing home care, and is assessed by an appropriate medical provider to be in need of institutional nursing home care. In addition, the veteran must meet the required VA eligibility criteria for nursing home care or the contract nursing home program and the eligibility criteria for the specific state veterans' home. For VA nursing home care, a veteran may be subject to a copayment depending on their financial status. VA social workers can help explain eligibility and copayment requirements.

Long-Term Care Services. In addition to nursing home care, VA offers a variety of other long-term care services either directly or by contract with community-based agencies. Such services include adult day health care, inpatient or outpatient respite care, inpatient or outpatient geriatric evaluation and management, hospice and palliative care, and home based primary care. Veterans receiving these services may be subject to a copayment.

READJUSTMENT COUNSELING

Readjustment counseling is provided at over 200 community-based Vet centers located in all 50 states, the District of Columbia, Guam, Puerto Rico, and the U.S. Virgin Islands, and is designed to help combat veterans in their readjustment to civilian life. Vet center staff provide group, individual and family counseling plus a wide range of other services to include medical referral, homeless veteran services, employment services, VA benefit referral, and the brokering of non-VA services.

Eligible veterans include those who served on active duty in a combat theater during:

• *Vietnam War* – February 28, 1961 to May 7, 1975;

• *Korean War* – June 27, 1950 to July 27, 1954 (eligible for the Korean Service Medal);

• *World War II* – Three eligible categories: European-African-Middle Eastern Campaign Medal (December 7, 1941 to November 8, 1945), Asiatic-Pacific Campaign Medal (December 7, 1941 to March 2, 1946), or American Campaign Medal (December 7, 1941 to March 2, 1946);

• *American Merchant Marines* – In oceangoing service during the period of armed conflict, December 7, 1941 to August 15, 1945;

• *Vietnam era veterans not in the war zone* – August 5, 1964 to May 7, 1975 (eligible until January 1, 2004, however, veterans who were Vet center clients prior to January 1, 2004, remain eligible);

- *Lebanon* – August 25, 1982 to February 26, 1984;
- *Grenada* – October 23, 1983 to November 21, 1983;
- *Panama* – December 20, 1989 to January 31, 1990;
- *Persian Gulf* – August 2, 1990 to – a date to be determined;
- *Somalia* – September 17, 1992 to – a date to be determined;
- *Operations Joint Endeavor, Joint Guard, & Joint Forge* – Veterans who participated in one or more of the three successive operations in the former Yugoslavia (Bosnia-Herzegovina and Croatia, aboard U.S. Naval vessels operating in the Adriatic Sea, or air spaces above those areas);
- *Global War On Terrorism* – Veterans who serve or have served in military expeditions to combat terrorism on or after September 11, 2001 and before a date yet to be established. Includes Operation Enduring Freedom and Operation Iraqi Freedom.

Vet centers also provide bereavement counseling to the families of military personnel killed in action and sexual trauma counseling to veterans who suffered sexual trauma while on active duty.

Readjustment difficulties can include Post-Traumatic Stress Disorder (PTSD) or any other problems that affect functioning within the family, work, school or other areas of everyday life. For additional information, contact the nearest vet center, or get more information at www.vetcenter.va.gov.

OUTPATIENT PHARMACY SERVICES

The VA provides free outpatient pharmacy services to: (1) veterans with a service-connected disability of 50 percent or more; (2) veterans receiving medication for service-connected conditions; (3) veterans whose annual income does not exceed the maximum annual VA pension rate; (4) veterans enrolled in Priority Group 6 who receive medication for service-connected conditions; (5) veterans receiving medication for conditions related to sexual trauma while serving on active duty; (6) certain veterans receiving medication for treatment of cancer of the head or neck; (7) veterans receiving medication for a VA approved research project; and (8) former POWs.

Other veterans are charged a copayment of $8 for each 30-day or less supply of medication. For veterans enrolled in Priority Groups 2 through 6, the VA sets a maximum annual copayment amount. The copayments apply to prescription and over-the-counter medications, such as aspirin, cough syrup or vitamins, dispensed by a VA pharmacy.

OUTPATIENT DENTAL TREATMENT

Outpatient dental treatment provided by the VA includes examinations and the full spectrum of diagnostic, surgical, restorative and preventive procedures. The following veterans may receive care:

1. Veterans having service-connected and compensable dental disabilities or conditions.
2. Former prisoners of war.
3. Veterans with service-connected, noncompensable dental conditions as a result of combat wounds or service injuries.
4. Veterans with nonservice-connected dental conditions determined by VA to be aggravating a service-connected medical problem.
5. Veterans having service-connected conditions rated as permanently and totally disabling, or rated 100 percent by reason of individual unemployability.

6. Veterans participating in a vocational rehabilitation program.
7. Certain enrolled homeless veterans.
8. Veterans with nonservice-connected dental conditions for which treatment was started while the veteran was an inpatient in a VA facility.
9. Veterans requiring treatment for dental conditions clinically determined to be complicating a medical condition currently under treatment.

In addition, veterans may receive one-time dental treatment for service-connected and noncompensable dental disabilities if the following conditions are met: the dental condition can be shown to have existed at time of discharge; the veteran served on active military duty for at least 90 days; the veteran applied to VA for dental care within 180 days of discharge or release from active duty, and the certificate of discharge does not include certification that all appropriate dental treatment had been rendered prior to discharge.

BENEFICIARY TRAVEL

Certain veterans may be reimbursed for travel costs to receive VA medical care. Effective November 17, 2008, reimbursement is paid at $.415 per mile. Scheduled appointments qualify for round-trip mileage. Unscheduled visits may only qualify for return mileage. The deductible, effective January 9, 2009 is $3.00 for one-way and $6.00 for round trip. The maximum monthly deductible is $18.00.

Beneficiary travel payments may be made to the following:

1. Veterans whose service-connected disabilities are rated 30 percent or more.
2. Veterans traveling for treatment of a service connected condition.
3. Veterans who receive a VA pension.
4. Veterans whose income does not exceed the maximum annual VA pension rate.
5. Veterans traveling for scheduled compensation or pension examinations.
6. Veterans participating in the Vocational Rehabilitation Program.

A veteran whose medical condition requires an ambulance or specially equipped van, may qualify for such transportation if they meet one of the qualifications in 1 to 4, above, and travel is preauthorized. Advance authorization is not required in an emergency if a delay would be hazardous to life or health.

MEDICAL CARE FOR DEPENDENTS AND SURVIVORS (CHAMPVA)

The Civilian Health and Medical Program of the Department of Veterans Affairs (CHAMPVA) is a comprehensive health care program in which the VA shares the cost of covered health care services and supplies with eligible beneficiaries. To be eligible for CHAMPVA, an individual cannot be eligible for TRICARE and must be one of the following:

1. The spouse or child of a veteran who VA has rated as 100 percent permanently and totally disabled for a service-connected disability.
2. The surviving spouse or child of a veteran who died from a VA rated service-connected disability, or who, at the time of death, was rated 100 percent permanently and totally disabled.
3. The surviving spouse or child of a military member who died in the line of duty, not due to misconduct. However, in most cases, these family members are eligible for TRICARE, not CHAMPVA.

A surviving spouse under age 55 who remarries loses CHAMPVA eligibility on midnight of the date of remarriage. However, eligibility may be reestablished if the remarriage is terminated by death, divorce or annulment effective the first day of the month after the termination of the remarriage. A CHAMPVA eligible surviving spouse who is 55 or older does not lose eligibility upon remarriage.

Individuals who have Medicare entitlement may also have CHAMPVA eligibility secondary to Medicare. Beneficiaries who are eligible for CHAMPVA and under age 65 must be enrolled in both Medicare Parts A and B. For benefits beyond age 65, beneficiaries must meet the following conditions: (1) if the beneficiary was 65 or older prior to June 5, 2001, and was otherwise eligible for CHAMPVA, and was entitled to Medicare Part A coverage, then the beneficiary is eligible for CHAMPVA without having to enroll in Medicare Part B; (2) if the beneficiary turned 65 on or before June 5, 2001, and had Medicare Parts A and B, they must keep both Parts to be eligible; (3) if the beneficiary turns 65 on or after June 5, 2001, they must be enrolled in Medicare Parts A and B to be eligible.

For more information, or to apply for CHAMPVA benefits, visit the CHAMPVA website www.va.gov/hac, call 1-800-733-8387, or contact the VA Health Administration Center, P.O. Box 469063, Denver, CO 80246-9063.

Many VA medical centers provide services to CHAMPVA beneficiaries under the CHAMPVA In-house Treatment Initiative (CITI) Program. CHAMPVA beneficiaries who use a CITI facility incur no cost for the services they receive. However, services provided under this program are on a space available basis, after the needs of veterans are met. Because of this, not all services are available at all times, and the same services may not be available every day. CHAMPVA beneficiaries with Medicare entitlement are not eligible to participate in the CITI program.

VA FACILITIES – WHERE TO GO FOR HELP

VA installations are listed by state. Information on VA benefits may be obtained from the following installations: Regional Offices; other offices; Regional Office and Insurance Centers in Philadelphia and St. Paul, and United States Veterans Assistance Centers; Medical Centers (Hospitals); Domiciliary Care; Nursing Home Care; Outpatient Clinic (independent); Outpatient Clinic (physically separated from hospital); and Outpatient Clinic Satellite.

Beneficiaries residing or traveling overseas requiring information or assistance relative to VA benefits should contact the nearest American Embassy or Consulate.

ALABAMA

Medical Centers:

Birmingham 35233 (700 S. 19th St., 205-933-8101)

Montgomery (West Campus) 36109 (215 Perry Hill Rd., 334-272-4670)

Tuscaloosa 35404 (3701 Loop Rd. E., 205-554-2000)

Tuskegee (East Campus) 36083 (2400 Hospital Rd., 334-727-0550)

Vet Centers:

Birmingham 35233 (1201 2nd Ave. S., 205-731-0550)

Mobile 36606 (2577 Government Blvd., 251-478-5906)

Montgomery (215 Perry Hill Rd., Bldg. 6, 334-272-4670 ext. 4483)

ALASKA

Medical Center:

Anchorage 99508 (2925 DeBarr Rd., 907-257-4700 or 1-888-353-7574)

Vet Centers:
Anchorage 99508 (4201 Tudor Centre Dr., Ste. 115, 907-563-6966)
Fairbanks 99701 (540 4th Ave., Ste. 100, 907-456-4238)
Soldotna 99669 (Red Diamond Ctr., Bldg. F, Ste. 4, 43335 Kalifornsky Beach Rd., 907-260-7640)
Wasilla 99654 (851 E. Westpoint Dr., Ste. 111, 907-376-4318 or 1-800-478-4318)

ARIZONA
Medical Centers:
Phoenix 85012 (650 E. Indian School Rd., 602-277-5551)
Prescott 86313 (500 Hwy. 89 N., 928-445-4860)
Tucson 85723 (3601 S. 6th Ave., 520-792-1450)
Vet Centers:
Chinle 86503 (Navajo Rt. 7, Old BIA Compound, Bldg. 59, 928-674-3682)
Hotevilla 86030 (1 Main St., 928-734-5166)
Mesa 85202 (1303 S. Longmore, Ste. 5, 480-610-6727)
Phoenix 85012 (77 E. Weldon Ave., Ste. 100, 602-640-2981)
Prescott 86305 (3180 Stillwater Dr., Ste. A, 928-778-3469)
Tucson 85719 (3055 N. 1st Ave., 520-882-0333)

ARKANSAS
Medical Centers:
Fayetteville 72703 (1100 N. College Ave., 479-443-4301)
Little Rock 72205 (4300 W. 7th St., 501-257-1000)
North Little Rock 72114 (2200 Fort Roots Dr., 501-257-1000)
Vet Center:
North Little Rock 72114 (201 W. Broadway, Ste. A, 501-324-6395)

CALIFORNIA
Medical Centers:
Fresno 93703 (2615 E. Clinton Ave., 559-225-6100)
Livermore 94550 (4951 Arroyo Rd., 925-373-4700)
Loma Linda 92357 (11201 Benton St., 909-825-7084 or 1-800-741-8387)
Long Beach 90822 (5901 E. 7th St., 562-826-8000)
Los Angeles 90073 (11301 Wilshire Blvd., 310-478-3711)
Mather 95655 (10535 Hospital Way, 916-366-5366)
Menlo Park 94025 (795 Willow Rd., 650-493-5000)
Palo Alto 94304 (3801 Miranda Ave., 650-493-5000)
San Diego 92161 (3350 La Jolla Village Dr., 858-552-8585)
San Francisco 94121 (4150 Clement St., 415-221-4810)
Vet Centers:
Capitola 95010 (1350 41st Ave., Ste. 102, 831-464-4575)
Chico 95926 (280 Cohasset Rd., Ste. 100, 530-899-8549)
Commerce 90022 (5400 E. Olympic Blvd., Ste. 140, 323-728-9966)
Concord 94520 (1899 Clayton Rd., Ste. 140, 925-680-4526)
Corona 92879 (800 Magnolia Ave, Ste. 110, 951-734-0525)
Eureka 95501 (2830 G St., Ste. A, 707-444-8271)
Fresno 93726 (3636 N. 1st St., Ste. 112, 559-487-5660)
Garden Grove 92840 (12453 Lewis St., Ste. 101, 714-776-0161)
Los Angeles 90230 (5730 Uplander Way, Ste. 100, Culver City, 310-641-0326)
Los Angeles 90247 (1045 W. Redondo Beach Blvd., Ste. 150, Gardena, 310-767-1221)
Modesto 95351 (1219 N. Carpenter Rd., Ste. 12, 209-569-0713)
Oakland 94612 (1504 Franklin St., Ste. 200, 510-763-3904)
Redwood City 94062 (2946 Broadway St., 650-299-0672)
Rohnert Park 94928 (6225 State Farm Dr., Ste. 101, 707-586-3295)
Sacramento 95825 (1111 Howe Ave., Ste. 390, 916-566-7430)
San Bernardino 92408 (155 W. Hospitality Ln., Ste. 140, 909-890-0797)

San Diego 92103 (2900 6th Ave., 619-294-2040)

San Francisco 94102 (505 Polk St., 415-441-5051)

San Jose 95112 (278 N. 2nd St., 408-993-0729)

San Marcos 92069 (1 Civic Center Dr., Ste. 150, 760-744-6914)

Sepulveda 91343 (9737 Hascle Ave., 818-892-9227)

Ventura 93001 (790 E. Santa Clara St., Ste. 100, 805-585-1860)

COLORADO

Medical Centers:

Denver 80220 (1055 Clermont St., 303-399-8020)

Grand Junction 81501 (2121 North Ave., 970-242-0731)

Vet Centers:

Boulder 80302 (2336 Canyon Blvd., Ste. 103, 303-440-7306)

Colorado Springs 80903 (602 S. Nevada Ave., 719-471-9992)

Denver 80230 (7465 E. First Ave., Ste. B, 303-326-0645)

Fort Collins 80524 (1100 Poudre River Dr., Lower Level, 970-221-5176)

Grand Junction 81505 (2472 F. Rd., Unit 16, 970-245-4156)

Pueblo 81001 (509 E. 13th St., Rm. 18, 719-546-6666, ext. 133)

CONNECTICUT

Medical Centers:

Newington Division 06111 (555 Willard Ave., 860-666-6951)

West Haven Division 06516 (950 Campbell Ave., 203-932-5711)

Vet Centers:

Rocky Hill 06067 (25 Elm St., Ste. A, Wethersfield, 860-563-8800)

Norwich 06360 (2 Cliff St., 860-887-1755)

West Haven 06516 (141 Captain Thomas Blvd., 203-932-9899)

DELAWARE

Medical Center:

Wilmington 19805 (1601 Kirkwood Hwy., 302-994-2511)

Vet Center:

Wilmington 19805 (1601 Kirkwood Hwy., Bldg. 3, 302-994-1660)

DISTRICT OF COLUMBIA

Medical Center:

Washington, DC 20422 (50 Irving St. N.W., 202-745-8000)

Vet Center:

Washington, DC 20011 (1250 Taylor St. N.W., 202-726-5212)

FLORIDA

Medical Centers:

Bay Pines 33744 (10000 Bay Pines Blvd., 727-398-6661)

Gainesville 32608 (1601 S.W. Archer Rd., 352-376-1611)

Lake City 32025 (619 S. Marion Ave., 386-755-3016)

Miami 33125 (1201 N.W. 16th St., 305-324-4455)

Orlando 32803 (5201 Raymond St., 407-629-1599)

Tampa 33612 (13000 Bruce B. Downs Blvd., 813-972-2000)

West Palm Beach 33410 (7305 N. Military Trl., 561-882-8262)

Vet Centers:

Ft. Lauderdale 33304 (713 N.E. 3rd Ave., 954-356-7926)

Ft. Myers 33916 (4110 Center Pointe Dr., Unit 204, 239-479-4401)

Gainesville 32607 (105 N.W. 75th St., Ste. 2, 352-331-1408)

Jacksonville 32202 (300 E. State St., Ste. J, 904-232-3621)

Melbourne 32935 (2098 Sarno Rd., 321-254-3410)

Miami 33122 (8280 N.W. 27th St., Ste. 511, 305-859-8387)

Orlando 32822 (5575 S. Semoran Blvd., Ste. 36, 407-857-2800)

Palm Beach 33461 (2311 10th Ave. N., Ste. 13, 561-585-0441)

Pensacola 32506 (4501 Twin Oaks Dr., Ste. 104, 850-456-5886)

Sarasota 34231 (4801 Swift Rd., Ste. 100, 941-927-8285)

St. Petersburg 33713 (2880 1st Ave. N., 727-893-3791)

Tallahassee 32303 (548 Bradford Rd., 850-942-8810)

Tampa 33612 (8900 N. Armenia Ave., Ste. 312, 813-228-2621)

GEORGIA

Medical Centers:

Augusta 30904 (1 Freedom Way, 706-733-0188)

Decatur 30033 (1670 Clairmont Rd., 404-321-6111)

Dublin 31021 (1826 Veterans Blvd., 478-272-1210)

Vet Centers:

Atlanta 30324 (1440 Dutch Valley Pl., Ste. 1100, 404-347-7264)

Lawrenceville 30043 (930 River Centre Pl., 770-963-1809)

Macon 31201 (750 Riverside Dr., 478-477-3813)

Savannah 31406 (308A Commercial Dr., 912-652-4097)

GUAM

Medical Center

Agana Heights 96919 (U.S. Naval Hospital, Wing E-200, Box 7608, 671-344-9200/9122)

Vet Center:

Hagatna 96910 (222 Chalan Santo Papa, Reflection Ctr., Ste. 201, 671-472-7160/7161)

HAWAII

Medical Center:

Honolulu 96819-1522 (459 Patterson Rd., 808-566-1000)

Vet Centers:

Hilo 96720 (126 Pu'uhonu Way #2, 808-969-3833)

Honolulu 96814 (1680 Kapiolani Blvd., Ste. F3, 808-973-8387)

Kailua-Kona 96740 (73-4976 Kamanu St., Ste. 207, 808-329-0574)

Lihue 96766 (3-3367 Kuhio Hwy., Ste. 101, 808-246-1163)

Wailuku 96793 (35 Lunalilo St., Ste. 101, 808-242-8557)

IDAHO

Medical Center:

Boise 83702 (500 W. Fort St., 208-422-1000)

Vet Centers:

Boise 83705 (5440 Franklin Rd., Ste. 100, 208-342-3612)

Pocatello 83201 (1800 Garrett Way, 208-232-0316)

ILLINOIS

Medical Centers:

Chicago 60612 (820 S. Damen Ave., 312-569-8387)

Danville 61832 (1900 E. Main St., 217-554-3000)

Hines 60141 (5th Ave. and Roosevelt Rd., 708-202-8387)

Marion 62959 (2401 W. Main St., 618-997-5311)

North Chicago 60064 (3001 Green Bay Rd., 847-688-1900)

Vet Centers:

Chicago 60620 (7731 S. Halsted St., 773-962-3740)

Chicago Heights 60411 (1600 S. Halsted St., 708-754-0340)

East St. Louis 62203 (1265 N. 89th St., Ste. 5, 618-397-6602)

Evanston 60202 (565 Howard St., 847-332-1019)

Moline 61265 (1529 46th Ave., Ste. 6, 309-762-6954)

Oak Park 60302 (155 S. Oak Park Ave., 708-383-3225)

Peoria 61603 (3310 N. Prospect Rd., Ste. 102, 309-688-2170)

Rockford 61108 (4960 E. State St., Ste. 3, 815-395-1276)

Springfield 62703 (1227 S. 9th St., 217-492-4955)

INDIANA

Medical Centers:

Indianapolis 46202 (1481 W. 10th St., 317-554-0000)

Fort Wayne 46805 (2121 Lake Ave., 260-426-5431)

Marion 46953 (1700 E. 38th St., 765-674-3321)

Vet Centers:
Evansville 47711 (311 N. Weinbach Ave., 812-473-5993)
Fort Wayne 46802 (528 W. Berry St., 260-460-1456)
Indianapolis 46208 (3833 N. Meridian St., Ste. 120, 317-988-1600)
Merrillville 46410 (6505 Broadway, 219-736-5633)

IOWA
Medical Centers:
Des Moines 50310 (3600 30th St., 515-699-5999 or 1-800-294-8387)
Knoxville 50138 (1515 W. Pleasant St., 641-842-3101 or 1-800-816-8878)
Iowa City 52246 (601 Hwy. 6 W., 319-338-0581)
Vet Centers:
Cedar Rapids 52402 (1642 42nd St. N.E., 319-378-0016)
Des Moines 50310 (2600 Martin Luther King Jr. Pkwy., 515-284-4929)
Sioux City 51104 (1551 Indian Hills Dr., Ste. 214, 712-255-3808)

KANSAS
Medical Centers:
Leavenworth 66048 (4101 S. 4th St. Trafficway, 913-682-2000)
Topeka 66622 (2200 S.W. Gage Blvd., 785-350-3111)
Wichita 67218 (5500 E. Kellogg, 316-685-2221)
Vet Centers:
Manhattan 66502 (205 S. 4th St., Ste. B, 785-587-8257)
Wichita 67202 (251 N. Water, 316-685-2221 ext. 41080)

KENTUCKY
Medical Centers:
Lexington 40502 (1101 Veterans Dr., 859-233-4511)
Lexington 40511 (2250 Leestown Rd., 888-824-3577)
Louisville 40206 (800 Zorn Ave., 502-895-3401)
Vet Centers:
Lexington 40507 (301 E. Vine St., 859-253-0717)
Louisville 40208 (1347 S. 3rd St., 502-634-1916)

LOUISIANA
Medical Centers:
New Orleans 70112 (1601 Perdido St., 504-412-3700)
Pineville 71360 (2495 Shreveport Hwy., 71 N., 318-473-0010)
Shreveport 71101 (510 E. Stoner Ave., 318-221-8411)
Vet Centers:
Baton Rouge 70809 (5207 Essen Ln., Ste. 3, 225-757-0042)
Kenner 70062 (2200 Veteran Blvd., Ste. 114, 504-464-4743)
Shreveport 71104 (2800 Youree Dr., Bldg. 1, Ste. 105, 318-861-1776)

MAINE
Medical Center:
Augusta 04330 (1 VA Ctr., 207-623-8411)
Vet Centers:
Bangor 04401 (368 Harlow St., In-Town Plaza, 207-947-3391)
Caribou 04736 (456 York St., 207-496-3900)
Lewiston 04240 (Parkway Complex, 29 Westminster St., 207-783-0068)
Portland 04103 (475 Stevens Ave., 207-780-3584)
Springvale 04083 (628 Main St., 207-490-1513)

MARYLAND
Medical Centers:
Baltimore 21201 (10 N. Green St., 410-605-7000 or 1-800-463-6295)
Perry Point 21902 (Perry Point Medical Ctr., 410-642-2411)
Vet Centers:
Aberdeen 21001 (223 W. Bel Air Ave., 410-272-6771)
Baltimore 21208 (1777 Reisterstown Rd., Ste. 199, 410-764-9400 or 1-888-792-8900)

Cambridge 21613 (830 Chesapeake Dr., 410-228-6305)

Elkton 21921 (103 Chesapeake Blvd., Ste. A, 410-392-4485)

Silver Spring 20910 (1015 Spring St., Ste. 101, 301-589-1073)

MASSACHUSETTS

Medical Centers:

Bedford 01730 (200 Springs Rd., 1-800-VETMED1 or 781-275-7500)

Boston 02130 (150 S. Huntington Ave., 617-232-9500)

Brockton 02301 (940 Belmont St., 508-583-4500)

Leeds 01053 (421 N. Main St., 413-584-4040)

West Roxbury 02132 (1400 VFW Pkwy., 617-323-7700)

Vet Centers:

Boston 02215 (665 Beacon St., Ste. 100, 617-424-0665)

Brockton 02301 (1041-L Pearl St., 508-580-2730)

Hyannis 02601 (474 W. Main St., 508-778-0124)

Lowell 01852 (73 E. Merrimack St., 978-453-1151)

New Bedford 02740 (468 North St., 508-999-6920)

Springfield 01103 (1985 Main St., Northgate Plaza, 413-737-5167)

Worcester 01604 (691 Grafton St., 508-753-7902)

MICHIGAN

Medical Centers:

Ann Arbor 48105 (2215 Fuller Rd., 734-769-7100)

Battle Creek 49015 (5500 Armstrong Rd., 269-966-5600)

Detroit 48201 (4646 John R St., 313-576-1000)

Iron Mountain 49801 (325 E. H St., 906-774-3300 or 1-800-215-8262 in MI and WI)

Saginaw 48602 (1500 Weiss St., 989-497-2500)

Vet Centers:

Dearborn 48124 (2881 Monroe St., Ste. 100, 313-277-1428)

Detroit 48101 (4161 Cass Ave., 313-831-6509)

Escanaba 49829 (3500 Ludington St., Ste. 100, 906-233-0244)

Grand Rapids 49546 (2050 Brenton Rd., S.E., 616-285-5795)

MINNESOTA

Medical Centers:

Minneapolis 55417 (One Veterans Dr., 612-725-2000)

St. Cloud 56303 (4801 Veterans Dr., 320-252-1670 or 1-800-247-1739)

Vet Centers:

Duluth 55802 (405 E. Superior St., Ste. 160, 218-722-8654)

New Brighton 55112 (550 County Rd. D, Ste. 10, 651-644-4022)

MISSISSIPPI

Medical Centers:

Biloxi 39531 (400 Veterans Ave., 228-523-5000)

Jackson 39216 (1500 E. Woodrow Wilson Dr., 601-362-4471)

Vet Centers:

Biloxi 39531 (288 Veterans Ave., 228-388-9938)

Jackson 39216 (1755 Lelia Dr., Ste. 104, 601-965-5727)

MISSOURI

Medical Centers:

Columbia 65201 (800 Hospital Dr., 573-814-6000)

Kansas City 64128 (4801 E. Linwood Blvd., 816-861-4700)

Poplar Bluff 63901 (1500 N. Westwood Blvd., 573-686-4151)

St. Louis 63106 (John Cochran Div., 915 N. Grand Blvd., 314-652-4100)

St. Louis 63125 (1 Jefferson Barracks Dr., 314-652-4100)

Vet Centers:

Kansas City 64111 (301 E. Armour Blvd., Ste. 305, 816-753-1866)

St. Louis 63103 (2901 Olive St., 314-531-5355)

St. Louis 63114 (2122 Kratky Rd., 314-426-5864)

MONTANA
Medical Center:
Fort Harrison 59636 (3687 Veterans Dr., 406-442-6410)
Vet Centers:
Billings 59102 (1234 Ave. C, 406-657-6071)
Missoula 59802 (500 N. Higgins Ave., 406-721-4918)

NEBRASKA
Medical Centers:
Grand Island 68803 (2201 N. Broadwell Ave., 308-382-3660)
Lincoln 68510 (600 S. 70th St., 402-489-3802)
Omaha 68105 (4101 Woolworth Ave., 402-346-8800)
Vet Centers:
Lincoln 68508 (920 L St., 402-437-5298)
Omaha 68131 (2428 Cuming St., 402-346-6735)

NEVADA
Medical Centers:
Nellis AFB 89191 (4700 Las Vegas Blvd. N., 702-653-2260)
Reno 89502 (1000 Locust St., 1-888-838-6256)
Vet Centers:
Las Vegas 89146 (1919 S. Jones Blvd., Ste. A, 702-251-7873)
Reno 89503 (1155 W. 4th St., Ste. 101, 775-323-1294)

NEW HAMPSHIRE
Medical Center:
Manchester 03104 (718 Smyth Rd., 603-624-4366 or 1-800-892-8384)
Vet Centers:
Auburn 03032 (15 Dartmouth Dr., Ste. 201, 603-623-4204)
Manchester 03104 (103 Liberty St., 603-668-7060)

NEW JERSEY
Medical Centers:
East Orange 07018 (385 Tremont Ave., 973-676-1000)
Lyons 07939 (151 Knollcroft Rd., 908-647-0180)
Vet Centers:
Bloomfield 07003 (2 Broad St., 973-748-0980)
Ewing 08618 (934 Parkway Ave., 609-882-5744)
Secaucus 07094 (110A Meadowlands Pkwy., Ste. 102, 201-223-7787)
Ventnor 08406 (6601 Ventnor Ave., Ste. 105, 609-487-8387)

NEW MEXICO
Medical Center:
Albuquerque 87108 (1501 San Pedro Dr. S.E., 505-265-1711)
Vet Centers:
Albuquerque 87104 (1600 Mountain Rd. N.W., 505-346-6562)
Farmington 87402 (4251 E. Main, Ste. C, 505-327-9684)
Las Cruces 88001 (230 S. Water St., 575-523-9826)
Santa Fe 87505 (2209 Brothers Rd., Ste. 110, 505-988-6562)

NEW YORK
Medical Centers:
Albany 12208 (113 Holland Ave., 518-626-5000)
Batavia 14020 (222 Richmond Ave., 585-297-1000)
Bath 14810 (76 Veterans Ave., 607-664-4000)
Bronx 10468 (130 W. Kingsbridge Rd., 718-584-9000)
Brooklyn 11209 (800 Poly Pl., 718-836-6600)
Buffalo 14215 (3495 Bailey Ave., 716-834-9200)
Canandaigua 14424 (400 Fort Hill Ave., 585-394-2000)
Castle Point 12511 (100 Rt. 9D Castle Point Rd., 845-831-2000)
Montrose 10548 (2094 Albany Post Rd., 914-737-4400)
Northport 11768 (79 Middleville Rd., 631-261-4400)
New York 10010 (423 E. 23rd St. and 1st Ave., 212-686-7500)
St. Albans 11425 (179-00 Linden Blvd., 718-526-1000)
Syracuse 13210 (800 Irving Ave., 315-425-4400)

Vet Centers:

Albany 12205 (17 Computer Dr., W., 518-626-5130)

Babylon 11702 (116 W. Main St., 631-661-3930)

Binghamton 13901 (53 Chenango St., 866-716-8213)

Bronx 10468 (2471 Morris Ave 1A, 718-367-3500)

Brooklyn 11201 (25 Chapel St., Ste. 604, 6th Fl., 718-624-2765)

Buffalo 14202 (564 Franklin St., 716-882-0505)

Harlem 10035 (2279 3rd Ave., 2nd Fl., 212-426-2200)

Manhattan 10004 (32 Broadway, 2nd Fl., 212-742-9591)

Middletown 10940 (726 E. Main St., Ste. 203, 845-342-9917)

Rochester 14620 (1867 Mt. Hope Ave., 585-232-5040)

Staten Island 10301 (150 Richmond Ter., 718-816-4499)

Syracuse 13210 (716 E. Washington St., Ste. 101, 315-478-7127)

Watertown 13601 (210 Court St., 866-610-0358)

White Plains 10601 (300 Hamilton Ave., 914-682-6251)

Woodhaven 11421 (75-10B 91st Ave.,718-296-2871)

NORTH CAROLINA

Medical Centers:

Asheville 28805 (1100 Tunnel Rd., 828-298-7911)

Durham 27705 (508 Fulton St., 919-286-0411)

Fayetteville 28301 (2300 Ramsey St., 910-488-2120)

Salisbury 28144 (1601 Brenner Ave., 704-638-9000)

Vet Centers:

Charlotte 28262 (2114 Ben Craig Dr., 704-333-6107)

Fayetteville 28311 (4140 Ramsey St., Ste. 110, 910-488-6252)

Greensboro 27406 (2009 S. Elm-Eugene St., 336-333-5366)

Greenville 27858 (1021 W.H. Smith Blvd., Ste. 100, 252-355-7920)

Raleigh 27604 (1649 Old Louisburg Rd., 919-856-4616)

NORTH DAKOTA

Medical Center:

Fargo 58102 (2101 N. Elm St., 701-232-3241)

Vet Centers:

Bismarck 58501 (1684 Capital Way, 701-224-9751)

Fargo 58103 (3310 Fiechtner Dr., Ste. 100, 701-237-0942)

Minot 58701 (1400 20th Ave. S.W., Ste. 2, 701-852-0177)

OHIO

Medical Centers:

Brecksville 44141 (10000 Brecksville Rd., 440-526-3030)

Chillicothe 45601 (17273 State Rt. 104, 740-773-1141)

Cincinnati 45220 (3200 Vine St., 513-861-3100)

Cleveland 44106 (10701 East Blvd., 216-791-3800)

Dayton 45428 (4100 W. 3rd St., 937-268-6511)

Vet Centers:

Cincinnati 45203 (801-B W. 8th St., Ste. 126, 513-763-3500)

Cleveland 44113 (4242 Lorain Ave., Ste. 201, 216-939-0784)

Cleveland Heights 44118 (2022 Lee Rd., 216-932-8471)

Columbus 43215 (30 Spruce St., 1st Fl., 614-257-5550)

Dayton 45408 (627 Edwin C. Moses Blvd., 6th Fl., 937-461-9150)

Parma 44129 (5700 Pearl Rd., Ste. 102, 440-845-5023

Toledo 43614 (1565 S. Byene Rd., Ste. 104, 419-213-7533)

OKLAHOMA

Medical Centers:

Muskogee 74401 (1011 Honor Heights Dr., 918-683-3261)

Oklahoma City 73104 (921 N.E. 13th St., 405-270-0501)

Vet Centers:

Oklahoma City 73118 (1024 N.W. 47th, Ste. B, 405-456-5184)

Tulsa 74112 (1408 S. Harvard, 918-748-5105)

OREGON

Medical Centers:

Portland 97239 (3710 S.W. U.S. Veterans Hospital Rd., 503-220-8262)

Roseburg 97471 (913 N.W. Garden Valley Blvd., 541-440-1000)

Vet Centers:

Eugene 97401 (1255 Pearl St., Ste. 200, 541-465-6918)

Grants Pass 97526 (211 S.E. 10th St., 541-479-6912)

Portland 97220 (8383 N.E. Sandy Blvd., Ste. 110, 503-273-5370)

Salem 97301 (617 Chemeketa St. N.E., Ste. 100, 503-362-9911)

PENNSYLVANIA

Medical Centers:

Altoona 16602 (2907 Pleasant Valley Blvd., 814-943-8164)

Aspinwall 15215 (100 Patroit Way, 412-688-6000 or 1-866-482-7488)

Butler 16001 (325 New Castle Rd., 724-287-4781 or 1-800-362-8262)

Coatesville 19320 (1400 Blackhorse Hill Rd., 610-384-7711)

Erie 16504 (135 E. 38th St., 814-868-8661)

Lebanon 17042 (1700 S. Lincoln Ave., 717-272-6621)

Philadelphia 19101 (3900 Woodland Ave., 215-823-5800 or 1-800-949-1001)

Pittsburgh 15219 (Brackenridge and Allequeippa St., 412-688-6000 or 1-866-482-7488)

Pittsburgh 15206 (7180 Highland Dr., 412-365-4900 or 1-866-482-7488)

Wilkes-Barre 18711 (1111 E. End Blvd., 570-824-3521 or 1-877-928-2621)

Vet Centers:

DuBois 15801 (100 Meadow Ln., Ste. 8, 814-372-2095)

Erie 16501 (1001 State St., Ste. 102, 814-453-7955)

Harrisburg 17102 (1500 N. 2nd St., 717-782-3954)

McKeesport 15131 (Oakpark Mall, 2001 Lincoln Way, 412-678-7704)

Philadelphia 19107 (801 Arch St., Ste. 102, 215-627-0238)

Philadelphia 19120 (101 E. Olney Ave., 215-924-4670)

Pittsburgh 15205 (2500 Baldwick Rd., Ste. 15, 412-920-1765)

Scranton 18505 (1002 Pittston Ave., 570-344-2676)

Williamsport 17701 (49 E. 4th St., Ste. 104, 570-327-5281)

PHILIPPINES

Regional Office:

Manila 0930 (1131 Roxas Blvd., 011-632-528-2500)

Clinic:

Manila 1300 (2201 Roxas Blvd., Pasay City, 011-632-833-4566)

PUERTO RICO

Medical Center:

San Juan 00921 (10 Calle Casia, 787-641-7582)

Vet Centers:

Arecibo 00612 (50 Gonzalo Marin St., 787-879-4510)

Ponce 00730 (35 Mayor St., Ste. 1, 787-841-3260)

San Juan 00921 (Condominio Medical Center Plaza, Ste. LC8-11, La Riviera, Rio Piedras, 787-749-4409)

RHODE ISLAND

Medical Center:

Providence 02908 (830 Chalkstone Ave., 401-273-7100)

Vet Center:

Warwick 02889 (2038 Warwick Ave., 401-739-0167)

SOUTH CAROLINA

Medical Centers:

Charleston 29401 (109 Bee St., 843-577-5011)

Columbia 29209 (6439 Garners Ferry Rd., 803-776-4000)

Vet Centers:

Columbia 29201 (1513 Pickens St., 803-765-9944)

Greenville 29601 (14 Lavinia Ave., 864-271-2711)

North Charleston 29406 (5603A Rivers Ave., 843-747-8387)

SOUTH DAKOTA

Medical Centers:

Fort Meade 57741 (113 Comanche Rd., 605-347-2511 or 1-800-743-1070)

Hot Springs 57747 (500 N. 5th St., 605-745-2000 or 1-800-764-5370)

Sioux Falls 57105 (2501 W. 22nd St., 605-336-3230)

Vet Centers:

Martin 57551 (105 E. Hwy. 18, 605-685-1300)

Rapid City 57701 (621 6th St., Ste. 101, 605-348-0077)

Sioux Falls 57104 (601 S. Cliff Ave., Ste. C, 605-332-0856)

TENNESSEE

Medical Centers:

Johnson City 37684 (Veterans Way and Lamont St., 423-926-1171)

Memphis 38104 (1030 Jefferson Ave., 901-523-8990)

Murfreesboro 37129 (3400 Lebanon Rd., 615-867-6000 or 1-800-876-7093)

Nashville 37212 (1310 24th Ave. S., 615-327-4751)

Vet Centers:

Chattanooga 37411 (951 Eastgate Loop Rd., Bldg. 5700, Ste. 300, 423-855-6570)

Johnson City 37604 (1615A W. Market St., 423-928-8387)

Knoxville 37914 (2817 E. Magnolia Ave., 865-545-4680)

Memphis 38104 (1407 Union Ave., Ste. 410, 901-544-0173)

Memphis 38127 (3461 Austin Peay Hwy., 901-261-4500)

Morristown 37813 (925 E. Morris Blvd., 423-586-9100)

Nashville 37217 (1420 Donelson Pike, Ste. A-5, 615-366-1220)

TEXAS

Medical Centers:

Amarillo 79106 (6010 Amarillo Blvd. W., 806-355-9703)

Big Spring 79720 (300 Veterans Blvd., 432-263-7361)

Bonham 75418 (1201 E. Ninth St., 1-800-924-8387)

Dallas 75216 (4500 S. Lancaster Rd., 1-800-849-3597)

Houston 77030 (2002 Holcombe Blvd., 713-791-1414)

Kerrville 78028 (3600 Memorial Blvd., 830-896-2020)

San Antonio 78229 (7400 Merton Minter Blvd., 210-617-5300)

Temple 76504 (1901 Veterans Memorial Dr., 254-778-4811)

Waco 76711 (4800 Memorial Dr., 254-752-6581)

Vet Centers:

Amarillo 79109 (3414 Olsen Blvd., Ste. E., 806-354-9779)

Austin 78745 (1110 W. William Cannon Dr., Ste. 301, 512-416-1314)

Corpus Christi 78411 (4646 Corona Dr., Ste. 250, 361-854-9961)

Dallas 75216 (4500 S. Lancaster Rd., Bldg. 69, 214-857-1254)

Dallas 75231 (10501 N. Central Expy., Ste. 213, 214-361-5896)

El Paso 79925 (1155 Westmoreland Dr., Ste. 121, 915-772-5368)

Fort Worth 76104 (1305 W. Magnolia, Ste. B, 817-921-9095)

Harker Heights 76548 (302 Millers Crossing, Ste. 4, 254-953-7100)

Houston 77024 (701 N. Post Oak Rd., Ste. 102, 713-682-2288)

Houston 77098 (2990 Richmond Ave., Ste. 325, 713-523-0884)

Laredo 78041 (6020 McPherson Rd., #1A, 956-723-4680)

Lubbock 79410 (3208 34th St., 806-792-9782)

McAllen 78504 (801 W. Nolana Loop, Ste. 140, 956-631-2147)

Midland 79703 (3404 W. Illinois, Ste. 1, 432-697-8222)

San Antonio 78212 (231 W. Cypress St., Ste. 100, 1st Fl., 210-472-4025)

UTAH

Medical Center:

Salt Lake City 84148 (500 Foothill Dr., 801-582-1565)

Vet Centers:
Provo 84604 (1807 N. 1120 W., 801-377-1117)
Salt Lake City 84106 (1354 E. 3300 S., 801-584-1294)

VERMONT
Medical Center:
White River Junction 05009 (215 N. Main St., 802-295-9363)
Vet Centers:
South Burlington 05403 (359 Dorset St., 802-862-1806)
White River Junction 05001 (222 Holiday Inn Dr., Office Complex #2, 802-295-2908 or 1-800-649-6603)

VIRGINIA
Medical Centers:
Hampton 23667 (100 Emancipation Dr., 757-722-9961)
Richmond 23249 (1201 Broad Rock Blvd., 804-675-5000)
Salem 24153 (1970 Roanoke Blvd., 540-982-2463)
Vet Centers:
Alexandria 22310 (6890 S. Kings Hwy., Ste. 204, 703-360-8633)
Norfolk 23504 (1711 Church St., Ste. A & B, 757-623-7584)
Richmond 23230 (4902 Fitzhugh Ave., 804-353-8958)
Roanoke 24016 (350 Albemarle Ave. S.W., 540-342-9726)

VIRGIN ISLANDS
Medical and Vet Centers:
Kings Hill 00850 (Box 10553, R.R. 02, The Village Mall, 340-778-5553)
St. Thomas 00802 (Upper Havensight Mall, Ste. 310, 340-774-6674)

WASHINGTON
Medical Centers:
Seattle 98108 (1660 S. Columbian Way, 206-762-1010)
Spokane 99205 (4815 N. Assembly St., 509-434-7000)
Tacoma 98493 (9600 Veterans Dr. S.W., American Lake Division, 253-582-8440)
Walla Walla 99362 (77 Wainwright Dr., 509-525-5200)
Vet Centers:
Bellingham 98229 (3800 Byron Ave., Ste. 124, 360-733-9226)
Everett 98201 (3311 Wetmore Ave., 425-252-9701)
Seattle 98121 (2030 9th Ave., Ste. 210, 206-553-2706)
Spokane 99206 (100 N. Mullan Rd., Ste. 102, 509-444-8387)
Tacoma 98409 (4916 Center St., Ste. E, 253-565-7038)
Yakima 98901 (1111 N. 1st St., Ste. 1, 509-457-2736)

WEST VIRGINIA
Medical Centers:
Beckley 25801 (200 Veterans Ave., 304-255-2121)
Clarksburg 26301 (1 Medical Center Dr., 304-623-3461)
Huntington 25704 (1540 Spring Valley Dr., 304-429-6741)
Martinsburg 25405 (510 Butler Ave., 304-263-0811 or 1-800-817-3807)
Vet Centers:
Beckley 25801 (1000 Johnstown Rd., 304-252-8220)
Charleston 25302 (521 Central Ave., 304-343-3825)
Henlawson 25624 (Outstation, 21 Veterans Ave., 304-752-4453)
Huntington 25701 (3135 16th St. Rd., Ste. 11, 304-523-8387)
Martinsburg 25401 (900 Winchester Ave., 304-263-6776)
Morgantown 26508 (1083 Greenbag Rd., 304-291-4303)
Princeton 24740 (905 Mercer St., 304-425-5653)
Wheeling 26003 (1206 Chaplin St., Ste. 175, 304-232-0587)

WISCONSIN
Medical Centers:
Madison 53705 (2500 Overlook Ter., 608-256-1901)
Milwaukee 53295 (5000 W. National Ave., 414-384-2000)
Tomah 54660 (500 E. Veterans St., 608-372-3971)

Vet Centers:
Madison 53703 (706 Williamson St., 608-264-5342)
Milwaukee 53218 (5401 N. 76th St., Ste. 100, 414-536-1301)

WYOMING
Medical Centers
Cheyenne 82001 (2360 E. Pershing Blvd., 307-778-7550)
Sheridan 82801 (1898 Fort Rd., 307-672-3473)
Vet Centers:
Casper 82601 (1030 N. Poplar St., Ste. B, 307-261-5355)
Cheyenne 82001 (3219 E. Pershing Blvd., 307-778-7370)

DEPENDENCY AND INDEMNITY COMPENSATION

Dependency and Indemnity Compensation (DIC) may be authorized for survivors of military personnel who die in the line of duty while on active duty or active duty for training. If death occurs during inactive duty for training, benefits are payable in some cases.

Compensation may be authorized when death occurs following service mentioned earlier if the death is due to a service-connected disability. DIC payments may be authorized for surviving spouses who have not remarried (or who remarry after age 57), unmarried children under 18 (as well as certain helpless children and those between 18 and 23 if attending a VA-approved school), and low income parents of deceased veterans. DIC may also be paid in certain cases involving a veteran's death unrelated to military service. Eligibility may also result from a veteran whose death resulted from a nonservice-related injury or disease, and who was receiving, or was entitled to receive, VA compensation for a service-connected disability that was rated as totally disabling: for at least 10 years immediately before death; or since the veteran's release from active duty and for at least five years immediately preceding death; or for at least one year before death if the veteran was a former prisoner of war who died after September 30, 1999.

Rates of Dependency and Indemnity Compensation

Spouses' Payments. Surviving spouses of veterans who died on or after January 1, 1993, receive $1,154 a month. That rate is increased by $246 per month if the veteran was totally disabled due to service-connected disabilities continuously for at least eight years prior to death and the surviving spouse was married to the veteran for those same eight years. For a spouse entitled to DIC based on the veteran's death prior to January 1, 1993, the amount paid is $1,154 or an amount based on the veteran's pay grade, whichever is greater. If the death occurred prior to January 1, 1993 and the veteran was totally disabled due to service-connected disabilities continuously for at least eight years prior to death, then the amount paid is the higher of the amount based on the pay grade or $1,154 plus $246. The following table shows the rates based on pay grades which correspond to military ranks. Payments are not affected by income that the spouse may receive from earnings or other sources. If the spouse receives SBP benefits under the Survivor Benefit Plan administered by the Department of Defense, the law requires offset of SBP payments against DIC payments. (See section on SURVIVOR BENEFIT PLAN.)

MONTHLY DIC PAYMENTS TO SURVIVING SPOUSES OF VETERANS WHO DIED BEFORE JANUARY 1, 1993

COMMISSIONED OFFICERS		WARRANT OFFICERS		ENLISTED PERSONS	
Pay Grade	Monthly Rate	Pay Grade	Monthly Rate	Pay Grade	Monthly Rate
O-1	$1,219	W-1	$1,219	E-1	$1,154
O-2	1,260	W-2	1,267	E-2	1,154
O-3	1,347	W-3	1,305	E-3	1,154
O-4	1,427	W-4	1,380	E-4	1,154
O-5	1,571			E-5	1,154
O-6	1,771			E-6	1,154
O-7	1,912			E-7	1,194
O-8	2,100			E-8	1,260
O-9	2,246			E-9	1,314
O-10	2,463			E-9*	1,419
O-10**	2,643				

*Rate for veteran who served as Senior Enlisted Member of Service.
**Rate for veteran who served as Service Chief.

NOTES:

1. Add $286 for each dependent child under age 18.
2. For pay grades E-7 – E-9; W-1 – W-4; and O-1 – O-3, the base rate is $1,400 if the veteran is rated totally disabled due to service-connected disabilities for at least eight years prior to death and the surviving spouse was married to the veteran for those same eight years.

Additional Amounts for Certain Surviving Spouses. The payment to a surviving spouse who has a child or children under 18 years of age is increased by $250 per month for up to two years after entitlement to DIC commences. This is in addition to any other DIC compensation the surviving spouse may receive (including payments for dependent children). Eligibility for this increased benefit ceases on the first month after the month the surviving spouse no longer has a dependent child under 18 years of age.

Effects of Remarriage on DIC Payments. Effective January 1, 2004, a surviving spouse who remarries after age 57 will not lose eligibility for DIC payments. A surviving spouse who remarries prior to age 57, normally loses eligibility for DIC. However, payments may be reinstated if the remarriage ends in either death or divorce.

Spouses will receive an additional DIC payment of $286 for each child under 18 who is in the spouse's custody. A spouse may qualify for special additional payments based on disability status. If the spouse is a patient in a nursing home or if the VA determines that the spouse needs regular health care and assistance from another person, the aid and attendance allowance of $286 will be paid. For less severe disability, if the spouse is substantially confined to the residence, VA will pay the housebound allowance of $135.

Children's Payments. Normally, payments for minor children (under age 18) will be included with a surviving spouse's payment. In 2009, the rate of payment for each minor child is $286 per month. Payments are made for older children under certain circumstances. Children between the ages of 18 and 23 are eligible for payments if they remain unmarried and are attending an educational program approved by VA. Such programs can include secondary school, college level programs and trade, technical, or business training programs. Payments may also be made for unmarried children over age 18 who became helpless, as determined by VA, prior to age 18 as a result of physical or mental disability.

When no spouse is entitled to receive payments, children will still be paid provided they remain unmarried and meet general requirements for age, school attendance, or helplessness.

Parents' Payments. Parents of deceased veterans may be paid DIC provided they meet certain income tests. The rate of payment will depend on whether both parents survive and whether a surviving parent is currently married. The income limit for two parents together is $18,087; the limit for one parent or two parents not together is $13,456.

Social Security Eligibility. Eligibility for DIC does not affect eligibility for Social Security, and eligibility for Social Security does not affect eligibility for DIC, for either surviving spouses or children.

Survivors may be eligible for Restored Entitlement Program for Survivors (REPS) payments if they are eligible for DIC and the veteran's death occurred in service prior to August 13, 1981, or resulted from disability related to military service and acquired before August 13, 1981. Eligible spouses may be paid REPS benefits only when Social Security eligibility ceases because the youngest child has reached age 16, and the spouse has a child between the ages of 16 and 18. Children over 18 may receive REPS payments to age 22 if they are unmarried and attending a post-secondary educational program which meets certain requirements, including full-time attendance.

DEATH PENSION

Pensions based on need are available for surviving spouses and unmarried children of deceased veterans with wartime service. Spouses must not have remarried and children must be under age 18, or under age 23 if attending a VA-approved school. Pension is not payable to those with estates large enough to provide maintenance. The veteran must have been discharged under conditions other than dishonorable and must have had 90 days or more of active military service, at least one day of which was during a period of war, or a service-connected disability justifying discharge for disability. If the veteran died in service but not in the line of duty, benefits may be payable if the veteran had completed at least two years of honorable service. Children who become incapable of self-support because of a disability before age 18 may be eligible for a pension as long as the condition exists, unless the child marries or the child's income exceeds the applicable limit.

The Aid and Attendance, or Housebound allowance may be payable to the surviving spouse of a wartime veteran who has in-home care or who lives in a nursing home or assisted-living facility.

Aid and Attendance. A surviving spouse may be eligible for the Aid and Attendance allowance when the surviving spouse:

1. requires the aid of another person in order to perform personal functions required in everyday living, such as bathing, feeding, dressing, attending to the wants of nature, adjusting prosthetic devices, or protecting the person from the hazards of his/her daily environment, or
2. is bedridden, in that his/her disability or disabilities requires that he/she remain in bed apart from any prescribed course of convalescence or treatment, or
3. is a patient in a nursing home due to mental or physical incapacity, or
4. is blind, or so nearly blind as to have corrected visual acuity of 5/200 or less in both eyes, or concentric contraction of the visual field to 5 degrees or less.

Housebound. A surviving spouse may be eligible for the Housebound allowance if the claimant is substantially confined to his/her immediate premises because of permanent disability, however, both Aid and Attendance and Housebound benefits cannot be paid at the same time.

The Improved Pension program provides a monthly payment to bring an eligible person's income to a support level established by law. The payment is reduced by the annual income from other sources such as Social Security paid to the surviving spouse or dependent children. Medical expenses may be deducted from the income ceiling.

Maximum Annual Pension Rates for Surviving Spouse:

Surviving spouse without dependent children $7,933
Surviving spouse with one dependent child $10,385
Surviving spouse in need of regular aid and attendance without dependent children $12,681
Surviving spouse in need of regular aid and attendance with one dependent child $15,128
Surviving spouse permanently housebound without dependent children $9,696
Surviving spouse permanently housebound with one dependent child $12,144
Increase for each additional dependent child $2,020

APPEALS

Veterans and other claimants for VA benefits have the right to appeal decisions made by a VA regional office or medical center. Typical issues appealed are disability compensation, pension, education benefits, recovery of overpayments, medication copayment debts, and reimbursement for medical services that were not authorized.

A claimant has one year from the date of the notification of a VA decision to file an appeal. The first step in the appeal process is for a claimant to file a written notice of disagreement with the VA regional office or medical center that made the decision. This is a written statement that a claimant disagrees with VA's decision. Following receipt of the written notice, VA will furnish the claimant a Statement of the Case describing what facts, laws and regulations were used in deciding the case. To complete the request for appeal, the claimant must file a Substantive Appeal within 60 days of the mailing of the Statement of the Case, or within one year from the date VA mailed its decision, whichever period ends later.

Board of Veterans Appeals. The Board of Veterans Appeals, located in Washington, DC, makes decisions on appeals on behalf of the Secretary of Veterans Affairs. Although it is not required, a veteran's service organization, an agent or an attorney may represent a claimant. Appellants may present their case in person to a member of the Board at a hearing in Washington, DC, at a VA regional office or by videoconference. The texts of appeal decisions made by the Board, as well as a plain language pamphlet, *"How Do I Appeal?"* can be found at www.va.gov/vbs/bva. This pamphlet may also be requested by writing to Hearings and Transcription Unit (014HRG), Board of Veterans' Appeals, 810 Vermont Avenue, N.W., Washington, DC 20420.

U.S. Court of Appeals for Veterans Claims. A final Board of Veterans Appeals decision that does not grant a claimant the benefits desired may be appealed to the U.S. Court of Appeals for Veterans Claims, an independent court, not part of the Department of Veterans Affairs.

Notice of an appeal must be received by the court with a postmark that is within 120 days after the date—stamped on the decision—on which the

Board of Veterans Appeals mailed its decision. The court reviews the record considered by the Board of Veterans Appeals.

It does not hold trials or receive new evidence. Appellants may represent themselves before the court or have lawyers or approved agents as representatives. Oral argument is held only at the direction of the court. Either party may appeal a decision of the court to the U.S. Court of Appeals for the Federal Circuit and may seek review in the Supreme Court of the United States.

The court's website, www.vetapp.gov, contains court decisions, case status information, rules and procedures, and other special announcements. For other questions, write to the Clerk of the Court, 625 Indiana Ave. N.W., Ste. 900, Washington, DC 20004, or call the clerk's office at (202) 501-5970.

BURIAL INFORMATION FOR RETIREES

Military retirees, their spouses, and minor children may be buried in national cemeteries. There is no charge for opening or closing of graves. Arrangements and expenses such as obtaining the services of a funeral director, selecting a funeral home, preparation of the remains, casketing, and transportation to the national cemetery are the responsibility of the survivors or the deceased's estate.

In the case of an active duty member who is retired from active service while hospitalized, and remains hospitalized until time of death, the military will provide for care and disposition of the remains including the assistance of the military services in making funeral arrangements. It is important to note that the retiree must be continuously hospitalized from admission to death for this entitlement.

When a retiree is eligible, and military authorities arrange for disposition of remains, services of preparation at the place of death, removal, embalming preparation and preservation, casket, and transportation to a common carrier are normally obtained under contract; however, when no contract is available, military authorities negotiate with local funeral directors to obtain these services. The military authorities also will provide the following as required.

1. Cremation (if cremation is requested in writing by the Person Authorized to Direct Disposition (PADD) of the remains).
2. A suitable urn for the cremated remains.
3. A U.S. flag to drape the casket.
4. Transportation of remains, accompanied by an escort, from the place of death to the place designated by the PADD.
5. Military honors, if requested, will be provided at the place of interment.

Military contract standards are high, and it is generally advantageous to leave arrangements for preparation of remains to military authorities. When the PADD makes private arrangements, the Government will reimburse up to $8,800 of expenses, depending on the type of arrangements made, for the member's burial. In addition, the Government will provide travel for the next-of-kin to the place of burial.

The Government will mark all graves in national and post cemeteries with regulation headstones and markers of the types authorized. When interment is made in a private or civilian cemetery, the PADD may obtain a regulation Government headstone or marker, without expense, to mark the grave of a deceased member of the Armed Forces or a former member of the Armed Forces who was discharged under conditions other than dishonorable. Applicants may select either an upright headstone of marble or a flat marker of marble, granite, or bronze, dependent upon the type permissible in the cemetery.

Special Requirement for Combat Theater Deaths

The law requires the Secretary concerned to provide transportation for the remains of a member of the Armed Forces who dies in a combat theater of operations and whose remains are returned through the mortuary facility at Dover AFB, Delaware, to the escorted remains destination. Transportation by air requires that the primary mission of the aircraft be transportation of such remains (although more than one set of remains may be transported on the same flight). In addition, the law requires that the remains be continuously escorted from Dover AFB to the escorted remains destination and that an appropriate Honor Guard detail be provided.

Prohibition on Eligibility for Burial Benefits

Notwithstanding the list of persons eligible for burial in Arlington National Cemetery or a National Cemetery, below, the following persons are not eligible for burial in Arlington National Cemetery or a National Cemetery: (1) A person who has been convicted of an offense under federal law for which a sentence of imprisonment for life or the death penalty may be imposed and whose conviction is final (other than a person whose sentence was commuted by the President); (2) A person who has been convicted under state law of the willful, deliberate, or premeditated unlawful killing of another human being for which a sentence of imprisonment for life or the death penalty may be imposed and whose conviction is final (other than a person whose sentence was commuted by the Governor of a state); or (3) A person who is found to have committed a federal capital crime or a state capital crime, but has not been convicted of such crime by reason of such person not being available for trial due to death or flight to avoid prosecution.

Burial in Arlington National Cemetery

The persons specified below are eligible for ground burial in Arlington National Cemetery. The last period of active duty of former members of the Armed Forces must have ended honorably. Interment may be casketed or cremated remains.

1. Any active duty member of the Armed Forces (except those members serving on active duty for training only).
2. Any veteran who is retired from active military service with the Armed Forces.
3. Any veteran who is retired from a Reserve Component, has reached age 60 and has started drawing retired pay, and who served a period of active duty (other than for training).
4. Any former member of the Armed Forces separated honorably prior to October 1, 1949 for medical reasons and who was rated at 30 percent or greater disabled effective on the day of discharge.
5. Any former member of the Armed Forces who has been awarded one of the following decorations: Medal of Honor; Distinguished Service Cross (Navy Cross or Air Force Cross); Distinguished Service Medal; Silver Star; Purple Heart.

6. The President of the United States or any former President of the United States.

7. Any former member of the Armed Forces who served on active duty (other than for training) and who held any of the following positions: an elective office of the U.S. Government; Office of the Chief Justice of the United States or of an Associate Justice of the Supreme Court of the United States; an office listed, at the time the person held the position, in 5 USC 5312 or 5313 (Levels I and II of the Executive Schedule); or the chief of a mission who was at any time during his/her tenure classified in Class I under the provisions of Section 411, Act of 13 August 1946, 60 Stat. 1002, as amended (22 USC 866) or as listed in State Department memorandum dated March 21, 1988.

8. Any former prisoner of war who, while a prisoner of war, served honorably in the active military, naval, or air service, whose last period of military, naval or air service terminated honorably, and who died on or after November 30, 1993.

9. The spouse, widow or widower, minor child, or permanently dependent child, and certain unmarried adult children of any of the above eligible veterans.

10. The widow or widower of a member of the Armed Forces who was lost or buried at sea or officially determined to be missing in action; a member of the Armed Forces who is interred in a U.S. military cemetery overseas that is maintained by the American Battle Monuments Commission; a member of the Armed Forces who is interred in Arlington National Cemetery as part of a group burial.

11. The surviving spouse, minor child, or permanently dependent child of any person already buried in Arlington National Cemetery.

12. The parents of a minor child, or permanently dependent child whose remains, based on the eligibility of a parent, are already buried in Arlington National Cemetery. A spouse divorced from the primary eligible, or widowed and remarried, is not eligible for interment.

13. Provided certain conditions are met, a former member of the Armed Forces may be buried in the same grave with a close relative who is already buried and is the primary eligible.

Servicemembers who do not meet these requirements may qualify for inurnment in the cemetery's columbarium complex, which houses cremated remains. Any honorably discharged veteran, spouse, or dependent children may be inurned in the same family niche.

Space assignment in Arlington National Cemetery is assigned only when there is need. No site can be reserved, and only one gravesite will be assigned per family. If a spouse or eligible child dies first, space will be assigned for the spouse's or child's interment, provided the servicemember agrees in writing to be buried in the same site.

There is no charge for a grave (or columbarium niche) in Arlington National Cemetery, for opening or closing a grave or for a Government headstone or marker. Except for active duty servicemembers, all other costs must be borne by the family.

For information on Arlington burials, write to the Superintendent, Arlington National Cemetery, Arlington, VA 22211, or call (703) 607-8000.

Burial in National Cemeteries

Basis for Eligibility. Members of the Armed Forces, including Reserve Component members, who die on active duty; Reserve Component members if they were entitled to retired pay at the time of death, or would have been entitled had they not been under age 60; and all veterans (including retirees) discharged under conditions other than dishonorable. Spouses, remarried surviving spouses and dependent children may also be eligible.

Burial in a VA national cemetery is available for eligible veterans, their spouses and dependents at no cost to the family and includes the gravesite, grave-liner, opening and closing of the grave, a head-stone or marker, and perpetual care as part of a national shrine. For veterans, benefits also include a burial flag (with case for active duty) and military funeral honors.

Burial options are limited to those available at a specific cemetery but may include in-ground casket, or interment of cremated remains in a columbarium, in ground or in a scatter garden. The funeral director or the next of kin makes interment arrangements by contacting the national cemetery in which burial is desired.

Surviving spouses of veterans who died on or after January 1, 2000, do not lose eligibility for burial in a national cemetery if they remarry. Burial of dependent children is limited to unmarried children under age 21, or under age 23 if a full-time student at an approved educational institution. Also eligible are unmarried adult children who become physically or mentally disabled and incapable of self-support before age 21, or age 23 if a full-time student.

For more information, call 1-800-697-6947, the VA National Cemetery System's Office of Memorial Programs in Washington, DC, Monday through Friday, 8:00 a.m. - 5:00 p.m. EST.

NATIONAL CEMETERIES

Only the physical address of each cemetery is listed. Call the number provided to obtain proper mailing address.

(As of October 2008)

ALABAMA

Fort Mitchell National Cemetery
553 Hwy. 165
Fort Mitchell, AL 36856
(334) 855-4731

Mobile National Cemetery (Closed)
1202 Virginia St.
Mobile, AL 36604
Contact: Barrancas National Cemetery, FL, (850) 453-4846

ALASKA

Fort Richardson National Cemetery
Bldg. #58-512, Davis Hwy.
Fort Richardson, AK 99505
(907) 384-7075

Sitka National Cemetery
803 Sawmill Creek Rd.
Sitka, AK 99835
Contact: Fort Richardson National Cemetery, AK, (907) 384-7075

ARIZONA

National Memorial Cemetery of Arizona
23029 N. Cave Creek Rd.
Phoenix, AZ 85024
(480) 513-3600

Prescott National Cemetery
VA Medical Center, 500 Hwy. 89 N.
Prescott, AZ 86301, Contact: National Memorial Cemetery of Arizona, AZ, (480) 513-3600

ARKANSAS

Fayetteville National Cemetery
700 Government Ave.
Fayetteville, AR 72701
(479) 444-5051

Fort Smith National Cemetery
522 Garland Ave. and S. 6th St.
Fort Smith, AR 72901
(479) 783-5345

Little Rock National Cemetery
(Closed)
2523 Confederate Blvd.
Little Rock, AR 72206
(501) 324-6401

CALIFORNIA

Fort Rosecrans National Cemetery
Cabrillo Memorial Dr., Point Loma,
San Diego, CA 92106
(619) 553-2084

Golden Gate National Cemetery
(Closed)
1300 Sneath Ln.
San Bruno, CA 94066
(650) 761-1646

Los Angeles National Cemetery
(Closed)
950 S. Sepulveda Blvd.
Los Angeles, CA 90049
(310) 268-4675

Riverside National Cemetery
22495 Van Buren Blvd.
Riverside, CA 92518
(951) 653-8417

Sacramento Valley
VA National Cemetery
5810 Midway Rd.
Dixon, CA 95620
(707) 693-2460

San Francisco National Cemetery
(Closed)
Presidio of San Francisco
1 Lincoln Blvd.
San Francisco, CA 94129
Contact: Golden Gate National
Cemetery, CA, (650) 761-1646

San Joaquin Valley National Cemetery
32053 W. McCabe Rd.
Santa Nella, CA 95322
(209) 854-1040

COLORADO

Fort Logan National Cemetery
3698 S. Sheridan Blvd.
Denver, CO 80235
(303) 761-0117

Fort Lyon National Cemetery
15700 Rd. HH
Las Animas, CO 81054
(719) 456-0165

FLORIDA

Barrancas National Cemetery
Naval Air Station
80 Hovey Rd.
Pensacola, FL 32508
(850) 453-4846

Bay Pines National Cemetery
10000 Bay Pines Blvd.
Bay Pines, FL 33708
(727) 398-9426

Florida National Cemetery
6502 S.W. 102nd Ave.
Bushnell, FL 33513
(352) 793-7740 or 1074

Sarasota VA National Cemetery
9810 State Hwy. 72
Sarasota, FL 34241
1-877-861-9840

South Florida VA National Cemetery
6501 S. State Rd. 7
Lake Worth, FL 33449
(561) 649-6489

St. Augustine National Cemetery
(Closed)
104 Marine St., St.
Augustine, FL 32084
Contact: Florida National Cemetery
(352) 793-7740

GEORGIA

Andersonville National Cemetery
496 Cemetery Rd.
Andersonville, GA 31711
(229) 924-0343

Georgia National Cemetery
2025 Mount Carmel Church Ln.
Canton, GA 30114
(770) 479-9300 or (866) 236-8159

Marietta National Cemetery (Closed)
500 Washington Ave.
Marietta, GA 30060
(866) 236-8159

HAWAII
National Memorial
Cemetery of the Pacific
2177 Puowaina Dr.
Honolulu, HI 96813
(808) 532-3720

ILLINOIS
Abraham Lincoln National Cemetery
20953 W. Hoff Rd.
Elwood, IL 60421
(815) 423-9958
Alton National Cemetery
600 Pearl St., Alton, IL 62003
Contact: Jefferson Barracks National Cemetery, MO, (314) 845-8320
Camp Butler National Cemetery
5063 Camp Butler Rd.
Springfield, IL 62707
(217) 492-4070
Danville National Cemetery
1900 E. Main St.
Danville, IL 61832, (217) 554-4550
Mound City National Cemetery
Junction Hwy 37 & Rt. 51
Mound City, IL 62963
Contact: Jefferson Barracks National Cemetery, MO, (314) 845-8320
Quincy National Cemetery (Closed)
36th and Maine St.
Quincy, IL 62301
Contact: Rock Island National Cemetery, IL, (309) 782-2094
Rock Island National Cemetery
Bldg. 118 Rock Island Arsenal
Rock Island, IL 61299
(309) 782-2094

INDIANA
Crown Hill National Cemetery (Closed)
700 W. 38th St.
Indianapolis, IN 46208
Contact: Marion National Cemetery, IN, (765) 674-0284
Marion National Cemetery
VA Medical Center
1700 E. 38th St.
Marion, IN 46953
(765) 674-0284
New Albany National Cemetery
1943 Ekin Ave.
New Albany, IN 47150
Contact: Zachary Taylor National Cemetery, KY, (502) 893-3852

IOWA
Keokuk National Cemetery
1701 J St.
Keokuk, IA 52632
Contact: Rock Island National Cemetery, IL, (309) 782-2094

KANSAS
Fort Leavenworth National Cemetery (Closed)
395 Biddle Blvd.
Leavenworth, KS 66027
Contact: Leavenworth National Cemetery, KS, (913) 758-4106
Fort Scott National Cemetery
900 E. National Ave.
Fort Scott, KS 66701
(620) 223-2840
Leavenworth National Cemetery
4101S 4th St. Trafficway
Leavenworth, KS 66048
(913) 758-4106

KENTUCKY
Camp Nelson National Cemetery
6980 Danville Rd.
Nicholasville, KY 40356
(859) 885-5727
Cave Hill National Cemetery (Closed)
701 Baxter Ave.,
Louisville, KY 40204
Contact: Zachary Taylor National Cemetery, KY, (502) 893-3852
Danville National Cemetery (Closed)
277 N. First St.
Danville, KY 40442
Contact: Camp Nelson National Cemetery, KY, (859) 885-5727
Lebanon National Cemetery
20 Hwy. 208
Lebanon, KY 40033
Contact: Zachary Taylor National Cemetery, KY, (502) 893-3852
Lexington National Cemetery (Closed)
833 W. Main St.
Lexington, KY 40508
Contact: Camp Nelson National Cemetery, KY, (859) 885-5727
Mill Springs National Cemetery
9044 W. Hwy. 80
Nancy, KY 42544
Contact: Camp Nelson National Cemetery, KY, (859) 885-5727

Zachary Taylor National Cemetery (Closed)
4701 Brownsboro Rd.
Louisville, KY 40207
(502) 893-3852

LOUISIANA

Alexandria National Cemetery (Closed)
209 E. Shamrock St.
Pineville, LA 71360
Contact: Natchez National Cemetery, MS, (601) 445-4981

Baton Rouge National Cemetery (Closed)
220 N. 19th St.
Baton Rouge, LA 70806
Contact: Port Hudson National Cemetery, LA, (225) 654-3767

Port Hudson National Cemetery
20978 Port Hickey Rd.
Zachary, LA 70791
(225) 654-3767

MAINE

Togus National Cemetery (Closed)
VA Medical and Regional Office Center
1 VA Center
Togus, ME 04330
Contact: Massachusetts National Cemetery, MA, (508) 563-7113

MARYLAND

Annapolis National Cemetery (Closed)
800 West St.
Annapolis, MD 21401
Contact: Baltimore National Cemetery, MD, (410) 644-9696

Baltimore National Cemetery
5501 Frederick Ave.
Baltimore, MD 21228
(410) 644-9696

Loudon Park National Cemetery (Closed)
3445 Frederick Ave.
Baltimore, MD 21228
Contact: Baltimore National Cemetery, MD, (410) 644-9696

MASSACHUSETTS

Massachusetts National Cemetery
Connery Ave.
Bourne, MA 02532
(508) 563-7113

MICHIGAN

Fort Custer National Cemetery
15501 Dickman Rd.
Augusta, MI 49012
(269) 731-4164

Great Lakes National Cemetery
4200 Belford Rd.
Holly, MI 48442
(248) 328-0495

MINNESOTA

Fort Snelling National Cemetery
7601 34th Ave. S.
Minneapolis, MN 55450
(612) 726-1127

MISSISSIPPI

Biloxi National Cemetery
400 Veterans Ave.
Biloxi, MS 39531, (228) 388-6668

Corinth National Cemetery
1551 Horton St.
Corinth, MS 38834
Contact: Memphis National Cemetery, TN, (901) 386-8311

Natchez National Cemetery
41 Cemetery Rd.
Natchez, MS 39120
(601) 445-4981

MISSOURI

Jefferson Barracks National Cemetery
2900 Sheridan Rd.
St. Louis, MO 63125
(314) 845-8320

Jefferson City National Cemetery (Closed)
1024 E. McCarty St.
Jefferson City, MO 65101
Contact: Jefferson Barracks National Cemetery, MO, (314) 845-8320

Springfield National Cemetery
1702 E. Seminole St.
Springfield, MO 65804
(417) 881-9499

NEBRASKA

Fort McPherson National Cemetery
12004 S. Spur 56A
Maxwell, NE 69151
(308) 582-4433

NEW JERSEY

Beverly National Cemetery
916 Bridgeboro Rd.
Beverly, NJ 08010
(609) 877-5460

Finn's Point National Cemetery
Fort Mott Rd.
Salem, NJ 08079
Contact: Beverly National Cemetery, NJ, (609) 877-5460

NEW MEXICO

Fort Bayard National Cemetery
300 Camino De Paz
Fort Bayard, NM 88036
Contact: Fort Bliss National Cemetery, TX
(915) 564-0201 or (505) 537-5794

Santa Fe National Cemetery
501 N. Guadalupe St.
Santa Fe, NM 87501
(505) 988-6400 or 1-877-353-6295

NEW YORK

Bath National Cemetery
San Juan Ave.
Bath, NY 14810
(607) 664-4853

Calverton National Cemetery
210 Princeton Blvd.
Calverton, NY 11933
(631) 727-5410

Cypress Hills National Cemetery (Closed)
625 Jamaica Ave.
Brooklyn, NY 11208
Contact: Long Island National Cemetery, NY, (631) 454-4949

Gerald B.H. Solomon Saratoga National Cemetery
200 Duell Rd.
Schuylerville, NY 12871
(518) 581-9128

Long Island National Cemetery (Closed)
2040 Wellwood Ave.
Farmingdale, NY 11735
(631) 454-4949

Woodlawn National Cemetery
1825 Davis St.
Elmira, NY 14901
(607) 732-5411

NORTH CAROLINA

New Bern National Cemetery (Closed)
1711 National Ave.
New Bern, NC 28560
(252) 637-2912

Raleigh National Cemetery (Closed)
501 Rock Quarry Rd.
Raleigh, NC 27610
Contact: New Bern National Cemetery, NC, (252) 637-2912

Salisbury National Cemetery
501 Statesville Blvd.
Salisbury, NC 28144
(704) 636-2661

Wilmington National Cemetery (Closed)
2011 Market St.
Wilmington, NC 28403
Contact: New Bern National Cemetery, NC, (252) 637-2912

OHIO

Dayton National Cemetery
VA Medical Center
4100 W. Third St.
Dayton, OH 45428
(937) 262-2115

Ohio Western Reserve National Cemetery
10175 Rawiga Rd.
Seville, OH 44273
(330) 335-3069

OKLAHOMA

Fort Gibson National Cemetery
1423 Cemetery Rd.
Fort Gibson, OK 74434
(918) 478-2334

Fort Sill National Cemetery
2648 N.E. Jake Dunn Rd.
Elgin, OK 73538
(580) 492-3200

OREGON

Eagle Point National Cemetery
2763 Riley Rd.
Eagle Point, OR 97524
(541) 826-2511

Roseburg National Cemetery
VA Medical Center
913 Garden Valley Blvd.
Roseburg, OR 97470
Contact: Eagle Point National Cemetery, OR, (541) 826-2511

Willamette National Cemetery
11800 S.E. Mt. Scott Blvd.
Portland, OR 97086
(503) 273-5250

PENNSYLVANIA

Indiantown Gap National Cemetery
RR 2, Box 484, Indiantown Gap Rd., (Interstate 81, Exit 85B)
Annville, PA 17003
(717) 865-5254

National Cemetery of the Alleghenies
1158 Morgan Rd.
Bridgeville, PA 15017
(724) 746-4363

Philadelphia National Cemetery (Closed)
Haines St. and Limekiln Pk.
Philadelphia, PA 19138
Contact: Beverly National Cemetery, NJ, (609) 877-5460

PUERTO RICO

Puerto Rico National Cemetery
50 National Cemetery Ave.
Bo. Hato Tejas
Bayamon, PR 00961
(787) 798-6720

SOUTH CAROLINA

Beaufort National Cemetery
1601 Boundary St.
Beaufort, SC 29902
(843) 524-3925

Florence National Cemetery
803 E. National Cemetery Rd.
Florence, SC 29506
(843) 669-8783

SOUTH DAKOTA

Black Hills National Cemetery
20901 Pleasant Valley Dr.
Sturgis, SD 57785
(605) 347-3830

Fort Meade National Cemetery (Closed)
Old Stone Rd.
Sturgis, SD 57785
Contact: Black Hills National Cemetery, SD, (605) 347-3830

Hot Springs National Cemetery (Closed)
500 N. Fifth St.
Hot Springs, SD 57747
Contact: Black Hills National Cemetery, SD, (605) 347-3830

TENNESSEE

Chattanooga National Cemetery
1200 Bailey Ave.
Chattanooga, TN 37404
(423) 855-6590

Knoxville National Cemetery (Closed)
939 Tyson St. N.W.
Knoxville, TN 37917
Contact: Chattanooga National Cemetery, TN, (423) 855-6590

Memphis National Cemetery
3568 Townes Ave.
Memphis, TN 38122
(901) 386-8311

Mountain Home National Cemetery
Veterans Way & Lamont St.
Mountain Home, TN 37684
(423) 979-3535

Nashville National Cemetery (Closed)
1420 Gallatin Rd. S.
Madison, TN 37115
(615) 860-0086

TEXAS

Dallas-Fort Worth National Cemetery
2000 Mountain Creek Pkwy.
Dallas, TX 75211
(214) 467-3374

Fort Bliss National Cemetery
5200 Fred Wilson Blvd.
Fort Bliss, TX 79906
(915) 564-0201

Fort Sam Houston National Cemetery
1520 Harry Wurzbach Rd.
San Antonio, TX 78209,
(210) 820-3891

Houston National Cemetery
10410 Veterans Memorial Dr.
Houston, TX 77038
(281) 447-8686

Kerrville National Cemetery (Closed)
VA Medical Center
3600 Memorial Blvd.
Kerrville, TX 78028
Contact: Fort Sam Houston National Cemetery, TX, (210) 820-3891

San Antonio National Cemetery (Closed)
517 Paso Hondo St.
San Antonio, TX 78202
Contact: Fort Sam Houston National Cemetery, TX
(210) 820-3891

VIRGINIA

Alexandria National Cemetery (Closed)
1450 Wilkes St.
Alexandria, VA 22314
Contact: Quantico National Cemetery, VA, (703) 221-2183 or (703) 221-9939

Balls Bluff National Cemetery (Closed)
Rt. 7, Leesburg, VA 22075
Contact: Culpeper National Cemetery, VA, (540) 825-0027

City Point National Cemetery (Closed)
10th Ave. and Davis St.
Hopewell, VA 23860
Contact: Fort Harrison National Cemetery, VA, (804) 795-2031

Cold Harbor National Cemetery (Closed)
6038 Cold Harbor Rd.
Mechanicsville, VA 23111
Contact: Fort Harrison National Cemetery, VA, (804) 795-2031

Culpeper National Cemetery
305 U.S. Ave.
Culpeper, VA 22701
(540) 825-0027

Danville National Cemetery (Closed)
721 Lee St.
Danville, VA 24541
Contact: Salisbury National Cemetery, NC, (704) 636-2661

Fort Harrison National Cemetery (Closed)
8620 Varina Rd.
Richmond, VA 23231
(804) 795-2031

Glendale National Cemetery (Closed)
8301 Willis Church Rd.
Richmond, VA 23231
Contact: Fort Harrison National Cemetery, VA, (804) 795-2031

Hampton National Cemetery (Closed)
Cemetery Rd. at Marshall Ave.
Hampton, VA 23669
(757) 723-7104

Quantico National Cemetery
18424 Joplin Rd. (Rt. 619)
Triangle, VA 22172
(703) 221-2183 or (703) 221-9939

Richmond National Cemetery (Closed)
1701 Williamsburg Rd.
Richmond, VA 23231
Contact: Fort Harrison National Cemetery, VA, (804) 795-2031

Seven Pines National Cemetery (Closed)
400 E. Williamsburg Rd.
Sandston, VA 23150
Contact: Fort Harrison National Cemetery, VA, (804) 795-2031

Staunton National Cemetery (Closed)
901 Richmond Ave.
Staunton, VA 24401
Contact: Culpeper National Cemetery, VA, (540) 825-0027

Winchester National Cemetery (Closed)
401 National Ave.
Winchester, VA 22601
Contact: Culpeper National Cemetery, VA, (540) 825-0027

WASHINGTON

Tahoma National Cemetery
18600 S.E. 240th St.
Kent, WA 98042
(425) 413-9614

WEST VIRGINIA

Grafton National Cemetery (Closed)
431 Walnut St.
Grafton, WV 26354
Contact: West Virginia National Cemetery, WV, (304) 265-2044

West Virginia National Cemetery
Rt. 2, Box 127
Grafton, WV 26354
(304) 265-2044

WISCONSIN

Wood National Cemetery (Closed)
Bldg. 1301
5000 W. National Ave.
Milwaukee, WI 53295
(414) 382-5300

Burial in State Veterans Cemeteries

Any member of the Armed Forces who died while in the active military, naval, or air service or was discharged under conditions other than dishonorable may be buried in a federally assisted cemetery.

Qualified individuals who desire interment in a state veterans cemetery must apply through their individual state veterans agency. Many state veterans cemeteries are filled to capacity and are closed even though some spaces are reserved for spouses. Some cemeteries can bury only cremated remains.

A plot allowance may be paid to a state or an agency or political subdivision of a state. To qualify: (1) burial must be in a cemetery or section of a cemetery which is reserved for persons entitled to be buried in a national cemetery; (2) there is no charge for the plot of interment in such cemetery; (3) the cemetery is owned and operated by a state or political subdivision or agency thereof; and (4) death occurs on or after October 1, 1978.

STATE VETERANS CEMETERIES

Only the physical address of each cemetery is listed. Call the number provided to obtain the proper mailing address. All state cemeteries (except Kentucky, Nevada, Utah, and Wyoming) restrict burial to state residents.

(As of October 2008)

ARIZONA

Southern Arizona Veterans Memorial Cemetery
1300 Buffalo Soldier Trl.
Sierra Vista, AZ 85635,
(520) 458-7144

ARKANSAS

Arkansas State Veterans Cemetery
1501 W. Maryland Ave.
North Little Rock, AR 72120
(501) 683-2259

CALIFORNIA

Northern California Veterans Cemetery
11800 Gas Point Rd.
Igo, CA 96047
(866) 777-4533

Veterans Memorial Grove Cemetery
Veterans Home of California
100 California Dr.
Yountville, CA 94599
(707) 944-4541

COLORADO

Colorado State Veterans Center
3749 Sherman Ave.
Monte Vista, CO 81144
(719) 852-5118

Veterans Memorial Cemetery of Western Colorado
2830 D Rd.
Grand Junction, CO 81501
(970) 263-8986 or (970) 263-0089

CONNECTICUT

Colonel Raymond F. Gates Memorial Cemetery
Veterans Home and Hospital
394 West St.
Rocky Hill, CT 06067
(860) 721-5838

Middletown Veterans Cemetery
317 Bow Ln.
Middletown, CT 06457
(860) 344-1961

Spring Grove Veterans Cemetery (Closed)
22 Hecker Ave.
Darien, CT 06820
(860) 721- 5838

DELAWARE

Delaware Veterans Memorial Cemetery
New Castle County
2465 Chesapeake City Rd.
Bear, DE 19701
(302) 834-8046

Delaware Veterans Memorial Cemetery
Sussex County
26669 Patriots Way
Millsboro, DE 19966
(302) 934-5653

GEORGIA

Georgia Veterans Cemetery
8819 U.S. Hwy. 301
Glennville, GA 30427
(912) 654-5398

Georgia Veterans Memorial Cemetery
2617 Vinson Hwy.
Milledgeville, GA 31061
(478) 445-3363

HAWAII

East Hawaii Veterans Cemetery - Area I
(Closed)
(Island of Hawaii), County of Hawaii
150 Kapiolani St.
Hilo, HI 96720
(808) 961-8311

East Hawaii Veterans Cemetery - Area II
(Island of Hawaii), County of Hawaii
110 Laimana St.
Hilo, HI 96720
(808) 961-8311

Hawaii State Veterans Cemetery
(Island of Oahu)
45-349 Kamehameha Hwy.
Kaneohe, HI 96744
(808) 233-3630

Kauai Veterans Cemetery
(Island of Kauai)
4331 Lele Rd.
Hanapepe, HI 96716
(808) 241-3348

Lanai Veterans Cemetery
(Island of Lanai)
Cemetery Rd.
Lanai City, HI 96763
(808) 565-7086

Maui Veterans Cemetery
(Island of Maui)
Baldwin Ave.
Makawao, HI 96768
(808) 572-7272

Molokai Veterans Cemetery
(Island of Molokai)
2725 Lihi Pali Ave.
Hoolehua, HI 96729
(808) 553-3221

West Hawaii State Veterans Cemetery
(Island of Hawaii)
72-3245 Queen Kaahumanu Hwy.
Mile Marker 88
Kailua Kona, HI 96740
(808) 961-8311

IDAHO

Idaho State Veterans Cemetery
10100 Horseshoe Bend Rd.
Boise, ID 83714
(208) 334-4796

ILLINOIS

Sunset Cemetery
Illinois Veterans Home
1707 N. 12th St.
Quincy, IL 62301
(217) 222-8641

INDIANA

Indiana Veterans Home Cemetery
3851 N. River Rd.
W. Lafayette, IN 47906-3765
(765) 463-1502

Indiana Veterans Memorial Cemetery
1415 North Gate Rd.
Madison, IN 47250
(812) 273-9220

IOWA

Iowa Veterans Home and Cemetery
1301 Summit St.
Marshalltown, IA 50158
(641) 752-1501

Iowa Veterans Cemetery
34024 Veterans Memorial Dr.
Adel, IA 50003
(515) 996-9048

KANSAS

Kansas Veterans Cemetery at Fort Dodge
and Kansas Soldiers Home Cemetery
11560 U.S. Hwy 400
Dodge City, KS 67801
(620) 338-8772

Kansas Veterans Cemetery at Fort Riley
(Location to open in early 2009)
Wildcat Creek Rd.
Fort Riley, KS 66442
(785) 296-3976

Kansas Veterans Cemetery at Wakeeney
403 S. 13th St.
Wakeeney, KS 67672
(785) 743-5685

Kansas Veterans Cemetery at Winfield
1208 N. College St.
Winfield, KS 67156
(620) 229-2287

KENTUCKY

Kentucky Veterans Cemetery-Central
2501 N. Dixie Blvd.
Radcliff, KY 40160
(502) 564-9281

Kentucky Veterans Cemetery-North
205 Eibeck Lane
Williamstown, KY 41097
(502) 564-9281

Kentucky Veterans Cemetery-West
5817 Fort Campbell Blvd.
Hopkinsville, KY 42240
(270) 889-6106

LOUISIANA

Northwest Louisana Veterans Cemetery
7970 Mike Clard Rd.
Keithville, LA 71047
(318) 925-0612

MAINE

Maine Veterans Memorial Cemetery
163 Mount Vernon Rd.
Augusta, ME 04330
(207) 287-3481

Maine Veterans Memorial Cemetery
(Closed)
Old Belgrade Rd. & Rte. 27
Civic Center Dr.
Augusta, ME 04330
(207) 287-3481

Maine Veterans Memorial
Cemetery - Caribou
37 Lombard Rd.
Caribou, ME 04736
(207) 492-1173

MARYLAND

Cheltenham Veterans Cemetery
11301 Crain Hwy.
Cheltenham, MD 20623
(301) 372-6398

Crownsville Veterans Cemetery
1122 Sunrise Beach Rd.
Crownsville, MD 21032
(410) 987-6320

Eastern Shore Veterans Cemetery
6827 E. New Market Ellwood Rd.
Hurlock, MD 21643
(410) 943-3420

Garrison Forest Veterans Cemetery
11501 Garrison Forest Rd.
Owings Mills, MD 21117
(410) 363-6090

Rocky Gap Veterans Cemetery
14205 Pleasant Valley Rd. N.E.
Flintstone, MD 21530
(301) 777-2185

MASSACHUSETTS

Massachusetts State Veterans Cemetery
111 Glenallen St.
Winchendon, MA 01475
(978) 297-9501

Massachusetts Veterans
Memorial Cemetery
1390 Main St.
Agawam, MA 01001
(413) 821-9500

MICHIGAN

Grand Rapids Home for Veterans Cemetery
3000 Monroe Ave. N.E.
Grand Rapids, MI 49505
(616) 364-5400

MINNESOTA

Minnesota State Veterans Cemetery
15550 Hwy. 115
Little Falls, MN 56345
(320) 616-2527

MISSOURI

Missouri State Veterans
Cemetery at Bloomfield
17357 Stars and Stripes Way
Bloomfield, MO 63825
(573) 568-3871

Missouri State Veterans
Cemetery at Higginsville
20109 Bus. Hwy. 13
Higginsville, MO 64037
(660) 584-5252

Missouri State Veterans
Cemetery at Jacksonville
1479 County Rd., 1675
Jacksonville, MO 65260
(660) 295-4237

Missouri State Veterans
Cemetery - Springfield
5201 S. Southwood Rd.
Springfield, MO 65804
(417) 823-3944

St. James Missouri Veterans Home Cemetery
20 N. Jefferson
St. James, MO 65559
(573) 265-3271

MONTANA

Eastern Montana State Veterans Cemetery
Hwy. 59, Cemetery Rd.
Miles City, MT 59301
(406) 951-1325

Montana State Veterans Cemetery
Heroes Rd.
Ft. Harrison, MT 59636
(406) 324-3742

Montana Veterans Home Cemetery
400 Veterans Dr.
Columbia Falls, MT 59912
(406) 892-3256

Western Montana State Veterans Cemetery
1911 Tower Rd.
Missoula, MT 59804
(406) 721-5778

NEBRASKA

Nebraska Veterans Home Cemetery
2300 W. Capital Ave.
Grand Island, NE 68803
(308) 385-6252

NEVADA

Northern Nevada Veterans Memorial Cemetery
14 Veterans Way
Fernley, NV 89408
(775) 575-4441

Southern Nevada Veterans Memorial Cemetery
1900 Buchanan Blvd.
Boulder City, NV 89005
(702) 486-5920

NEW HAMPSHIRE

New Hampshire State Veterans Cemetery
110 Daniel Webster Hwy.
Boscawen, NH 03303
(603) 796-2026

NEW JERSEY

Brigadier General William C. Doyle Veterans Memorial Cemetery
350 Provenceline Rd.
Wrightstown, NJ 08562
(609) 758-7250

New Jersey Veterans Memorial Cemetery (Closed)
524 N.W. Boulevard
Vineland, NJ 08360
(856) 405-4200

NORTH CAROLINA

Coastal Carolina State Veterans Cemetery
110 Montford Point Rd.
Jacksonville, NC 28541
(910) 347-4550

Sandhills State Veterans Cemetery
400 Murchison Rd.
Spring Lake, NC 28390
(910) 436-5630

Western Carolina State Veterans Cemetery
962 Old US 70 Hwy.
Black Mountain, NC 28711
(828) 669-0684

NORTH DAKOTA

North Dakota Veterans Cemetery
1825 46th St.
Mandan, ND 58554
(701) 667-1418

OHIO

Ohio Veterans Home Cemetery
3416 Columbus Ave.
Sandusky, OH 44870
(419) 625-2454

OKLAHOMA

Oklahoma Veterans Cemetery (Closed)
2312 N.E. 36th.
Oklahoma City, OK 73111
(405) 228-5038

PENNSYLVANIA

Pennsylvania Soldiers and Sailors Home Cemetery
560 E. Third St.
Erie, PA 16507
(814) 871-4531

RHODE ISLAND

Rhode Island Veterans Memorial Cemetery
301 S. County Trl.
Exeter, RI 02822
(401) 268-3088

SOUTH CAROLINA

M.J. Dolly Cooper Veterans Cemetery
140 Inway Dr.
Anderson, SC 29621
(864) 332-8022

SOUTH DAKOTA

South Dakota State Veterans Home Cemetery
2500 Minnekahta Ave.
Hot Springs, SD 57747
(605) 745-5127

TENNESSEE

East Tennessee State Veterans Cemetery
5901 Lyons View Pike
Knoxville, TN 37919
(865) 594-6776

Middle Tennessee State Veterans Cemetery
7931 McCrory Ln.
Nashville, TN 37221
(615) 532-2238

West Tennessee State Veterans Cemetery
4000 Forest Hill Irene Rd.
Memphis, TN 38125
(901) 543-7005

TEXAS

Central Texas State Veterans Cemetery
11463 State Hwy. 195
Killeen, TX 76542
(254) 616-1770

Rio Grande Valley State Veterans Cemetery
2520 S. Inspiration Rd.
Mission, TX 78572
(956) 583-7227

Texas State Veterans Cemetery at Abilene
(May 2009)
7457 W. Lake Rd.
Abilene, TX 99601
(512) 463-5977

UTAH

Utah State Veterans Cemetery
Utah National Guard
17111 S. Camp Williams Rd.
Bluffdale, UT 84065
(801) 254-9036

VERMONT

Vermont Veterans Home War Memorial Cemetery
325 North St.
Bennington, VT 05201
(802) 447-6510/6528

Vermont Veterans Memorial Cemetery
487 Furnace Rd.
Randolph Center, VT 05061
(802) 828-3379

VIRGINIA

Virginia Veterans Cemetery at Amelia
10300 Pridesville Rd.
Amelia, VA 23002
(804) 561-1475

Virginia Veterans Cemetery at Suffolk
5310 Milners Rd.
Suffolk, VA 23434
(757) 255-7217

WASHINGTON

Washington Soldiers Home Colony and Cemetery
1301 Orting-Kapowsin Hwy.
Orting, WA 98360
(360) 893-4500

Washington Veterans Home Cemetery
1141 Beach Dr. E.
Port Orchard, WA 98366
(360) 895-4700

WISCONSIN

Northern Wisconsin Veterans Memorial Cemetery
N4063 Veterans Way
Spooner, WI 54801
(715) 635-5360

Southern Wisconsin Veterans Memorial Cemetery
21731 Spring St.
Union Grove, WI 53182
(262) 878-5660

Central Wisconsin Veterans Memorial Cemetery
Wisconsin Veterans Home
N2665 County Rd. QQ
King, WI 54946
(715) 258-5586, ext. 2203

WYOMING

Oregon Trail State Veterans Cemetery
80 Veterans Rd.
Evansville, WY 82636
(307) 235-6673

TERRITORIES

Commonwealth of the Northern Mariana Islands Veterans Cemetery
Marpi, Saipan, MP 96950
(670) 664-2650

Guam Veterans Cemetery
150 Spruance Dr.
Piti, GU 96915
(671) 477-4013

FUNERAL HONORS

By law, all eligible veterans are entitled to military funeral honors. Military members are eligible if they are serving on active duty or served on active duty and were discharged under other than dishonorable conditions. Members of the Selected Reserve are also eligible. In addition, former members of the Selected Reserve are eligible if they served at least one term of enlistment or period of initial obligated service and were discharged under other than dishonorable conditions, or were discharged due to a disability incurred or aggravated in the line of duty. However, a person is not eligible if they are convicted of a federal or state capital crime and are described in Section 2411(b) of Title 38, United States Code, or is otherwise eligible, but the circumstances are such that to provide military honors at the funeral or burial of the person would bring discredit upon the person's service (or former service).

Upon request, two servicemembers will fold and present the American flag to surviving family members, and a bugler will sound Taps. If a bugler is not available, a high-quality CD will be used.

At least one member of the funeral detail will be from the deceased veteran's parent military service. The other may be from the same service or another military service. Other authorized providers, such as members of a veteran's organization, may be used to augment the military detail. No particular rank is specified in the law, but the services by tradition have ensured the person presenting the flag to the family is at least the grade of the deceased veteran.

Department of Defense policy requires funeral directors, rather than families, to contact the military. Military funeral honors must be requested–they are not provided automatically. A DoD website explaining the funeral honors process is at www.militaryfuneralhonors.osd.mil. In addition, funeral directors may call 1-877-645-4667 to coordinate ceremonies.

REMAINS OF RETIREES AND FAMILY MEMBERS

The remains of retirees and authorized dependents who die while an inpatient in a Government medical treatment facility may be transported at Government expense from the place of death to the place of interment. The amount of reimbursement may not exceed the cost of transporting the remains from place of death to a point no further distant than the last permanent residence (not home of record) within the 50 states, Commonwealth of Puerto Rico, and the possessions and territories of the United States.

When death of a dependent occurs in an overseas command in which an Armed Forces mortuary is operated, remains of such dependents may be prepared (embalmed and certain supplies furnished) at such mortuaries, on a reimbursable basis, as a convenience to the sponsor. The sponsor, however, must bear the costs of preparation, casket, and interment.

The remains of retirees or their family members who die while overseas may be transported to CONUS aboard military aircraft or DoD contracted aircraft from the overseas point of debarkation to the nearest point of embarkation in the CONUS at no cost, on a space available basis. Additionally, mortuary services for the remains may be obtained from a military overseas mortuary on a reimbursable space available basis. The person making arrangements for the remains should contact the American Citizen's

Services Assistance Office at (202) 647-5226 for information and assistance in obtaining mortuary services and transportation.

Surviving spouses of deceased military retirees are permitted to accompany the remains aboard a DoD aircraft to a U.S. port of entry in the CONUS for burial. Once in the states, surviving spouses must pay for transporting the deceased and themselves to the burial location. The policy also includes a provision to allow spouses to remain in the states for up to 12 months while still retaining authority to return overseas aboard a Government flight.

UNITED STATES NAVY MORTUARY AFFAIRS BURIAL AT SEA PROGRAM

Burial at Sea is a means of final disposition of remains that is performed on United States Navy vessels. The committal ceremony is performed while the ship is deployed. Therefore, family members are not allowed to be present. The commanding officer of the ship assigned to perform the ceremony will notify the family of the date, time, and, longitude and latitude once the committal service has been completed.

The following are eligible for the program: (1) active duty members of the uniformed services; (2) retirees and veterans who were honorably discharged. (3) U.S. civilian Military Sealift Command marine personnel; and (4) dependent family members of uniformed services active duty personnel, retirees, and veterans.

After the death of the individual for whom the request for Burial at Sea is being made, the Person Authorized to Direct Disposition (PADD) should contact the United States Navy Mortuary Affairs office toll-free at 1-866-787-0081.

REIMBURSEMENT OF BURIAL EXPENSE (Department of Veterans Affairs)

Service-connected Deaths. The VA will pay a burial allowance up to $2,000 if the veteran's death is service-connected. VA will also pay the cost of transporting the remains of a service-disabled veteran to the national cemetery nearest the home of the deceased that has available gravesites. In such cases, the person who bore the veteran's burial expenses may claim reimbursement from the VA.

Nonservice-connected Deaths. The VA will pay a $300 burial and funeral expense allowance for veterans who at the time of death, were entitled to receive pension or compensation or would have been entitled to compensation but for receipt of military retirement pay.

The VA will also pay the actual cost (not to exceed $300) of the burial and funeral expense when the veteran dies in a VA facility, a nursing home under VA contract, or a state VA nursing home. Additional costs of transportation of the remains may be reimbursed in those cases.

In addition, for nonservice-connected deaths, the VA will pay a $300 plot allowance for a veteran who is eligible for burial in a national cemetery, but not buried in a national cemetery or other cemetery that is under U.S. Government jurisdiction. If the veteran is eligible for a burial allowance, or was discharged from active duty because of a disability incurred or aggravated in the line of duty, and is buried without charge for the cost of a plot or interment in a state-owned cemetery reserved solely for veteran burials, the $300 plot allowance may be paid to the state. If the veteran is buried in a commercial cemetery, the VA may pay the $300 plot allowance to an individual responsible for the disposition of the remains. In the event a former employer of the deceased or a state agency paid the burial expenses, the plot allowance reimbursement is limited to the difference between the entire amount of the expenses and the amount paid or assumed by those on behalf of the veteran.

BURIAL FLAGS (Department of Veterans Affairs)

In general, the VA provides a United States burial flag to memorialize veterans who have an other than dishonorable discharge. Also eligible are veterans who were entitled to retired pay for service in the Reserve Components, or would have been entitled to such pay but not for being under 60 years of age; and members or former members of the Selected Reserve who served at least one enlistment or, in the case of an officer, the period of initial obligation, or were discharged for disability incurred or aggravated in line of duty, or died while a member of the Selected Reserve. Specific eligibility requirements are listed on the application form.

To apply, complete VA Form 21-2008, "*Application for United States Flag for Burial Purposes.*" Flags may be obtained at any VA regional office or U.S. Post Office.

HEADSTONE AND MARKERS (Department of Veterans Affairs)

The VA furnishes upon request, at no charge to the applicant, a government headstone or marker for the grave of any deceased eligible veteran in any cemetery around the world. For all deaths occurring before September 11, 2001, the VA may provide a headstone or marker only for graves that are not marked with a private headstone. Spouses and dependents buried in a private cemetery are not eligible for a government-provided headstone or marker.

Flat bronze, granite or marble markers and upright granite and marble headstones are available. In the case of national cemeteries, the style chosen must be consistent with existing monuments at the place of burial. Niche markers also are available to mark columbaria used for inurnment of cremated remains.

Government-furnished headstones and markers must be inscribed with the name of the deceased, branch of service, and the year of birth and death, in this order. Headstones and markers also may be inscribed with other items, including an authorized emblem of belief and, space permitting, additional text including military grade, rate or rank, war service such as "World War II," complete dates of birth and death, military awards, military organizations and civilian or veteran affiliations. When burial or memorialization is in a national, state or military veterans cemetery, the headstone or marker must be ordered through cemetery officials. To apply and to obtain specific information on available styles, contact the cemetery where the headstone or marker is to be placed.

When burial occurs in a private cemetery, submit an application for a Government-furnished headstone or marker to the VA. The Government will ship the headstone or marker free of charge, but will not pay for its placement. Headstones and markers previously provided by the Government may be replaced at Government expense if badly deteriorated, illegible, stolen or vandalized.

VA also provides memorial headstones and markers, bearing the inscription "In Memory of" as their first line, to memorialize eligible veterans whose remains were not recovered or identified, were buried at sea, donated to science or cremated and scattered. Eligible dependents may be memorialized only in national or state veterans cemeteries. Dependents do not need to outlive the veteran from whom their eligibility is based to be memorialized. Memorial headstones or markers must be placed in national, state veterans, local or private cemeteries. VA supplies and ships memorial headstones and markers free of charge for placement in state, local and private cemeteries, but does not pay for their plots or placement.

To apply, submit a completed VA Form 40-1330, "*Application for Standard Government Headstone or Marker for Installation in a Private or State Veterans Cemetery*," along with a copy of the veteran's military service discharge document to Memorial Programs Service (41A1), Department of Veterans Affairs, 5109 Russell Rd. Quantico, VA 22134-3903. Documents may also be faxed to 1-800-455-7143. Do not send original discharge documents, as they will not be returned. For more information and application instructions, visit www.cem.va.gov.

DEATH GRATUITY (Department of Defense)

The Department of Defense provides a tax-free payment of $100,000 to assist the survivors of certain deceased servicemembers in meeting their financial needs immediately following the member's death and before other survivor benefits, if any, become available.

The death gratuity is paid in the event of death while a member of the Armed Forces is on active duty, active duty for training, or inactive duty training. It is also payable if a member or former member dies of a service-connected cause within 120 days after his or her discharge or release from active duty, or active duty for training. For inactive duty training, the gratuity is payable if death occurs within 120 days and is the result of injury received during that training.

A member may designate named beneficiaries to receive a portion (in 10 percent increments) of the death gratuity payable upon that member's death. Such an election requires the Secretary concerned to notify the spouse, if any, if a member names someone other than the spouse to receive a portion of, or the entire death gratuity. In the event the member designates only a portion or none of the death gratuity, payment is made to the first available person(s) using the following order of precedence: surviving spouse (if any); surviving children; surviving parents; the duly appointed executor or administrator of the member's estate; to other next of kin of the member entitled under the laws of domicile of the member at the time of death.

PRESIDENTIAL MEMORIAL CERTIFICATES (Department of Veterans Affairs)

A Presidential Memorial Certificate (PMC) is signed by the current President to honor the military service of honorably discharged deceased veterans.

The VA administers the PMC program by preparing the certificates that bear the President's signature expressing the country's grateful recognition of the veteran's service in the United States Armed Forces.

Eligible recipients include the deceased veteran's next of kin and loved ones. More than one certificate may be provided. Application for the certificate may be made in person at any VA regional office, by mail or FAX. The form can be found at: www.va.gov/vaforms/va/pdf/VA40-0247.pdf. Please be sure to include a copy of the veteran's discharge and death certificate. Send requests to: Presidential Memorial Certificates (41A1C), National Cemetery Administration, 5109 Russell Rd., Quantico, VA 22134. Send FAX requests to: 1-800-455-7143. Additional information and a sample certificate can be found at www.cem.va.gov/CEM/pmc.asp.

PERSONAL AFFAIRS FOR SURVIVORS

General. Retired personnel are concerned with the future well being of their surviving family members. This section outlines survivor entitlements *(many of these entitlements are covered in this almanac)* and suggests things the retired member can do now to prepare their families for the future.

One of the most important points to keep in mind is that retired pay ceases upon death of the member. (See RETIREES CASUALTY ASSISTANCE SERVICES section which follows.) There is no form of compensation or pension that is automatically paid to the survivors of a retired member. In order to obtain any benefits to which the family members and survivors may be entitled, a claim must be properly filed with the appropriate agency.

The surviving spouse of a retiree is entitled to burial in a national military or state cemetery. In such cases, the VA will provide a headstone or marker. No headstones or markers are provided if burial takes place in a private cemetery.

Wills. Of all documents, perhaps the most important is a will. Regardless of how financially stable you left your family members, the absence of a will makes it very difficult for your survivors to receive all that you intended. If you die without a will, the state laws in which you reside will determine how your property will be distributed. Military legal assistance offices can prepare a will for you. If a legal assistance officer is not available, a civilian attorney can assist you. Most attorneys charge a nominal fee. If you have made a will, you should review it periodically particularly if you have moved from the state in which it was originally drawn or if significant changes have occurred.

Power of Attorney. A retired military member may need to designate a person to have power of attorney. The power of attorney is a written document given by one person to another authorizing the latter to legally act for the former. Since this document is binding, it should be executed with the advice of an attorney. The two categories of power of attorney are general and special. *The general power of attorney* authorization allows a designated person to execute any and all legal acts in the principal's name. *The special power of attorney* authorization permits a person to do only specific acts designated by the principal, such as: (1) Medical power of attorney—this document would allow your designee to give permission for emergency medical treatment for your child while you and your spouse are outside the U.S. on vacation; (2) Permission for your spouse to register your car or to sell your home when you cannot be present due to illness. Consult an attorney before deciding to use the power of attorney designation.

Casualty Assistance. When a retiree dies, the Services may offer assistance to the surviving spouse through their Casualty Assistance Offices. Additionally, the survivors should notify the closest Casualty Area Command (CAC) or Casualty Assistance Office (CAO), preferably of the service to which the retiree belonged. Contact should also be made with the Retirement Services/Activities Office for assistance and guidance since the Service CAO/CAC may not be able to provide casualty assistance to the surviving family members. See GUIDE TO MILITARY INSTALLATION FACILITIES in this edition for listing of bases where Casualty Assistance Offices are located. The Casualty Assistance Officer will notify the Defense Finance and Accounting Service. Although the survivor is faced with many pressing responsibilities after the retiree's death, notifying the CAC should be one of the priorities at the top of the list. This early notification may prevent financial hardship.

For example, retired pay should be stopped as soon as possible. Since entitlement to retired pay ends with the death of the retiree, early notification of death prevents an overpayment of retired pay which the survivor would have to pay back to DFAS. Early notification starts the processing for any benefits, as well as funeral support. The retiree can also help prevent financial hardship for the survivors by making sure that family members know where the important documents relating to marriage, birth, divorce, etc. are kept.

Within 24 hours of notification, the CAO or Retired Activities Office (RAO)/Retirement Services Officer (RSO) will contact the survivor. If the survivor wishes, the CAO or RAO/RSO helps with planning the funeral arrangements. The CAO or RAO/RSO can determine whether transportation support is available for the funeral in that area. After the funeral, the survivor and the CAO or RAO/RSO should discuss benefits such as ID card, Social Security, Veterans Administration and Survivor Benefits, as well as other topics such as a new will, taxes, and installation vehicle registration. The survivor should have the following documents on hand—survivor's birth certificate, retiree's discharge certificate(s), their marriage certificate, and if applicable, divorce certificate(s). During this meeting, the CAO or RAO/RSO and survivor may take the opportunity to either make appointments to visit agencies such as the VA, Social Security, and Installation Legal Assistance Offices or they may visit them if no appointments are needed.

Consultation and advice on all civil legal matters are provided to the retiree's survivors on a space available basis. Eligible family members may need to obtain legal advice and assistance in connection with settling the estate of the deceased member, making a new will for the surviving spouse, tax questions, or other related legal matters. Legal assistance officers can help in preparing the initial income tax return but can only give preliminary advice on probate and settlement of the estate, court appearances, preparation of inheritance tax returns, or in regard to civilian employment and business matters. Since these matters are outside the scope of the Legal Assistance Program, it may be necessary for survivors to obtain the services of civilian counsel. The legal assistance officer may provide a referral to such counsel through the local bar association. Government agencies, such as the VA, Social Security Administration, and DFAS will help to process claims for benefits and entitlements. Such matters are expeditiously handled without the need for civilian counsel.

Notes: (1) DFAS may take 30-60 days after submission of your completed packet before Survivor Benefit Plan benefits start arriving. (2) VA benefits may take two to six months from the time of your application until the benefit checks arrive. (3) Social Security benefits should start in 30 days. (4) Be sure to take notes or make copies of all applications you file to remind yourself of what items have been completed.

RETIREES CASUALTY ASSISTANCE SERVICES

When a retiree dies, the next of kin should immediately call the Defense Finance and Accounting Service (DFAS)—Cleveland Center. The Retired Casualty Section at DFAS can be reached toll-free, at 1-800-321-1080, from anywhere in the United States. DFAS will review and close the retiree's pay account. They will also mail a survivor benefits packet to the decedent's next-of-kin. The packet typically consists of forms for applying for the retiree's unpaid pay and the Survivor Benefit Plan annuity, if applicable. These forms must be completed as soon as possible and returned to DFAS with a photocopy of the death certificate.

Local Retired Activities Offices may also provide assistance. This includes providing information, counseling on eligibility for various survivor benefits, referral services and answering other questions you may have.

The best way to take care of your survivors after you are gone is to leave them the information they will need.

Consider the following charts and take time to fill out all that applies to you. Make sure your survivors know the location of this valuable information.

Things to know and plan for upon death of a retiree Retiree

- Obtain information required for death certificate (date/place of birth, father's name, mother's maiden name, etc.)
- Collect information for obituary notice (names, relation and locations of appropriate relatives, etc.)
- Assure that your survivors have a current military ID card in order to use military medical facilities, commissary, exchange, etc.
- Outline contents of your safety deposit box
- Keep current list of beneficiaries of all insurance policies
- Update will, living will and any letters of instruction

Survivor

- Make at least eight copies of death certificate (It may take several copies to clear estates.)
- Contact insurance companies as appropriate
- Make necessary changes in your DEERS enrollment (TRICARE, etc.)
- Change direct deposit of Social Security benefits & military retirement payments (entitlements) as necessary
- Verify entitlement to burial benefits (headstone, payments, etc.)
- Contact VA for Presidential Memorial Certificate
- Obtain an American flag if desired (contact VA)
- Update will
- Destroy or cancel extra credit cards
- Make appropriate changes to all joint ownerships
- Be prepared to turn in Retirees ID card (where and when required)
- Arrange for Military Funeral Honors

Important Offices/Organizations	Phone Numbers
Casualty Assistance Office (immediate assistance upon retiree's death)	Army: 1-800-626-3317 Navy: 1-800-368-3202 MC: 1-800-847-1597 AF: 1-800-433-0048 CG: 1-800-772-8724
Mortuary Affairs	
Hospital (Military)	
Legal Office (Military)	
Retiree Activities Office	
VA Hotline	1-800-827-1000
Social Security Hotline	1-800-772-1213
ID Card/DEERS Updates	1-800-538-9552
Family Support Center	
Finance (DFAS - Cleveland)	1-800-321-1080
SBP/RC-SPB	
VA Insurance	1-800-669-8477

<table>
<tr><th colspan="3">RETIREES CASUALTY ASSISTANCE</th></tr>
<tr><td colspan="3">Name of Retiree</td></tr>
<tr><td colspan="3">Address</td></tr>
<tr><td>City</td><td>State</td><td>Zip</td></tr>
<tr><td>SSN</td><td rowspan="2">Years of Service</td><td rowspan="2">Military Grade</td></tr>
<tr><td>Branch of Service</td></tr>
<tr><td>Date of Birth</td><td colspan="2">Place of Birth</td></tr>
<tr><td>Date of Marriage</td><td colspan="2">Place of Marriage</td></tr>
<tr><th colspan="3">Survivor Annuity Information</th></tr>
<tr><td colspan="3">Enrolled in RSFPP, SBP (circle all that apply)</td></tr>
<tr><td colspan="3">Did you disenroll from this plan? Yes or No</td></tr>
<tr><td colspan="3">Survivor Coverage Information</td></tr>
<tr><td colspan="2">Coverage type: spouse only, etc.</td><td>Monthly Cost</td></tr>
<tr><td rowspan="2">Annuity Base Amount</td><td colspan="2">55 percent SBP annuity amount:
$</td></tr>
<tr><td colspan="2">RSFPP annuity:
$</td></tr>
<tr><td colspan="3">Veterans Administration Data</td></tr>
<tr><td colspan="2">Eligible to draw VA disability compensation (even if not in receipt now): Yes No</td><td rowspan="2">Disability Claim number</td></tr>
<tr><td colspan="2">Veteran's claim number(s)</td></tr>
<tr><td colspan="3">VA Insurance policy number(s)</td></tr>
<tr><td>Type</td><td colspan="2">Amount</td></tr>
<tr><td colspan="2">Any known paid-up add'l VA Insurance $</td><td>As of Date</td></tr>
<tr><td colspan="3">Social Security Data</td></tr>
<tr><td>Receiving Social Security:
Yes No</td><td colspan="2">If yes, age and date at which first received:</td></tr>
<tr><td colspan="2">Social Security Claim Number</td><td>Month Filed</td></tr>
<tr><td colspan="2">Type of Benefit(s)</td><td rowspan="3">Beginning month of entitlement</td></tr>
<tr><td colspan="2">Monthly Amount</td></tr>
<tr><td colspan="2">Bank and account number</td></tr>
</table>

Retirement Pay Data	
Retiree gross and net pay data: as of date	
Gross Pay	Net Pay
Deduction Amount	For
Deduction Amount	For
Deduction Amount	For
Taxable Income Amount	
Burial Information	
Who should be notified of your death? (Make a separate list including name, address, phone number and relationship.)	
Do you want to be buried or cremated?	
Name of cemetery where you want to be buried.	
Do you want to be buried in uniform?	
Do you want a memorial service?	
Have you purchased a burial plot?	
Do you have a preference of funeral homes?	
Do you want a military honor guard?	
Document	**Location**
Copies of report(s) of separation from active duty (DD Form 214, etc.)	
Copies of retirement orders	
Copies of any legal certificates: birth, death, marriage, divorce, adoption and naturalization papers (if applicable)	
Social Security data	
VA Insurance data	
Updated will and letter of instructions	
Names of banks, credit unions, etc. (account numbers)	
Updated list of assets and liabilities	
Insurance policies, numbers, instructions, payments, etc.	
Military Identification Cards	
Safe Deposit Box (List Contents)	
Vehicle Title and Registration	
Real Estate Deeds	
Tax Returns	

VA AND MILITARY INSURANCE PROGRAMS

The Office of Servicemembers' Group Life Insurance (OSGLI) is an administrative office created by Prudential Insurance Company of America to administer the day-to-day operations of the SGLI and VGLI programs. Applications for VGLI should be sent to OSGLI, P.O. Box 41618, Philadelphia, PA 19176-9913. All other correspondence should be directed to OSGLI at 80 Livingston Ave., Roseland, NJ 07068-1733. OSGLI's toll-free customer service telephone number is 1-800-419-1473.

For inquiries regarding other VA insurance programs administered by VA, contact the Department of Veterans Affairs, Regional Office and Insurance Center, P.O. Box 42954, Philadelphia, PA 19101. If the policy number is unknown, send any available information, such as the veteran's VA claim number, Social Security number, date of birth, military serial number, or military service branch and dates of service.

In addition, the Department of Veterans Affairs has a national toll-free line for veterans seeking information about Government Life Insurance other than SGLI and VGLI. Policyholders and beneficiaries can call 1-800-669-8477 to obtain immediate service on address or policy changes, dividend information and general insurance information, or to report the death of a policyholder. Specialists are available Monday to Friday between 8:30 am - 6:00 pm EST. After hours, callers may leave a recorded message that generally will be answered on the next workday.

The VA Insurance website, www.insurance.va.gov can answer many questions regarding all Government Life Insurance programs.

VA SUPERVISED PROGRAMS

Servicemembers' Group Life Insurance (SGLI)

Member Coverage. All members of the uniformed services, including most Reserve Component and National Guardmembers, commissioned officers of the Public Health Service and the National Oceanic and Atmospheric Administration, and cadets and midshipmen of the service academies, are automatically insured under Servicemembers' Group Life Insurance (SGLI) for the (statutory) maximum of $400,000 unless an election is filed reducing the insurance by $50,000 increments or canceling it entirely. Reserve Component and National Guardmembers who are mobilized and called to active duty are automatically insured for $400,000 SGLI coverage upon mobilization even if the member had previously elected not to have any SGLI or to have less than the $400,000 maximum. Those Reserve Component and National Guardmembers who do not want SGLI coverage or want less than the $400,000 SGLI when they are mobilized must elect in writing to reduce or decline coverage at the time they are mobilized.

All members who wish to elect less than maximum insurance, or who decline coverage entirely, must complete form SGLV 8286, *Servicemembers' Group Life Insurance Election and Certificate*. If a member later wishes to obtain or increase coverage (for other than a statutory automatic increase in coverage), the member must complete form SGLV 8285 and provide proof of good health.

The cost for this coverage is $3.25 per $50,000, the smallest increment issued, or $26.00 per month for the full $400,000 coverage (an additional premium of $1.00 is charged for a traumatic injury protection rider described below). For members with full-time coverage, SGLI coverage is in effect 24 hours a day, 365 days a year, regardless of duty status.

The Department of Defense reimburses all SGLI premiums paid by members serving in the Operation Iraqi Freedom or Operation Enduring Freedom theaters of operations.

Eligibility of Certain Individual Ready Reservists for SGLI. Certain members of the Individual Ready Reserve (IRR) who are subject to **involuntary call-up authority** are automatically eligible for full-time coverage under the SGLI program. Members of the IRR who are not in the special category subject to involuntary call-up are eligible for Veterans' Group Life Insurance and are only eligible for SGLI when called to active duty.

The SGLI coverage may be converted, upon release from active duty, active duty for training, initial active duty for training or upon separation from the Ready Reserve, to VGLI or, if desired, to a commercial life insurance policy effective at the end of the free 120 day SGLI coverage period.

In the event a servicemember is declared totally disabled at the time of separation from active duty, full-time SGLI coverage will be extended for as long as the disability continues, up to a maximum of two years. The extension is not automatic. Application must be made to the Office of Servicemembers' Group Life Insurance for the extension. A member who is granted a SGLI disability extension is automatically converted to VGLI coverage at the end of the extension, without evidence of good health, provided the required premiums are paid.

Requirement for Spousal Notification. The member's service is required to provide written notification to the spouse of a member who, upon submitting an SGLI election, (1) declines coverage; (2) elects less than maximum coverage, for either an initial election or subsequent election; and (3) elects a beneficiary who is not the spouse or child of that member, unless the spouse was previously notified of such a beneficiary election and immediately prior to the election, the spouse is not the designated beneficiary regardless of the amount of coverage.

The requirement for spousal notification is satisfied upon written notification being mailed to the spouse's last known address.

Traumatic Injury Protection. This coverage provides insurance protection for severe losses resulting from a traumatic injury. This program is known as SGLI Traumatic Injury Protection or TSGLI. All servicemembers who have SGLI coverage are automatically insured under this program unless they make a valid election to decline SGLI coverage. Coverage is in effect so long as SGLI coverage is in effect at the time the traumatic injury occurs and the member is not separated from service. TSGLI coverage ends at midnight on the day of separation from service. In addition, any servicemembers who, on or between October 7, 2001 and November 30, 2005 suffered a loss as a result of a traumatic injury that was incurred in Operation Enduring Freedom or Operation Iraqi Freedom are also eligible for the TSGLI benefit. The maximum amount of coverage is $100,000. Payments to insured members who suffer a traumatic injury range from $25,000 to $100,000, depending on the nature of the traumatic injury as prescribed in regulations by the Department of Veterans Affairs. The premium for this coverage is $1.00 per month for active duty members and Reserve Component members with full time SGLI coverage; $1.00 per year for Reserve Component members with part time SGLI coverage; and is provided at no cost to Reserve Component members who otherwise are not covered while performing funeral honors duty and one day muster duty.

Spouse and Dependent Child Coverage. Family Servicemembers' Group Life Insurance (FSGLI) is a program extended to the spouses and dependent children of members insured under the SGLI program. FSGLI provides up to a maximum of $100,000 of insurance coverage for spouses, not to exceed the amount of SGLI the insured member has in force, and $10,000 for dependent children. FSGLI is the servicemember's benefit, and the member is the beneficiary of the policy.

Spousal coverage is issued in increments of $10,000 up to the $100,000 maximum. FSGLI coverage is automatic. Spouse premiums are age-based and are deducted based on the spouse's information in the member's DEERS records. Therefore, it is important that the DEERS record accurately reflect the member's marital status and dependent information. The member is responsible for updating DEERS. The result of incorrect information is the wrong premiums being deducted and the potential for a large debt of back premiums to account for the automatic coverage.

MONTHLY RATES FOR SPOUSE COVERAGE

Age of Spouse	Monthly Premium Rates for $100,000 of Coverage	Age of Spouse	Monthly Premium Rates for $100,000 of Coverage
Under 35	$5.50	50-54	$27.00
35-39	$7.00	55-59	$40.00
40-44	$9.00	60 and Over	$52.00
45-49	$14.00		

Eligible dependent children are automatically insured for $10,000 each, as long as the servicemember maintains a minimum of $10,000 SGLI coverage. Effective October 10, 2008, dependent child coverage also extends to a member's stillborn child. No child can be insured under the coverage of more than one servicemember. The coverage is provided at no cost to the servicemember and cannot be reduced or declined.

A spouse may elect to convert his or her coverage to a commercial policy with a participating company within 120 days following a "termination of family coverage" event. These include: the date the servicemember elects in writing to terminate their spouse's coverage or their own coverage, the date the servicemember separates from service, the date of the servicemember's death, and the date of the servicemember's divorce from the spouse. There is no conversion option available to children.

Veterans' Group Life Insurance (VGLI)

Veterans' Group Life Insurance (VGLI) is a term insurance policy available, upon application, to all members being separated or released from active duty, active duty for training, initial active duty for training or upon separation from the Ready Reserve, who were covered under SGLI. VGLI is issued in multiples of $10,000 up to a maximum current coverage of $400,000. VGLI may not be issued in an amount greater than the amount of SGLI carried at the time of separation from service. VGLI is renewable for life and may be converted to a commercial insurance policy at any time.

Members wishing to continue coverage under the VGLI program must submit application and payment of first premium within 120 days of separation. If application (unless totally disabled) is not made within the

120-day period, application may be made within one year after SGLI coverage terminates if sufficient evidence of good health is provided. If the member is totally disabled on the date of separation, and approved for the SGLI extension (two years from the date of separation), once the SGLI expires the member will automatically be converted to VGLI for the same level of coverage the member had under SGLI, unless the member chooses a lesser amount. However, if the member decides for some reason that they do not want the VGLI, the member should simply not pay the premium and the policy will be cancelled. The effective date of VGLI will be the day following the end of the two-year period or the day following the date the total disability ends, whichever is earlier.

The Office of Servicemembers' Group Life Insurance notifies servicemembers who separate from active duty by mailing them computer-produced VGLI applications shortly after they separate. VGLI applications are mailed to eligible members three times: generally within 60 days after separation; just prior to 120 days after separation when the "no health questions" period ends; and again shortly before the end of the 16-month application period.

OSGLI also notifies separating Reservists to inform them of their entitlement to VGLI. Sometimes there is a delay between the time the Reservists separate and the receipt of the records in OSGLI. Reservists who have not received a computer-produced VGLI application within three weeks of separation should contact OSGLI to request a VGLI application. The VGLI application is also available on the VA website: www.insurance.va.gov. When submitting the generic application, Reservists must include a copy of their separation orders.

Full-time VGLI coverage is also available to members of the Individual Ready Reserve (IRR) and the Inactive National Guard (ING).

The VGLI premium rate structure contains 11 age brackets. The monthly premiums rates are listed in the following table. Rates are effective July 1, 2008. Different premium rates may apply for policies with initial or renewal dates prior to July 1, 2008.

Monthly Premiums for $1,000 of VGLI Coverage (Effective July 1, 2008)											
AGE GROUP	29 and under	30-34	35-39	40-44	45-49	50-54	55-59	60-64	65-69	70-74	75 and over
PREMIUM	$.08	$.10	$.13	$.17	$.22	$.36	$.67	$1.08	$1.50	$2.25	$4.50

Prior to October 1, 2002, when a veteran enrolled in VGLI, the same premium rate would apply for five years, even if the insured reached the next age group during that five-year period. For all VGLI policies that become effective or renew on or after October 1, 2002, the veteran will begin paying the premium rate corresponding to his or her age at time of enrollment or renewal. However, when the insured reaches the next age group, the new premium for that age group will apply.

For example, under the old structure, if a veteran enrolled in VGLI at age 38, the veteran would pay the premium rate for age group 35-39 for five years until the policy renewed, and the veteran was age 43. At renewal, the veteran would pay the premium rate for age group 40-44 for the next five years. Under the new structure, this same veteran will pay the premium rate for age group 35-39 until reaching age 40, at which time the premium rate for age group 40-44 will apply.

The law allows a terminally ill person insured under either SGLI or VGLI to elect to receive a portion of the face value of the policy (up to 50 percent) in a lump-sum as an accelerated death benefit, discounted by a small amount to reflect the loss of interest for the nine-month period of terminal illness. To receive the accelerated benefit, the member or veteran must have a medical prognosis of a life expectancy of nine months or less. The portion of the face value of insurance which is not paid in a lump-sum as an accelerated death benefit would remain payable to the beneficiary following the death of the insured.

VA ADMINISTERED PROGRAMS

National Service Life Insurance (NSLI). Policies issued under this insurance program have a prefix of "V", "RS/W", "J", "JR" or "JS". The maximum face value of an NSLI policy is $10,000. Term premiums in the "V" and "RS" programs are capped at the age 70 rates. Dividends are paid annually on all NSLI policies. The NSLI program is closed and no new policies are being issued.

Service-Disabled Veterans Insurance (S-DVI). Veterans who are granted a service-connected disability by the VA, but are otherwise in good health, may apply to VA for up to $10,000 in life insurance coverage at standard insurance rates. This insurance is limited to veterans who left service on or after April 25, 1951, and must be applied for within two years of the date they are granted a new service-connected disability. Veterans who are totally disabled may apply for a waiver of premiums. For those veterans who are eligible for this waiver, an additional policy of up to $20,000 is available. Premiums, however, cannot be waived on the additional insurance. Term premiums for the S-DVI program are also capped at age 70 rates.

Policy Provisions for NSLI and S-DVI.

Waiver of Premiums. All NSLI and S-DVI policies (except for supplemental S-DVI policies) contain a provision for waiver of payment of premiums if the insured becomes totally disabled for at least six consecutive months prior to the insured person's 65th birthday. Total disability is defined as any impairment of mind or body that prevents someone from performing substantially gainful employment.

Automatic Renewal. A five-year term policy, which is not lapsed at the end of the term period, is automatically renewed for an additional five-year period, at higher premium rates. However, term rates are "capped" at the age 70 rate and will never go higher.

Reinstatement of Lapsed Insurance. Lapsed term policies may be reinstated within five years from the date of lapse. However, for term policies that have lapsed for six months or more, evidence of good health is required in order to reinstate the policy. Permanent plans of insurance can also be reinstated, but all back premiums with interest must be paid. Contact the Insurance Center for details.

Converting Term Policies. A term policy that is in force may be converted to a permanent plan. Upon reaching renewal at age 70 or older, NSLI and S-DVI term policies on total disability premium waiver are automatically converted to a permanent plan of insurance, which provides cash and loan values as well as higher annual dividends. S-DVI policies do not participate in dividends.

Policy Loan. Government Life Insurance policyholders with permanent plans of insurance can take out a loan against the cash surrender value of their policy. The policy cannot be lapsed and must be in-force for at least one year before a policy has loan value. Policy loans may be granted up to 94 percent of the cash surrender value less any indebtedness.

Policy loans have a variable interest rate with a minimum rate of five percent and a maximum rate of 12 percent. The current interest rate for these adjustable rate loans is five percent. Each October 1, the interest rate may be increased or decreased depending on the 10-year constant maturity, U.S. Treasury Securities index. The current interest rate is available at any VA office or by calling 1-800-669-8477.

Modified Life Policy. A "modified life at age 65" plan of insurance is available to NSLI and S-DVI policyholders. The premium rates for this plan remain the same throughout the premium-paying period, while the face value reduces by 50 percent at age 65. The reduced amount may be replaced with a "special ordinary life" plan, for an additional premium. A "modified life at age 70" plan is also available, which is like the modified life at age 65 plan except that face value reduction does not occur until age 70. The premium rate is only slightly higher than for the modified life at age 65 plan.

Veterans' Mortgage Life Insurance (VMLI). The maximum amount of mortgage life insurance available for veterans who have received a grant from VA for specially adapted housing is $90,000. Protection is available upon submission of an application and information about the insured's mortgage. Premiums are automatically deducted from VA compensation payments, or paid directly, if the veteran does not receive VA compensation, and will continue until the mortgage has been liquidated, the home is sold, or the veteran elects to discontinue the coverage. If the property for which the mortgage is insured is sold, VMLI may be obtained on the mortgage of a subsequent home. Effective October 2005, premiums were reduced for all policyholders by an average of 35 percent.

Dividends. Policies with prefixes of "K", "V", "RS/W", "J", "JR" or "JS" pay dividends. The dividends are paid automatically on the policy anniversary date. Dividends are not payable to individuals insured under SGLI or VGLI. These are group policies and therefore, do not pay dividends. In addition, no dividends are payable in the S-DVI and VMLI programs.

EMPLOYMENT OF RETIRED MILITARY PERSONNEL

The law governing the employment of retired military personnel in federal civilian jobs contains the following major provisions:

With certain exceptions, military retirees are not entitled to placement in the veterans preference subgroup of their tenure group for reduction-in-force purposes, and they get credit only for length of military service performed during a war or in any campaign or expedition for which a campaign badge has been issued. This provision does not apply to: (1) military personnel whose retirement is based on a disability incurred in the line of duty as a result of armed conflict or caused by an instrumentality of war; (2) those whose retirement is based on less than 20 years of active service and; (3) those employed on November 30, 1964, in positions in the federal civilian service to which the laws on veterans preference apply and who have not had a break in service of more than 30 days since then. Military retirees below the rank of major or lieutenant commander (O-4) get veterans preference in examinations and appointments even though they do not get a preference in Reductions-in-Force (RIF).

Credit for military service of military retirees for annual leave purposes is limited to actual service during a war declared by Congress; while participating in a campaign or expedition for which a campaign badge is authorized; or all active duty when retirement was based on a disability received as a direct result of armed conflict or caused by an instrumentality of war and incurred in the line of duty for a period of war during World War II, the Korean conflict, the Vietnam era, or the Persian Gulf War.

Military retirees must wait at least 180 days after retirement before taking a civilian position in the DoD unless prior approval is obtained from the Service Secretary. This restriction may be waived for shortage category jobs or in a national emergency.

2009 PAY SCHEDULE FOR GENERAL SCHEDULE EMPLOYEES*

Longevity Steps	1	2	3	4	5	6	7	8	9	10
Grade GS										
1	17,540	18,126	18,709	19,290	19,873	20,216	20,792	21,373	21,396	21,944
2	19,721	20,190	20,842	21,396	21,635	22,271	22,907	23,543	24,179	24,815
3	21,517	22,234	22,951	23,668	24,385	25,102	25,819	26,536	27,253	27,970
4	24,156	24,961	25,766	26,571	27,376	28,181	28,986	29,791	30,596	31,401
5	27,026	27,927	28,828	29,729	30,630	31,531	32,432	33,333	34,234	35,135
6	30,125	31,129	32,133	33,137	34,141	35,145	36,149	37,153	38,157	39,161
7	33,477	34,593	35,709	36,825	37,941	39,057	40,173	41,289	42,405	43,521
8	37,075	38,311	39,547	40,783	42,019	43,255	44,491	45,727	46,963	48,199
9	40,949	42,314	43,679	45,044	46,409	47,774	49,139	50,504	51,869	53,234
10	45,095	46,598	48,101	49,604	51,107	52,610	54,113	55,616	57,119	58,622
11	49,544	51,195	52,846	54,497	56,148	57,799	59,450	61,101	62,752	64,403
12	59,383	61,362	63,341	65,320	67,299	69,278	71,257	73,236	75,215	77,194
13	70,615	72,969	75,323	77,677	80,031	82,385	84,739	87,093	89,447	91,801
14	83,445	86,227	89,009	91,791	94,573	97,355	100,137	102,919	105,701	108,483
15	98,156	101,428	104,700	107,972	111,244	114,516	117,788	121,060	124,332	127,604

*Rates shown do not include locality payments payable in the 48 contiguous states and the District of Columbia. There are 32 separate locality pay areas that receive between 13.86 to 34.35 percent locality pay. The range of increases for all locality areas in 2009 is 3.39 to 4.78 percent.

CREDIT DEPOSITS FOR POST-1956 MILITARY SERVICE

The Omnibus Budget Reconciliation Act of 1982, P.L. 97-253, established requirements for credit for Post-1956 military service for civil service retirement purposes. The law distinguishes between employees employed before, and on or after October 1, 1982.

Employed before October 1, 1982. Individuals who first became employed in a position under Civil Service Retirement System (CSRS) before October 1, 1982, have the option of either: (1) making the deposit for post-1956 military service or (2) receiving credit as in the past (without making the deposit) and having their annuity recomputed at age 62 to eliminate post-1956 military service if they are eligible for Social Security old-age or survivor benefits. Included are individuals who were, previous to October 1, 1982 covered under the CSRS and again employed under CSRS on or after October 1, 1982.

Employed on or after October 1, 1982. Individuals who first became employed under the CSRS on or after October 1, 1982, or after January 1, 1984, who are subject to the Federal Employee Retirement System (FERS), receive credit for their post-1956 military service only if a deposit for the military service is made.

Amount of Deposit. The deposit under CSRS is seven percent of basic military pay received, plus interest. For individuals subject to FERS, the deposit is three percent of basic military pay. An interest-free grace period is available until two years after the initial appointment.

Payment Options. Employees making service credit deposits for their post-1956 service may elect to make such deposits in either a lump sum payment, installment payments, or through payroll deductions. Employees interested in making such a deposit should contact their personnel office for additional instructions.

Q. *Is military service creditable under the CSRS or FERS?*

A. As a general rule, military service is creditable provided it was active service, was terminated under honorable conditions, and was performed before separation from a civilian position under the Retirement System. For exceptions to this general rule, see the questions immediately following.

Q. Does the receipt of pension or compensation under laws administered by the Veterans' Administration bar the crediting of military service?

A. If an individual is receiving military retired or retainer pay, no retirement credit can be given for the military service (including pre-1957 military service) unless the retired pay is waived or was (1) awarded because of a service-connected disability incurred either in combat with an enemy of the United States or caused by an instrumentality of war and incurred in the line of duty during a period of war, or (2) awarded under Reserve retirement provisions (10 USC, Chapter 1223).

An applicant for disability retirement who is receiving military retired or retainer pay (or Veterans' Administration pension or compensation in lieu of such military retired or retainer pay) under conditions requiring that it be waived in order to make the military service available for civil service retirement credit is not eligible for computation under the guaranteed minimum provisions of the law unless he or she waives the military retired or retainer pay (and renounces the Veterans' Administration pension or compensation, if any).

Q. Can military retired pay be waived so that the military service will be credited under the CSRS or FERS?

A. Yes.

SURVIVOR OPTIONS

Q. May the survivor of a deceased federal employee waive the military retired pay the employee was receiving at time of death?

A. No.

Q. What choice does an employee's survivor (widow, widower or dependent child) have between crediting military service toward retirement or towards social security?

A. With regard to military service before January 1, 1957, there is a choice. The survivor can choose to have military service used under the CSRS or FERS or credited toward the Social Security benefit. When a surviving spouse appears eligible for a survivor annuity, and the deceased employee had post-1956 military service for which deposit was not made or completed, the surviving spouse may elect to make or complete the deposit in order to have the military service credited under the CSRS or FERS. The employing agency is responsible for advising the survivor of his or her right to make the deposit and for counseling the survivor about the effect of the decision to make or not make the deposit. If the survivor decides to make (or complete) the deposit, the agency must collect the amount due in one lump sum.

Q. What is the effect of an election by a survivor to credit military service performed before January 1, 1957 toward Social Security rather than using it for retirement, and vice versa?

A. The survivor cannot receive any credit for the deceased employee's military service in any annuity under the retirement system, if he or she elects to credit such military service toward Social Security. The survivor may still be eligible to receive Social Security even though the election was made to use the military service for retirement if there is sufficient other covered employment, but no credit for the military service will be allowed in computing the Social Security benefit.

VETERANS' PREFERENCE IN GOVERNMENT EMPLOYMENT

Since the Civil War, veterans of the Armed Forces have been given some degree of preference in appointments to federal jobs. Recognizing their sacrifices, Congress enacted laws to prevent veterans seeking federal employment from being penalized for their time in military service. Veterans' preference recognizes the economic loss suffered by citizens who have served their country in uniform, restores veterans to a favorable competitive position for Government employment, and acknowledges the larger obligation owed to disabled veterans.

Veterans' preference is not so much a reward for being in uniform as it is a way to help make up for the economic loss suffered by those who answered the nation's call to arms. Historically, Congress has reserved preference for those who were either disabled or who served in combat areas. Eligible veterans receive many advantages in federal employment, including preference for initial employment and a higher retention standing in the event of layoffs. However, the veterans' preference laws do not guarantee the veteran a job, nor do they give veterans' preference in internal agency actions such as promotion, transfer, reassignment, and reinstatement.

Veterans' preference in its present form comes from the Veterans' Preference Act of 1944, as amended, and is now codified in various provisions of Title 5, United States Code. By law, veterans who are disabled or who served on active duty in the Armed Forces during certain specified time periods or in military campaigns are entitled to preference over others in hiring from competitive lists of eligibles and also in retention during reductions in force. Preference applies to hiring for virtually all jobs, whether in the competitive or excepted service.

PREFERENCE IN EMPLOYMENT AND RETENTION

Veterans' Preference in Hiring. Five-point preference is given to those honorably separated veterans (this means an honorable or general discharge) who served on active duty (not active duty for training) in the Armed Forces: (1) during any war (this means a war declared by Congress, the last of which was World War II); (2) during the period April 28, 1952, through July 1, 1955; (3) for more than 180 consecutive days, any part of which occurred after January 31, 1955, and before October 15, 1976; (4) during the Gulf War period beginning August 2, 1990, and ending January 2, 1992; (5) in a campaign or expedition for which a campaign medal has been authorized, such as El Salvador, Lebanon, Granada, Panama, Southwest Asia, Somalia, Haiti, Afghanistan, and Iraq; or (6) for more than 180 consecutive days, any part of which occurred during the period beginning on September 11, 2001, and ending on the date prescribed by Presidential proclamation or by law as the last date of Operation Iraqi Freedom.

Campaign medal holders and Gulf War veterans who originally enlisted after September 7, 1980, or began on active duty on or after October 14, 1982, without having previously completed 24 months of continuous active duty, must have served continuously for 24 months or the full period called or ordered to active duty.

Military retirees at or above the rank of major or lieutenant commander are not entitled to preference unless they qualify as disabled veterans.

Ten-point preference is given to those honorably separated veterans who:

1. qualify as disabled veterans because they have served on active duty in the Armed Forces at any time and have a present service-connected disability or are receiving compensation, disability retirement benefits, or pension from the military or the Department of Veterans Affairs;
2. are Purple Heart recipients;
3. are the spouse of a veteran unable to work because of a service-connected disability;
4. are the unmarried widow of certain deceased veterans; and
5. are the mother of a veteran who died in service or who is permanently and totally disabled.

When applying for federal jobs, eligible veterans should claim preference on their application or resume. Applicants claiming ten-point preference must complete form SF-15, *"Application for Ten-Point Veteran Preference."* Veterans who are still in the service may be granted five-point tentative preference on the basis of information contained in their applications, but they must produce a DD Form 214 prior to appointment to document entitlement to preference.

Note: Reservists who are retired from the Reserves but are not receiving retired pay are not considered "retired military" for purposes of veterans' preference.

The Department of Labor's Office of the Assistant Secretary for Policy and Veterans Employment and Training Service developed an "expert system" to help veterans receive the preferences to which they are entitled. This system helps veterans determine the type of preference to which they are entitled, the benefits associated with the preference and the steps necessary to file a complaint due to the failure of a federal agency to provide those benefits. To find out whether you qualify for veterans' preference, visit the veterans' preference e-laws website at www.dol.gov/elaws/vetspref.htm.

The Veterans Employment Opportunities Act of 1998 gives veterans access to federal job opportunities that might otherwise be closed to them. The law requires that: (1) agencies allow eligible veterans to compete for vacancies advertised under the agency's merit promotion procedures ***when the agency is seeking applications from individuals outside its own workforce;*** (2) all merit promotion announcements open to applicants outside an agency's workforce include a statement that these eligible veterans may apply; and (3) veterans who are selected be given career conditional appointments.

There is a redress system for preference eligibles and the law makes it a prohibited personnel practice for an agency to knowingly take or fail to take a personnel action if that action or failure to act would violate a statutory or regulatory veterans' preference requirement.

Preference in Retention During Layoffs. In Government reduction in force programs brought about by the economy or other factors, Congress has given employees with veterans' preference (except for certain retired military personnel) in the federal service job priority rights over certain nonveterans.

Career civil service preference-eligible veterans have job retention rights over all other federal workers in the same occupational series and grade at their place of employment.

Preference-eligible veterans with career-conditional tenure do not have job retention rights over nonveterans who have career civil service tenure. However, they do have retention rights over nonveteran workers with the same tenure. Preference-eligible veterans with Temporary Appointment Pending Establishment of a Register (TAPER) appointments have retention rights over nonveteran TAPERs.

Thus, between two persons who are doing similar work in identical positions and serving under the same tenure, the preference-eligible veteran is retained over the nonveteran. Preference-eligible veterans who are subject to an involuntarily reduction-in-force also have a wider range of retention rights than nonveterans.

Veterans' Preference Complaints and Redress. Veterans who believe that they have not been properly accorded their rights have several different avenues for complaints, depending upon the nature of the complaint and the individual's veteran status.

The Veterans Employment Opportunities Act of 1998, as amended, allows preference/other eligibles to complain to the Department of Labor's Veterans Employment and Training Service (VETS) when the person believes an agency has violated his or her rights under any statute or regulation relating to veterans' preference.

Veterans Recruitment Appointment (VRA)

The VRA is a special authority by which agencies can appoint an eligible veteran without competition. The VRA is an excepted appointment to a position that is otherwise in the competitive service. After two years of satisfactory service, the veteran is converted to a career-conditional appointment in the competitive service. **Note:** A veteran may be given a noncompetitive temporary or term appointment based on VRA eligibility. These appointments do not lead to career jobs.

Eligibility. To be eligible for a VRA appointment, a veteran must have served on active duty in the Armed Forces for more than 180 days and have been separated under conditions other than dishonorable. The 180-day requirement does not apply to veterans released from active duty because of a service-connected disability, or to members of a Reserve Component ordered to active duty during a period of war or in a campaign or expedition for which a campaign or expeditionary medal is authorized.

Public Law 107-288 changed the eligibility to include all qualified covered veterans. The term "covered veterans" means any of the following veterans: (1) disabled veterans; (2) veterans who served on active duty in the Armed Forces during a war or in a campaign or expedition for which a campaign badge has been authorized (for VRA eligibility, the term "period of war" includes the Vietnam era and the Persian Gulf War beginning August 2, 1990 and ending November 30, 1995, but does not include other operations such as Panama and Somalia); (3) veterans who, while serving on active duty in the Armed Forces, participated in a United States military operation for which an Armed Forces Service Medal was awarded pursuant to Executive Order 12985; and (4) recently separated veterans. It also removed the 10 year limit for covered veterans.

Terms and conditions of employment. VRA eligibles may be appointed to any position for which qualified up to GS-11 or equivalent (the promotion potential of the position is not a factor). The veteran must meet the qualification requirements for the position. (Any military service is considered qualifying for GS-3 or equivalent.) After two years of substantial continuous service in a permanent position under a VRA, the appointment will be converted to a career or career conditional appointment in the competitive service, providing performance has been satisfactory. Once hired, VRAs are treated like any other competitive service employee and may be promoted, reassigned, or transferred. VRA appointees with less than 15 years of education must complete a training program established by the agency.

VETERANS' EMPLOYMENT AND TRAINING

The U.S. Department of Labor (DOL) is committed to providing veterans and transitioning military personnel with the resources and services they need to succeed in the 21st century workforce. Special emphasis is placed on assisting disabled, recently separated, and veterans with barriers to employment so that they may find meaningful employment. Veterans receive priority in all DOL funded employment and training programs.

Veterans' Employment and Training Service (VETS)

The Veterans' Employment and Training Service is an agency within the DOL whose mission is to maximize employment opportunities for veterans and transitioning service members; to protect their employment rights, and to meet labor-market demands with qualified veterans.

A Director for Veterans' Employment and Training (DVET) is located in each state, Puerto Rico, and the District of Columbia to provide technical assistance and oversight of employment services offered to veterans throughout the One Stop Career Center System in each state. Disabled

Veterans' Outreach Program (DVOP) specialists and Local Veterans' Employment Representatives (LVER) staff provide direct services to veterans seeking employment. Questions can be addressed to the DVET located in each state whose telephone numbers can be located in the blue pages of local telephone directories, displayed under United States Government, Labor Department, Veterans' Employment and Training Service, or in the online directory available through the VETS website at: www.dol.gov/vets.

Employment Assistance. Most One Stop Career Center System offices provide veterans with priority of service in employment and training opportunities, including assessments, the provision of labor market information, job development, and referrals to job opportunities as needed. Special consideration is given to veterans with special needs whom may choose services tailored to their needs on a one-to-one basis. These offices can assist veterans in finding employment, job opportunities or unemployment compensation, and may be located in the blue pages of local telephone directories under State Government, Labor Department, Employment Security, Commerce Department or a related Department listing.

Disabled Veterans' Outreach Program (DVOP) specialist. DVOP specialists are usually located in the One Stop Career Centers. They are specialists who provide intensive services to veterans with special needs and whose primary focus is to assist eligible veterans who have employment barriers and may benefit from their specialized training and assistance.

Local Veterans' Employment Representative (LVER) staff. LVER staff are usually located in One Stop Career Center System offices as well. However, their primary function is to conduct outreach to employers and to promote veteran participation in the local civilian labor market. In addition, they facilitate the employment and training of veterans by conducting job search workshops, forming job search groups and by establishing a network of employers who value the discipline and training that veterans acquire as a result of their military service.

Veterans' Workforce Investment Program (VWIP). The statutory intent of VWIP is to support employment and training programs that meet the needs of veterans with service-connected disabilities, veterans who have significant barriers to employment, veterans who served on active duty in the Armed Forces during a war or in a campaign or expedition for which a campaign badge has been authorized, and to servicemembers who recently separated from military service (within 48 months).

The ultimate goal of this program is to assist veterans in obtaining gainful employment. VWIP allows for specialized employment, training to include licensing, credentialing, and various local, state, and federal educational resources that are customized and tailored to meet the specific needs of each veteran participant. In some cases veterans who are female, minority, economically disadvantaged, homeless, and/or disabled veterans may be targeted to receive these specialized resources.

VWIP grant funds are competitively awarded. There are currently 12 states in which VWIP grantees operate. These programs complement services generally provided by the One Stop Career Center System. Information about VWIP programs and links to job and training opportunities offered through the One-Stop Career Centers can be found in telephone directories under State Government, and the websites of each State's Public Labor Exchange or State Workforce Investment Board office.

Homeless Veterans' Reintegration Program (HVRP). This program focuses on providing customized employment and training services to homeless veterans with the goal of assisting their successful reintegration into the workforce and thereby breaking the cyclical patterns of homelessness and associated problems. HVRP addresses the problems facing homeless veterans by stimulating the development of effective service delivery systems addressing the complex issues of homelessness. HVRP grant funds are competitively awarded to mostly non-profit community organizations to provide job search, counseling, job placement assistance, remedial education, classroom and on-the-job training, and referral to support services. HVRP is designed to allow flexibility so that grantees may identify and address particular barriers that impact the participation of homeless individuals in the local workforce.

Veterans Affirmative Action. States' public employment service offices provide veterans priority in referrals to jobs for federal contractor job openings. Employment service staffs provide assistance to employers in meeting these program obligations. Employers with federal contracts of $25,000 or above are mandated to list job openings with the State Workforce Agency. In addition, Public Law 107-288, requires employers with federal contracts of at least $100,000 to take affirmative action on behalf of disabled veterans; veterans who served on active duty in the Armed Forces during a war or in a campaign or expedition for which a campaign badge has been authorized; veterans who, while serving on active duty in the Armed Forces, participated in a U.S. military operation for which an Armed Forces service medal was awarded pursuant to Executive Order 12985; and veterans who have separated from the U.S. military service within 36 months.

Transition Assistance Program (TAP). The Department of Labor, in conjunction with the Departments of Defense, Homeland Security, and Veterans Affairs, provides TAP Employment Workshops to transitioning military personnel and their spouses who are within 12 months of separation from active duty, or within 24 months of retirement from the military. These workshops provide information on job search skills, the labor market, employment and training opportunities and veterans' benefits. TAP is now available to servicemembers at most military bases in Europe, Japan, Korea and Guam. For additional information, contact your military installation's Family Support Center or the Transition Assistance office.

Work Opportunity Tax Credit (WOTC). The WOTC provides a tax credit to employers who hire individuals in nine targeted groups, which include qualified veterans and certain disabled veterans. For purposes of qualifying for the tax credit, employers should know the prospective employee is not considered a member of one of the targeted groups unless the individual is certified as such by a Public Labor Exchange or employment service agency. Applicants must submit IRS Form 8850, "*Work Opportunity Credit Pre-screening Notice and Certification Request*," and ETA Form 9061, "*Individual Characteristics Form.*" Information regarding the WOTC can be obtained from any One-Stop Career Center and at the Department of Labor, Employment and Training Administration (DOL/ETA) website at www.doleta.gov.

Recovery & Employment Assistance Lifelines (REALifelines). REALifelines is a comprehensive new program to provide individualized job training, counseling and re-employment services to each and every veteran seriously injured or wounded in the Global War on Terrorism. The initiative is a joint project of the U.S. Department of Labor and the Department of Defense. It creates a seamless, personalized assistance network to ensure that seriously wounded and injured servicemembers who cannot return to active duty are trained for rewarding new careers in the private sector. Dedicated professionals help each person discover his or her special interests and unique talents to find the right career path to suit his or her special needs. In addition to assisting wounded veterans, job training and employment services are also available to spouses in families that have suffered an active duty casualty, as well as to family members who have temporarily left their jobs to be with their loved ones during recovery.

e-VETS. VETS maintains a website which serves as an Internet gateway for hundreds of links dedicated to veterans' issues, programs, and veterans' benefits. The e-VETS website is www.dol.gov/elaws/evets.htm and is dedicated to assist veterans and their family members in sorting through the wide range of services and information available via the Internet. Search menus link to websites providing information on career assessment tools, job search skills and techniques, resume preparation tips, entrepreneurship, financial assistance, job opportunities, veteran's benefits, cost of living indexes, relocation, and many others. The e-VETS website is updated every 90 days to ensure the links are active and the information provided remains current and viable.

UNEMPLOYMENT COMPENSATION

Military personnel about to complete an active duty tour or career should be aware of the availability of an unemployment compensation program for ex-servicemembers. In the United States, the unemployment insurance program is based on a dual program of federal and state statutes. The Federal Social Security Act established the program in 1935. Much of the federal program is implemented through the Federal Unemployment Tax Act. Each state administers a separate unemployment insurance program within minimum guidelines established by federal statute.

A combination of federal and state taxes are levied upon employers to support the unemployment compensation systems. The proceeds from the unemployment taxes are deposited in an Unemployment Trust Fund. Each state has a separate account in the Fund to which deposits are made. The Federal Government provides funding for benefits for unemployed federal employees and ex-military personnel.

Unemployment Compensation for Ex-Servicemembers (UCX) is the benefit program for ex-military personnel to provide weekly income to meet basic needs while searching for employment.

Those who were on active duty within a branch of the U.S. military may be entitled to unemployment benefits based on that service. The military wages are assigned to the state where they first file a new claim after separation from active duty. They must meet the following requirements: the claimant must have been separated under honorable conditions, and must have completed a full term of service; or if released early, it must have been for a qualifying reason; or the claimant must have served on active duty as a member of a National Guard or Reserve Component continuously for 90 or more days.

UCX benefits are paid under the same conditions as benefits based on other employment. However, for claims purposes, military wages are determined by pay grade at the time of separation. A wage table furnished by the Federal Government, which shows the equivalent civilian wage for each military pay grade, is used for the determination.

Information the military furnishes about length of service and the reason for separation is considered final and binding. If any of this information is incorrect on the DD Form 214 or other military documents, it is the claimant's responsibility to contact the service or the Department of Veterans Affairs to have the information reviewed. The Unemployment Insurance office can help claimants with the process.

Conceptually, unemployment compensation is designed to provide benefits to most workers out of work due to no fault of their own for periods between jobs. Except in a few states where there are small employee payments, the system is financed by a payroll tax on employers. Most states now pay a maximum of 26 weeks.

In periods of very high unemployment in individual states, benefits may be extended. These extended benefits are funded on a shared basis, approximately half from state funds and half from federal sources. Almost all wage and salary workers are now covered by the federal-state system. State agencies take applications and administer payments.

Unemployment insurance pays benefits to qualified workers who are unemployed and looking for work. Unemployment compensation payments are intended to provide an unemployed worker time to find a new job equivalent to the one lost without major financial distress. Benefits are paid as a matter of right and are not based on need.

Military benefits cannot be paid for any week for which the claimant receives allowances from the Department of Veterans Affairs under Chapters 31 or 35 of Title 38, United States Code. (The chapters provide subsistence allowances for vocational rehabilitation training and educational assistance allowances to war orphans and widows, or, in some cases, children and spouses of permanent, totally disabled veterans.)

Benefits may be denied if, after military service, you quit a job without good cause, are fired for misconduct connected with your work or refuse a suitable job without good cause. Fines and/or imprisonment may result if intentional misinformation or falsification of facts is provided to state authorities for the purposes of obtaining benefits. If you discover any errors made when applying for benefits, notify the appropriate office immediately. Benefits may also be reduced or denied because of pay for unused leave, pension, severance pay, or retirement pay.

SIGNIFICANT STATE UNEMPLOYMENT INSURANCE PROVISIONS(1)
(As of July 2008)

State	Initial Waiting Period (weeks)	Weekly benefit amount for total unemployment		Benefit weeks of total unemployment	
		Minimum	Maximum	Minimum	Maximum(2)
Alabama	0	$45	$235	15	26
Alaska	1	44-68	248-320	16	26
Arizona	1	60	240	12	26
Arkansas	1	77	431	9	26
California	1	40	450	14	26
Colorado	1	25	431-475	13	26
Connecticut	0	15-30	501-576	26	26
Delaware	0	20	330	24	26
District of Columbia	1	50	359	19	26
Florida	1	32	275	9	26
Georgia	0	44	330	6	26
Hawaii	1	5	523	26	26
Idaho	1	58	364	10	26
Illinois	1	51-70	369-511	26	26
Indiana	1	50	390	8	26
Iowa	0	53-65	361-443	9	26
Kansas	1	105	423	10	26
Kentucky	0	39	415	15	26
Louisiana	1	10	284	26	26
Maine	1	60-90	344-516	14	26
Maryland	0	25-65	380	26	26
Massachusetts	1	32-48	600-900	10	30
Michigan	0	113-143	362	14	26
Minnesota	1	38	538	10	26
Mississippi	1	30	230	13	26
Missouri	1	35	320	8	26
Montana	1	120	407	8	28
Nebraska	1	30	298	14	26
Nevada	0	16	393	12	26
New Hampshire	0	32	427	26	26
New Jersey	0	85-97	560	1	26
New Mexico	1	66-99	355-455	26	26
New York	1	64	405	26	26
North Carolina	1	39	457	13	26
North Dakota	1	43	406	12	26
Ohio	1	103	365-493	20	26
Oklahoma	1	16	392	18	26
Oregon	1	113	482	3	26
Pennsylvania	1	35-43	539-547	16	26
Puerto Rico	1	7	133	26	26
Rhode Island	1	68-118	528-660	8	26
South Carolina	1	20	326	26	26
South Dakota	1	28	298	15	26
Tennessee	1	30	275	13	26
Texas	1	57	378	10	26
Utah	1	26	427	10	26
Vermont	0	64	425	26	26
Virginia	1	54	378	12	26
Virgin Islands	1	33	454	13	26
Washington	1	129	541	1	26
West Virginia	1	24	424	26	26
Wisconsin	0	53	355	12	26
Wyoming	0	30	415	11	26

(1) State laws may require that regular, as well as disability retirement payments to military personnel based on military service performed be deducted from unemployment compensation claims paid.
(2) Maximum based on military service is as shown above in state of residence.

NOTE: Some states may prorate or deny unemployment compensation for ex-servicemembers if receiving terminal or accrued leave pay or severance pay.

PART V

SURVIVOR BENEFIT PLAN

The Survivor Benefit Plan (SBP) is a Department of Defense insurance program that allows retired servicemembers to provide, upon death, a portion of their retired pay, in the form of an annuity, to a survivor. The program also provides automatic coverage for servicemembers who die in the line of duty, and certain servicemembers who die while eligible, but prior to making a program election. The SBP program applies to active duty and retired members of the Uniformed Services, including the U.S. Coast Guard, Public Health Service, and the National Oceanic and Atmospheric Administration.

In addition, within SBP is a program of coverage for Reserve Component members. The coverage—Reserve Component SBP (RC-SBP)—provides a Reserve Component member the opportunity to protect a Reserve Component annuity during the period of receiving notification of 20 years of qualifying service for a Reserve retirement (commonly referred to as the 20 year letter) and becoming eligible to receive retired pay at age 60 (or lower age, if eligible). Like SBP, there is a provision in RC-SBP to provide automatic coverage for certain servicemembers who die while performing inactive duty training and for Reserve Component members who die after completing the service required to receive Reserve Component retired pay, but before making an RC-SBP election.

Automatic Coverage. An active duty member who has a spouse or dependent child and is entitled to retired pay based on active service is automatically covered at the maximum level unless the member, with spouse concurrence if married, elects a lesser amount of coverage or no coverage. Such an election must be made prior to the first day the member becomes entitled to retired pay.

A Reserve Component member, upon being notified of eligibility to receive retired pay (20 year letter), is automatically covered at the maximum level unless the member, with spouse concurrence if married, elects a lesser amount of coverage or no coverage. Such an election must be made within 90 days of receiving the 20 year letter.

Annuities for Certain Deaths.

Active Duty Deaths. An annuity is payable to the survivors of a member who dies in the line of duty while serving on active duty. The annuity would also be payable if the death is not in the line of duty while serving on active duty, and the member had at least 20 years of active service.

The Secretary concerned may provide an optional annuity for child only SBP for an eligible active duty death that occurs after October 7, 2001.

The order of precedence for payment of a death on active duty annuity is as follows:

- to the member's former spouse if the member was required by court order or agreement to provide an annuity to the former spouse, or if the member voluntarily elected to provide an annuity to the former spouse.
- to the surviving spouse; or, if the Secretary concerned determines it to be more appropriate, to the surviving dependent children with respect to a death that occurs after October 7, 2001.
- to the surviving dependent children in the event there is no surviving spouse; the surviving spouse subsequently dies; or if the surviving spouse

loses eligibility due to remarrying prior to age 55, with respect to a death that occurs on or after September 10, 2001.

• to an insurable interest person, as determined by the Service Secretary, with respect to a death that occurs on or after November 24, 2003.

Reserve Component Deaths. An immediate annuity is payable to the survivors of a Reserve Component member who dies after completing the service required to receive retired pay (20 year letter) but before making an RC-SBP election; or with respect to deaths that occur on or after September 10, 2001, to the survivors of a member who dies from an illness or injury incurred or aggravated in the line of duty while performing inactive duty training.

The order of precedence for payment of an RC-SBP annuity for a reason described above is as follows:

• to the member's former spouse if the member was required by court order or agreement to provide an annuity to the former spouse, or if the member voluntarily elected to provide an annuity to the former spouse.
• to the surviving spouse.
• to the surviving dependent children in the event there is no surviving spouse or if the surviving spouse subsequently dies.

Eligibility to Provide an Annuity.

Active Duty Members. A member is eligible to elect SBP coverage (standard annuity) for a spouse, former spouse or dependent child upon retiring from active duty.

Reserve Component Members. A Reserve Component member is eligible to elect RC-SBP (Reserve Component annuity) for a spouse, former spouse, or dependent child upon notification of eligibility to receive retired pay (20 year letter). The member must elect one of the following options during the 90 day period after receiving the 20 year letter: (A) decline to make an election until attaining age 60 (or lower age, if eligible) and becoming entitled to receive retired pay; (B) elect coverage for an annuity to begin upon the member's death or upon date the member would become entitled to receive retired pay (age 60 or lower, if eligible), whichever is later; or (C) elect coverage for an annuity to begin upon the member's death, regardless of the member's age when death occurs.

Effective with notifications of eligibility to elect RC-SBP dated on or after January 1, 2001, a Reservist is automatically enrolled in option (C) at the maximum amount unless the member, with written spousal concurrence (if required), elects reduced coverage or not to participate in the plan within the 90 day period beginning on the date that the member receives notification of eligibility. A member who elected not to participate in the RC-SBP upon receiving the 20 year letter may elect SBP coverage upon becoming entitled to receive retired pay (age 60 or lower, if eligible).

A member, who does not have a spouse or dependent child when initially eligible to elect coverage, may elect coverage for a former spouse, or a person who has an insurable interest in the member.

Also eligible to elect coverage is: a member on the Temporary Disabled Retired List until the member is no longer eligible for disability retired pay; a member within the one year period after acquiring a spouse or dependent child in the case of an otherwise eligible member, but who does not have a spouse or child when originally eligible to elect coverage; any member who qualifies during an open enrollment period.

Coverage Available. There are several categories of beneficiaries that a member may choose from to provide an annuity under SBP and RC-SBP. The categories are:

Spouse Only. Eligibility for this category requires that a surviving spouse be a widow or widower who was married to a retiree at the time of his or her enrollment; or, if not married at the time of enrollment, was married to the deceased retiree for at least one year prior to the retiree's death; or, if not married at time of enrollment and was not married to the deceased retiree for at least one year prior to death, was the parent of issue by that marriage. Spouse coverage applies not only to the spouse a member has at time of enrollment, but also automatically to any subsequent spouse the member might acquire, unless the member elects to decline coverage for a subsequent spouse within one year of the date of marriage (concurrence of the subsequent spouse is not required, but that spouse will be notified of the member's declination).

Spouse and Child. SBP protection under this category is expanded to cover an eligible child or children if there is no surviving spouse, or if a surviving spouse subsequently dies or becomes ineligible to receive benefits due to remarriage before the age of 55. Thus, if there is a divorce or if the spouse dies before the retiree, the full SBP annuity will be paid to the eligible surviving child or children in the same manner as if the member had elected "child only" coverage.

Child Only. This option provides an annuity only for dependent children regardless of whether a member is married or not at time of enrollment (although a married member's spouse must concur with a child only election). Children remain beneficiaries until age 18, or age 22 if a full-time, unmarried student. Children mentally or physically incapable of self-support remain eligible, while unmarried, for as long as the incapacitation exists. A member with no dependent children at time of eligibility to elect coverage may elect coverage for the first child subsequently acquired, but the child must be added within one year of being acquired (born, adopted, etc.).

Former Spouse. A member who has a former spouse upon becoming eligible to elect a survivor annuity may elect coverage for a former spouse. If the member has more than one former spouse, the member must specify which former spouse is being covered. An election for a former spouse prevents payment of an annuity to a current spouse. A former spouse who was not a member's former spouse on the date a member became eligible to participate in SBP must have been married to the member for at least one year in order to be named as a former spouse beneficiary.

Former Spouse and Child. This category may be elected if the member and former spouse had a child as a result of their marriage. The same child eligibility rules for spouse and child coverage apply.

Person with an Insurable Interest. A member who does not have a spouse or dependent child when eligible to make a program election may elect to provide coverage for a person with an insurable interest in the member. The Department of Defense defines an insurable interest as "a natural person with an insurable interest who has a reasonable and lawful expectation of financial benefit from the continued life of the participating member, or any individual having a reasonable and lawful basis, founded upon the relation of parties to each other, either financial or of blood or affinity, to expect some benefit or advantage from the continuance of the life of the retired member." If the election is for a person who is more nearly related than a cousin, no proof of financial expectation is required.

An election for insurable interest coverage, for other than a dependent (as described in 10 USC 1072(2)), by a member who dies within one year after being retired due to a cause related to the disability for which retired, shall be voided and any premiums paid for that coverage will be paid to the person to whom the annuity would have been paid.

A member who discontinues coverage for an insurable interest beneficiary due to the death of that beneficiary, may elect a new insurable interest beneficiary within 180 days of the death of the previous beneficiary.

In addition to monthly premiums for the new beneficiary, the member must pay the difference between premiums already paid and what would have been paid had the new beneficiary been covered during the entire period of the coverage for the previous beneficiary up to the date of the new election, plus interest.

In addition, the member must live for at least two years from the date of the new election or the election is voided and any premiums paid for that coverage will be paid to the person to whom the annuity would have been paid.

Cost (Premiums). The cost for coverage varies depending on the category of coverage and level of coverage selected. Upon election, a member chooses a base amount. The base amount is a dollar amount selected by a member at the time of enrollment from which the member's monthly premium and the survivor's monthly annuity will be computed. A base amount may be any amount between a $300 minimum and the member's full gross retired pay. The base amount selected increases each year by the same percentage as the COLA for retired personnel. Effective October 1, 2008, a paid-up SBP provision becomes effective, and there will be no premiums due from a member who is at least 70 years of age and has paid for SBP coverage for at least 30 years.

Spouse/Former Spouse Coverage. The following table shows the cost for spouse or former spouse coverage for a standard annuity. The premium charge depends on certain factors, such as when the member joined the military (see table footnotes for detailed information). However, premiums for all base amounts $1,503 and greater (as of January 1, 2009) are always 6.5 percent of the base amount.

Spouse and Child/Former Spouse and Child Coverage. A member's premium for spouse and child coverage or former spouse and child coverage is calculated at the same rate as spouse or former spouse coverage, plus an additional charge for the child portion of the coverage. The amount of this additional charge is calculated using an actuarial factor determined by the combination of the ages of the member, the member's spouse, and the member's youngest child at time of enrollment. Typical premiums are about one-half of one percent of the base amount on average, but could be higher or lower depending on the difference in the age span between the parents and the youngest child. The additional premium charge for the child portion of the coverage terminates on the date the youngest child beneficiary loses eligibility to receive the annuity.

Base Amount of Retired Pay $	Monthly Payment for Surviving Spouse	SBP Premium* $	
	55%	2.5% of $701 plus 10% of remaining amount	6.5%
300.00	165.00	7.50	19.50
600.00	330.00	15.00	39.00
701.00	385.55	17.53	45.57
700.00	385.00	17.43	45.50
800.00	440.00	27.43	52.00
900.00	495.00	37.43	58.50
1000.00	550.00	47.43	65.00
1100.00	605.00	57.43	71.50
1200.00	660.00	67.43	78.00
1300.00	715.00	77.43	84.50
1400.00	770.00	87.43	91.00
1500.00	825.00	97.43	97.50
1502.14	826.18	97.64	97.64
1600.00	880.00		104.00
2000.00	1100.00		130.00
2500.00	1375.00		162.50
3000.00	1650.00		195.00
3500.00	1925.00		227.50

(Effective January 1, 2009)

*Premium amount in left column applies to members who entered service before March 1, 1990, any member receiving a disability retirement regardless of date entered, and any member providing for a Reserve Component annuity. Premiums for members with base amounts of $1,503 or greater retiring on or after January 1, 2009, are computed at 6.5 percent regardless of when they entered service.

Child Only Coverage. Premiums for child only coverage are calculated using an actuarial factor determined by a combination of the ages of the member and the member's youngest child at time of enrollment. Typical premiums are about 2.5 percent of the base amount, but could differ depending on the age span between the member and the youngest child. As with spouse and child/former spouse and child coverage, premium reductions in retired pay terminate on the date the youngest child beneficiary loses eligibility to receive the annuity.

Insurable Interest Coverage. Premiums for insurable interest coverage are 10 percent of the member's gross retired pay, plus an additional five percent for each full five years the beneficiary is younger than the member. The maximum premium is 40 percent of the member's retired pay. Premium reductions in retired pay continue as long as the beneficiary remains alive or until the member changes the election to cover a spouse and/or a child later acquired or requests termination of the insurable interest coverage. The annuity is 55 percent of retired pay, less the SBP premium.

Reserve Component Annuities. A member who elects a Reserve Component annuity does not make premium payments until age 60 (or lower age, if eligible) and entitlement for receipt of retired pay. At that time, the member pays an additional premium, in addition to the normal SBP costs, for the specified coverage that accounts for the period of coverage from the date of election until age 60 (or lower age, if eligible) when the member is entitled to receive retired pay. The additional premium does not terminate if the beneficiary loses eligibility.

Amount of Annuity. The amount of an annuity depends on several factors, such as whether the annuity is a standard annuity, Reserve Component annuity, or an annuity based on an active duty death, and whether the annuity is for a spouse or former spouse, dependent child, or insurable interest. Annuities increase each year at the same rate that the military member's retired pay would have increased. The following chart depicts the different annuity calculations:

		Spouse, Former Spouse, or Dependent Child 55 Percent of:	Insurable Interest
Active Duty Deaths	For death in the line of duty	the retired pay the member would have been entitled to on the day the member died based on 100 percent total disability	55 percent of the difference between the retired pay the member would have been entitled to on the day the member died based on 100 percent total disability and the sum of 10 percent of the retired pay plus an additional five percent of the retired pay for each full five years the designated beneficiary is younger than the member (maximum premium is 40 percent)
	For death not in the line of duty	the retired pay the member would have been entitled to on the day the member died based on years of creditable service for computation of retired pay	NOT APPLICABLE
Standard Annuity		the base amount selected by member	55 percent of the difference between the member's retired pay and the sum of 10 percent of the retired pay plus an additional five percent of the retired pay for each full five years the designated beneficiary is younger than the member (maximum premium is 40 percent)
Reserve Component Annuity	For death in the line of duty during inactive duty training (regardless of any previous election the member may have made), or during the 90 day period after receipt of 20 year letter	the retired pay the member would have been entitled to on the day the member died based on years of creditable service for computation of retired pay	NOT APPLICABLE
	For death after member elects RC-SBP coverage and after the 90 day period following receipt of 20 year letter	the difference between the base amount selected by member and the additional cost for Reserve Component coverage	A percentage of the retired pay that is less than 55 percent and which takes into account factors such as: age of the member when electing coverage; difference between age of member and elected beneficiary; whether the election was for an immediate or deferred annuity; actuarial tables, etc. The maximum reduction in retired pay is 40 percent plus the cost for Reserve Component coverage

Discontinuing Coverage. A member may elect to discontinue participation as follows:

Standard Annuity. During the period that is more than two years but less than three years after the first date of entitlement to receive retired pay. The member must submit the request to DFAS no earlier than the first day of the 25th month and no later than the last day of the 36th month from the date of entitlement to retired pay, with spouse concurrence (if applicable).

Reserve Component Annuity. During the period that begins on the second anniversary of commencement of retired pay and ends on the day before the third anniversary of commencement of retired pay. The member must submit the request to DFAS no earlier than the second anniversary and no later than the third anniversary, with spouse concurrence (if applicable).

A member who discontinues coverage for a Reserve Component Annuity continues to pay charges for the coverage provided, but not paid for, prior to becoming age 60 (or lower age, if eligible). Those costs continue until the earlier of: (1) the first day of the month in which a spouse or former spouse beneficiary is no longer eligible due to death, divorce, annulment or remarriage before age 55; (2) the first day of the month in which the insurable interest beneficiary dies; or (3) the first day of the month following the month in which the member's retired pay has been reduced for a total of 360 months. A member who elected child-only coverage continues to pay premiums until the retired pay has been reduced for a total of 360 months.

Insurable Interest Coverage. A member who has elected coverage for an insurable interest (not for a former spouse) may discontinue coverage at any time.

Members electing to terminate coverage are not eligible for continuation in the program; however, the member has 30 days after submitting a request to discontinue participation, to revoke such request. The effective date of the termination of coverage is the first day of the first month following the month in which the Secretary concerned receives the election.

Totally Disabled Members. A member who the VA has granted a service-connected, totally disabled rating and the member has suffered from that disability while so rated continuously for either 10 or more years, or at least five years from the member's date of last discharge or release from active duty may discontinue SBP or RC-SBP coverage. If the VA changes the rating of the member to less than totally disabled, the member may elect to resume coverage within the one year period immediately following the VA notifying the member of such a change. Coverage is limited to the type and amount of coverage initially elected, and premiums begin on the effective date that coverage is resumed.

REMARRIAGE AND ITS EFFECT ON THE SURVIVOR BENEFIT PLAN (SBP)

A retiree initially participating in SBP with either spouse or spouse and child coverage whose marriage ends in death, divorce, or annulment, has several options upon remarriage.

Upon notification to the finance service center of a change in marital status, SBP participation and costs are suspended, not terminated. If coverage was not converted to cover the former spouse (if applicable), then under the law, upon remarriage, the new spouse is automatically covered by SBP one year after the remarriage with the same level of coverage as the prior spouse. (An earlier date applies if a child is born of that marriage, or if the member remarries the former spouse.)

The options upon remarriage are: (1) resume spouse coverage at the same level and cost as was in effect for the first spouse (the "automatic" provision of the law); (2) elect not to resume spouse coverage, provided an election to do so is received by the finance center within one year after the remarriage (a member with spouse and child coverage who elects not to cover the new spouse will continue to have child only coverage and child only premiums will apply as long as there is a dependent child); and (3) increase the SBP base amount, if the previous election was for reduced coverage (the member must pay any difference, retroactive to the initial election, plus interest prior to the first year anniversary of the remarriage).

The important thing to remember is that the member must notify the finance center upon remarriage. If the finance center is not notified, the new spouse will be automatically covered under SBP, and the member will be responsible for SBP costs retroactive to one year after the remarriage.

DIC OFFSET

When a member of the uniformed services dies as a result of a service-connected cause, certain members of his or her family may be eligible for monthly DIC payments. (See section on DEPENDENCY AND INDEMNITY COMPENSATION.) Whenever a surviving spouse of an SBP participant is also entitled to DIC, the spouse's monthly SBP annuity is reduced by the amount of the DIC payment. The DIC payment is not taxable by the Federal Government. When the SBP annuity is greater than the amount of DIC, the total of the two payments, DIC and SBP, will be equal to the full amount of the SBP annuity. If the surviving spouse remarries and becomes ineligible for DIC, the full amount of SBP may be paid if the spouse is over age 55 when the remarriage occurs. When DIC is paid to the surviving spouse of an SBP participant, the dollar amount of the participant's SBP premiums that corresponds to the portion of the SBP annuity not payable are refunded to the beneficiary. This refund must be repaid for SBP eligibility to be restored if the surviving spouse loses DIC eligibility.

Special Survivor Indemnity Allowance. Beginning October 1, 2008, and ending on February 28, 2016, a surviving spouse who is entitled to DIC; is an SBP annuitant (although the SBP annuity may not be payable due to receipt of DIC); and whose SBP annuity is affected due to the receipt of DIC will receive a Survivor Indemnity Allowance as follows: for months during FY 2009: $50; for months during FY 2010: $60; for months during FY 2011: $70; for months during FY 2012: $80; for months during FY 2013: $90; and for all months beginning in FY 2014 and ending on February 28, 2016: $100. However, the maximum monthly payment of this allowance may not exceed the amount of the reduction in the SBP annuity due to receipt of DIC.

COLA ADJUSTMENTS

SBP premiums and survivor annuities are adjusted at the same time and by the same percentage as military retired pay. Military retired pay is adjusted annually at the same rate as the annual increase in Social Security Benefits. This is a significant advantage of SBP that is generally not offered by alternative insurance programs.

ESTIMATED SBP COSTS TO RETIREES AND BENEFIT PAYMENTS TO ANNUITANTS FOR FISCAL YEARS 1973-2008

FY*	Average Number of Retirees Making SBP Payments	Fiscal Year Cost to Retirees ($ in thousands)	Cumulative Cost to Retirees ($ in thousands)	Average Number of Beneficiaries Receiving Annuities	Fiscal Year Payments to Families ($ in thousands)	Cumulative Payments to Families ($ in thousands)
73	129,963	36,145	36,145	1,950	5,700	5,700
74	367,134	111,974	148,119	6,725	19,661	25,361
75	496,368	177,593	325,712	14,281	41,751	67,112
76	533,507	226,726	552,438	20,852	66,899	134,011
77	558,052	338,736	891,174	29,907	99,676	233,687
78	576,464	311,088	1,202,262	35,017	128,863	362,550
79	601,544	362,604	1,564,866	41,581	167,469	530,019
80	624,463	429,662	1,994,528	48,797	223,165	753,184
81	638,168	500,754	2,495,282	56,765	292,913	1,046,097
82	668,872	570,505	3,065,787	62,947	350,311	1,396,408
83	755,159	652,536	3,718,323	68,930	406,887	1,803,295
84	761,427	664,993	4,383,316	77,284	460,955	2,264,250
85	767,623	686,997	5,070,313	87,329	515,788	2,780,038
86	772,355	686,555	5,756,868	97,525	573,940	3,353,978
87	775,581	691,442	6,448,310	108,395	627,986	3,981,964
88	781,323	717,655	7,165,965	119,885	704,853	4,686,817
89	792,765	746,627	7,912,592	130,455	791,170	5,477,987
90	801,809	708,000	8,620,592	141,119	883,457	6,361,444
91	809,955	691,816	9,312,408	151,583	989,282	7,350,726
92	824,037	722,157	10,034,565	162,139	1,083,409	8,434,135
93	863,204	822,955	10,857,520	172,425	1,177,185	9,611,320
94	886,662	878,499	11,736,019	182,484	1,272,172	10,883,492
95	899,868	897,274	12,633,293	192,480	1,365,805	12,249,297
96	911,124	916,559	13,549,852	203,149	1,466,733	13,716,030
97	916,860	936,764	14,486,616	212,726	1,570,632	15,286,662
98	922,183	951,813	15,438,429	221,154	1,652,227	16,938,889
99	923,481	955,618	16,394,047	229,292	1,720,042	18,658,931
00	919,121	968,357	17,362,404	237,242	1,802,503	20,461,434
01	920,934	994,743	18,357,147	241,609	1,917,434	22,378,868
02	919,401	1,012,177	19,369,324	254,332	2,024,308	24,403,176
03	921,117	1,054,718	20,424,042	254,124	2,050,771	26,453,947
04	926,081	1,077,290	21,501,332	264,239	2,145,596	28,599,543
05	935,824	1,101,562	22,602,894	269,831	2,224,890	30,824,433
06	946,732	1,158,982	23,761,876	275,633	2,560,532	33,384,965
07	964,414	1,207,162	24,969,038	279,947	2,886,828	36,271,793
08	979,777	1,253,157	26,222,195	283,318	3,239,325	39,511,118

*Fiscal Years 1973-76 are July to June. Fiscal Year 77 is July to September (15 months). Fiscal Years 1978-2003 are October to September.

NOTES:

For Fiscal Years 73-79, only end month data was available. The estimate is the average of the end month and the previous years end month multiplied by 12.

For Fiscal Years 80-89 and 91-02, costs and counts of people are estimated by multiplying the midpoint of the Fiscal Year (March) by 12.

For Fiscal Year 90, legislation significantly reducing SBP costs to retirees became effective March 1, 1990. Figures for Fiscal Year 90 were estimated by multiplying the December numbers by 5 and the June numbers by 7 and adding.

TAX TREATMENT

Since SBP premiums are paid in the form of reductions in a participant's retired pay, they are not counted as taxable income. This tax break is especially important when comparing SBP to life insurance alternatives. For example, if a member's SBP premium is $100 a month and the member is in the 28 percent tax bracket, the member's actual out-of-pocket cost for SBP coverage is only $72, i.e., if the member elected to decline SBP participation and receive the $100 instead, the member would only get to keep $72 after paying taxes on the $100. Therefore, $72 is the amount that one should use in comparing the cost of SBP to the cost of a comparable amount of life insurance. Studies show that, except in unusually rare circumstances, no life insurance policy can guarantee equal or greater protection for equal or less cost than SBP.

A recipients SBP annuity is considered taxable income for federal income tax purposes. Some states exempt SBP annuities in whole or in part from taxable income for state income tax purposes. (See STATE INCOME TAX section.)

SOCIAL SECURITY

There are more than 1,300 Social Security offices located throughout the 50 states, Puerto Rico and the District of Columbia. Any one of these offices can give you information on Social Security matters when you visit or call by telephone. You can get the street address and telephone number of your nearest Social Security office from a telephone directory (listed under U.S. Government, Social Security Administration), from the Social Security website (www.ssa.gov), or from the U.S. Post Office. Calling the Social Security office before visiting can save a lot of time. You may be able to transact your business completely by phone. The Social Security Administration has a nationwide toll-free number, 1-800-772-1213. Social Security representatives are available from 7:00 a.m. - 7:00 p.m. on workdays. The best times to call are between 7:00 a.m. - 9:00 a.m. and between 5:00 p.m. - 7:00 p.m.

For help in planning your retirement or learning about benefits that may be available for you and your family, go to the benefits planner at www.ssa.gov/planners. You can also apply for retirement benefits on the Internet by going to www.ssa.gov/applytoretire.

Before you or your family can receive any Social Security benefits, you must apply for them. Do not delay in filing a claim if you are retiring or you are a spouse or surviving spouse. Benefits cannot be paid retroactively in most cases, but if retroactive benefits are available, the limit is 12 months. Contact any Social Security office if:

1. You are unable to work because of an illness or injury that has lasted or is expected to last a year or longer, or is expected to result in death.
2. You are within three months of age 62 or older and plan to retire.
3. You are within three months of age 65 even if you don't plan to retire. A delay in applying for monthly benefits can cause loss of some benefits.
4. A worker in your immediate family dies.
5. You or a dependent needs dialysis treatments or a kidney transplant because of a permanent kidney failure.

Make sure that your spouse or another member of your family knows to contact the Social Security office should any of these incidents occur.

PAYMENTS INTO SOCIAL SECURITY TRUST FUNDS

Since 1957, if you had military earnings for active duty service (including active duty for training), you paid Social Security taxes on those earnings and they are on your pay record. Inactive duty service in the Reserve Components (such as weekend drills) has been covered by Social Security since 1988. People who served in the military before 1957 did not pay into Social Security directly, but when they apply for benefits, their records are credited with special earnings for Social Security purposes that count toward any benefits that may be payable.

Under certain circumstances, special earnings for periods of active duty after 1957 can be credited to your military pay record for Social Security purposes. These extra earnings may help you qualify for Social Security or increase the amount of your Social Security benefit.

SPECIAL MILITARY SERVICE CREDITS

Prior to January 2002, special extra earnings were credited to military servicemembers. Military service in calendar year 2002 and future years no longer qualifies for these special extra earnings. Following is how military service credits apply for periods prior to January 2002.

Service from 1978 through 2001. For every $300 in active duty basic pay, you are credited with an additional $100 in earnings up to a maximum of $1,200 a year. If you enlisted after September 7, 1980, and did not complete at least 24 months of active duty or your full tour, you may not be able to receive the additional earnings. Check with the Social Security office for details.

Credits for Service in 1957 through 1977. You are credited with $300 in additional earnings for each calendar quarter in which you received active duty basic pay.

Service in 1940 through 1956. If you were in the military during this period, including attendance at a service academy, you did not pay Social Security taxes. However, under the following circumstances, your Social Security record may be credited with $160 a month in earnings for military service from September 16, 1940 through December 31, 1956:

- you were honorably discharged after 90 or more days of service, or you were released because of a disability or injury received in the line of duty; or
- you are still on active duty; or
- you are applying for survivors benefits and the veteran died while on active duty.

You cannot receive these special earnings credits if you're already receiving a federal benefit based on the same years of service. But there is one exception to this rule: if you were on active duty after 1956, you can still get the special earnings for 1951 through 1956, even if you're receiving a military retirement based on service during that period.

The maximum wage base on which Social Security taxes are withheld is $106,800 in 2009. There is no wage base limitation for Medicare tax; all covered wages are subject to Medicare tax. The Social Security tax is 6.2 percent. The Medicare tax is 1.45 percent. IRS Form W-2 (Withholding Tax Statement), prepared for each servicemember at the end of each calendar year, shows the amount of FICA tax withheld during the year. Where a nonpay status extends to any full calendar month, no FICA tax is deducted for that month.

Since military chaplains are considered employees of the Federal Government and not self-employed ministers, FICA tax is withheld from their base pay in the same manner as from base pay of other servicemembers. In addition, drill pay earned while on inactive duty training as a member of the Reserve Forces is subject to FICA taxes. Reserve and National Guard drill pay is subject to the same 7.65 percent withholding as for active duty members.

Social Security and Medicare taxes are not withheld from military allowances such as those for subsistence, uniforms, and housing, nor are these allowances counted as earnings for Social Security purposes. Hazardous duty and other special pays also do not count for FICA tax or earnings purposes.

Table 1 shows the FICA tax rates in effect from 1976 to present. Like civilian employers, the DoD as the servicemember's employer, matches the taxes contributed by its employees.

TABLE 1. FICA TAX CONTRIBUTION RATE SCHEDULE

Percent of Covered Earnings			
Years	Retirement, Survivors, and Disability Insurance Benefits	For Hospital Insurance	Total %
1976-77	4.95	0.90	5.85
1978	5.05	1.00	6.05
1979-80	5.08	1.05	6.13
1981	5.35	1.30	6.65
1982-83	5.40	1.30	6.70
1984	5.70	1.30	7.00
1985	5.70	1.35	7.05
1986-87	5.70	1.45	7.15
1988-89	6.06	1.45	7.51
1990 - Present	6.20	1.45	7.65

Military earnings and taxes were reported quarterly by the services through 1977 but are now reported annually. Periodically, funds to cover Social Security and Medicare taxes on the base pay of all military personnel are transferred by DoD to three trust funds set up for financing the retirement, survivors, disability and health insurance programs. One fund is used to finance the retirement and survivors insurance programs; a second is used to finance the disability insurance program; and a third, the Hospital Insurance Trust Fund, which began in 1966. The Social Security Administration administers the federal old age, survivors, and disability insurance programs. The health insurance program is administered by the Health Care Financing Administration, under the Department of Health and Human Services.

SOCIAL SECURITY BENEFITS AVAILABLE TO YOU

Old-Age, Survivors, and Disability Insurance (OASDI) benefits are paid to: (1) a retired insured worker age 62 or over; (2) eligible survivors and dependents of retired or disabled workers; and (3) a disabled insured worker under full retirement age. In addition, a lump sum death payment may be payable upon the death of the insured worker.

The purpose of Social Security benefits is to provide a partial replacement of earnings for the servicemember, or civilian worker, and his or her family when family income is reduced or stops because of the servicemember's or worker's retirement, disability, or death. The amount of monthly benefits is related to earnings covered under Social Security of the insured person and provides replacement for family income lost when one of these events happens.

The provisions of the Social Security law are of vital interest to servicemembers and veterans. This is true for both the servicemembers and veterans who have

Social Security insurance protection through taxes withheld from military pay after 1956 and for veterans who earned Social Security wage credits during active duty in the U.S. Armed Forces before and after World War II.

MEDICARE

Medicare is the federal health insurance program for Social Security recipients. There are two parts of the program that pay for different services. Part A (Hospital Insurance) helps pay for inpatient care in a hospital or skilled-nursing home and for home health and hospice care. Protection is also available for people who have chronic renal (kidney related) disease and to those receiving or are entitled to receive cash benefits for 24 months under the Social Security program because they are disabled. However, disabled beneficiaries with Amyotrophic Lateral Sclerosis (ALS) do not have to receive benefits for 24 months. If you are working, Part A can supplement your employer's health plan. Part B (Medical Insurance) helps pay for doctor's services, outpatient hospital care, and a number of other medical services and supplies.

Although most people do not pay a monthly premium for Part A coverage, enrollees do pay deductible and co-insurance amounts. Part B enrollees pay a monthly premium plus the deductible and co-insurance amounts. The premium amounts are set each year by law. The Government provides more than two-thirds of the cost of the program. This makes this plan affordable for older members. Usually, the premium increases if enrollment is delayed beyond age 65. Contact your local Social Security office or call toll-free 1-800-772-1213 for more details.

SOCIAL SECURITY COVERAGE

To be eligible for Old-Age, Survivors, Disability and Health Insurance (OASDHI) benefits, a person must have "insured status." "Insured" means that the person involved must have worked and earned Social Security credits from military service or in civilian wage-employment, or self-employment, covered by the Social Security Act for a specified amount of time, depending on the individual's age. The Social Security Administration determines if you are fully insured, currently insured, or insured for disability.

CREDITS NECESSARY FOR RETIREMENT AND SURVIVOR BENEFITS

To be fully insured, the servicemember must have at least one Social Security credit for each calendar year after 1950, or in each year after the year he or she attains age 21 if later, and prior to the year of attainment of age 62, death or disability. Table 2 shows the number of Social Security credits necessary for fully insured status.

Certain benefits can be paid when the servicemember is "currently insured," but not fully insured. You will be currently insured if you have at least six Social Security credits—through either military service or civilian work— during the 13 quarters (three years and three months) period ending with the quarter of death. If you have earned the necessary Social Security credits, your spouse and dependent children are eligible to get monthly benefits following your death regardless of your age at death.

If you have credit for at least 20 quarters within the 10 years ending with the quarter of disability and are fully insured, you are insured for Social Security disability benefits if you should become disabled at any time before you reach age 64½. Workers disabled before age 31 need less credit, and workers disabled by blindness need only to be fully insured.

Once a person has earned 40 Social Security credits—from military service, covered civilian work, or a combination of both—he or she is "permanently insured" for retirement and survivors insurance, but for disability, the current coverage as described above is also needed.

TABLE 2. CREDITS (QUARTERS OF COVERAGE) NEEDED TO BE FULLY INSURED

Year in which a worker reaches age 62 or dies	Will need this number of Social Security credits	Year in which a worker reaches age 62 or dies	Will need this number of Social Security credits
1975	24	1984	33
1976	25	1985	34
1977	26	1986	35
1978	27	1987	36
1979	28	1988	37
1980	29	1989	38
1981	30	1990	39
1982	31	1991 & later	40
1983	32		

DISABILITY BENEFITS

Extent of Disability and Amount of Work Needed

If you become so severely disabled that you are retired from the service and are not able to do other substantial work, you and your dependents may receive disability Social Security benefits if you qualify.

To be eligible for disability benefits, a person must have a disability that is so severe it, in the words of the law, makes him or her unable to "engage in any substantial gainful activity." It must be a physical or mental impairment that will show up in medical tests and examinations and one that has continued or is expected to continue for at least 12 consecutive months or result in death. A waiting period of five months after the disability began is required before benefits begin. However, a disability claim may be filed as soon as the disability occurs.

Servicemembers should bear in mind that disability has a specific meaning under the Social Security law. Thus, the fact that a servicemember is entitled to payments for "total disability" from another Government agency does not mean that, in every case, the person will be found eligible for disability benefits under Social Security. These benefits, however, may be received by eligible persons in addition to military disability payments based on service since 1956.

Disabled Widow or Widower

A disabled widow or widower of an eligible servicemember including a widow or widower who is divorced may begin to receive benefits as early as age 50.

Disabled Child

When a former servicemember receives retirement or disability insurance benefits or when he or she dies, Social Security benefits may also be paid to a child age 18 or over and continue to any age if the child was disabled before he or she reached age 22. See Table 3 for the amount of work needed and dependents who can receive disability benefits.

TABLE 3. SOCIAL SECURITY BENEFITS AND "INSURED STATUS" NEEDED

RETIREMENT PAYMENTS	
Monthly payments to—	If you are—
You as a retired worker	Fully insured.
And monthly payments to your—	
Spouse (or divorced spouse, if married for at least 10 years) 62 or over	Fully insured.
Dependent child (under 18 or any age if disabled before age 22)	Fully insured.
Student child 18 - 19 if full time student at a secondary school	Fully insured.
Spouse (regardless of age) if caring for entitled child under age 16 or disabled.	Fully insured.
SURVIVOR PAYMENTS	
Monthly payments to your—	If at death you were—
Widow or widower 60 or over	Fully insured.
Widow or widower or divorced spouse (regardless of age) if caring for child	Either fully or currently insured.
Disabled widow or widower 50-59	Fully insured.
Dependent child (under 18 or any age if disabled before age 22)	Either fully or currently insured.
Dependent parent (mother or father 62)	Fully insured.
Divorced spouse 60 or older, or spouse 50-59 and disabled (if married for at least 10 years)	Fully insured.
Lump-sum payment to your widow or widower, if living with you in the same household; otherwise can go to a spouse or child who was eligible for monthly benefits for the month of death.	Either fully or currently insured.
DISABILITY PAYMENTS	
Monthly payments to—	If you are—
You and your dependents* if you are totally disabled for work	Fully insured and have 20 credits out of the 40 calendar quarters ending with the one in which you became disabled.**

*Children under 18; children over 18, up to 22 in some cases, if a full time student at a secondary school; a disabled child 18 or older (who became disabled before 22nd birthday); a spouse at any age if caring for child or children under 16 or disabled and entitled to benefits; a husband or wife at age 62 whether or not a child entitled to benefits is in care.

**If you are under 31, you need fewer than 20 quarters of credit, depending on when your disability began.

HOW MEMBERS EARN SOCIAL SECURITY CREDIT

As you work and pay Social Security taxes, you earn up to a maximum of four "credits" for each year.

Credits determine whether you have at least the minimum amount of work needed to qualify for Social Security benefits. You earn credits by working in jobs covered by Social Security or by operating your own business as a self-employed person.

Before 1978, when employers reported employee wages every three months, the Social Security Administration called credits "quarters of coverage." At that time, employees received a quarter of coverage or credit if they earned at least $50 in wages in a three month calendar quarter.

Since 1978, employers have been reporting earnings on an annual basis. Credits are now based on the total wages and self-employment income during the year, no matter when the employee actually works. The employee may work all year to earn four credits, or may earn enough for all four in a much shorter length of time.

The amount of earnings it now takes to earn a credit changes each year. In 2008, employees earned one credit for each $1,050 of earnings. Employees who earned at least $4,200 during the year in 2008 are credited with the maximum four credits. In 2009, one credit is earned for each $1,090 of earnings.

Most people need 40 credits (10 years of work) to qualify for benefits. Younger people need fewer credits to be eligible for disability benefits or for their family to receive survivor benefits if they die.

An employee will probably earn more credits in their lifetime than the minimum number needed to be eligible for benefits. These extra credits do not increase the Social Security benefit amount; it is the average earnings over the employee's working years that determine how much the monthly benefit payment will be.

AMOUNTS OF SOCIAL SECURITY BENEFITS

Amounts of benefits payable on your Social Security record are determined from your average yearly taxable earnings over a certain period of time. Social Security benefits are computed by two methods of averaging earnings depending on when you are first eligible or the date you die. Usually, the basic amount payable is determined on the basis of the earnings adjusted to reflect earlier earnings in terms of the current dollar value. The Social Security Administration will explain how the earnings are indexed; the averaging period and how benefits are computed. On the Social Security records of most people, this period is from January 1, 1951 (or January 1 of the year of the 22nd birthday for those who have reached 21 since 1950), through the year before the person reaches 62 (with some adjustment for men reaching 62 before 1975), becomes disabled, or dies. Amounts of benefits payable on your record will depend on your average earnings over a certain number of these years. The lowest five years of earnings in the number of years used are excluded in figuring the average.

TABLE 4. PERCENTAGES FOR DEPENDENT AND SURVIVOR PAYMENTS

When you retire, become disabled, or die, monthly benefits can be paid to your—	This % times your monthly amount (before any reduction for age)*
Spouse or divorced spouse 62 or older, or spouse at any age with your entitled child (under 18 or disabled) in care[1 2]	50%
Child under 18 or 18-19 who qualifies as a student, or who is over 18 and is disabled before age 22	50% (if you are alive) 75% (if you are deceased)
Widow(er) or surviving divorced spouse who is at least age 60, or disabled and age 50-59	100%[1 2]
Child(ren)'s mother or father or surviving divorced mother or father	75%
First dependent parent	82½%
Second dependent parent (if two parents are entitled at the same time, each would receive 75%)	75%
[1] A divorced spouse must be 62 even with a child in care	[1]Reduced for months of entitlement before age 65; not reduced below 75%
[2] A divorced spouse must have been married to the worker for at least 10 years	[2]If the widow(er) or surviving divorced spouse has your entitled child in care. *Limited by a maximum amount payable to the family.

In every case, the benefit amount due to the insured person is figured when he or she applies for Social Security retirement or disability benefits, or when his or her family members file claims for survivor benefits. All payments to dependents or survivors are based on this amount. Table 4 shows the percentages used in figuring various dependent and survivor payments.

Since military personnel usually complete active service prior to age 62, it will probably be several years before you file for Social Security retirement benefits. Shortly before you reach retirement age, when you file your Social Security claim, the Social Security Administration will figure the exact amount of your benefit, based on all applicable earnings and the benefit schedule at that time. You can estimate the amounts of disability and survivor benefits that could be paid to you based on your Social Security record.

If you want to know the approximate amounts of benefits on your record, see ESTIMATING RETIREMENT BENEFITS.

REDUCED BENEFITS AS EARLY AS AGE 62

The Social Security law allows both men and women to apply for Social Security retirement, spouse, and divorced spouse benefits as early as age 62, (60 for widows, surviving divorced spouses, and widowers, or age 50 for surviving spouses and surviving divorced spouses if disabled).

When retirement benefits based on his or her Social Security record are paid to a former servicemember before Full Retirement Age (FRA), the amount of the benefit is permanently reduced. Benefits paid to a spouse of a former servicemember before he or she reaches FRA are also reduced, unless that spouse has care of the servicemember's entitled child under age 16, or a child disabled before age 22.

For a person whose FRA is 65, the reduction at age 62 is 20 percent, 13.33 percent at age 63, and 6.66 percent at age 64. (A retiree entitled to $8,500 a year at FRA would receive $6,800 a year at age 62.)

To reach an informed decision on when to retire, you should determine:

1. Whether you are fully insured.
2. How much earnings have been credited to your Social Security record.
3. What the Social Security Administration projects your estimated Primary Insurance Amount (PIA) will be at FRA. Your PIA is the monthly Social Security Payment that you would be entitled to at FRA based on your total earnings history.

To obtain a benefit estimate, you can call toll-free 1-800-772-1213 and request SSA Form 7004. After you complete the SSA Form 7004 and return it to Social Security, you should receive a detailed Social Security Statement within four to six weeks. Table 5 shows full retirement age and reductions that apply for accepting early benefits.

As a general rule, early retirement will give you about the same total Social Security benefits over your lifetime, but in smaller amounts to take into account the longer period you will receive them.

There are advantages and disadvantages to taking your benefit before your full retirement age. The advantage is that you collect benefits for a longer period of time. The disadvantage is that your benefit is permanently reduced. Each person's situation is different, so make sure you contact Social Security before you decide to retire.

TABLE 5

Social Security Full Retirement and Reductions by Age				
			At Age 62**	
Year of Birth*	Full Retirement Age	Age 62 Reduction Months	A $1,000 retirement benefit would be reduced to	The retirement benefit is reduced by***
1943-1954	66	48	$750	25.00%
1955	66 and 2 months	50	$741	25.83%
1956	66 and 4 months	52	$733	26.67%
1957	66 and 6 months	54	$725	27.50%
1958	66 and 8 months	56	$716	28.33%
1959	66 and 10 months	58	$708	29.17%
1960 and later	67	60	$700	30.00%

*If you were born on January 1, refer to the previous year.
**If you were born on the 1st of the month, figure the benefit as if your birthday was in the previous month. You must be at least 62 for the entire month to receive benefits.
***Percentages are approximate due to rounding.

BENEFITS MAY BE TAXABLE

Some people who receive Social Security benefits may have to pay income taxes on them. This will apply to individuals who have other substantial income in addition to Social Security benefits (for example, wages, self-employment, interest, dividends and other taxable income that you have to report on your tax return). No one pays taxes on more than 85 percent of his or her Social Security benefits and some pay on a smaller amount, based on these IRS rules:

If you file a federal tax return as an "individual" and your combined income is between $25,000 and $34,000, you may have to pay income tax on 50 percent of your Social Security benefits. If your combined income is above $34,000, up to 85 percent of your Social Security benefits is subject to income tax. On your 1040 tax return, your "combined income" is the sum of your adjusted gross income, plus nontaxable interest, plus one-half of your Social Security benefits.

If you file a joint return, you may have to pay taxes on 50 percent of your benefits if you and your spouse have a combined income that is between $32,000 and $44,000. If your combined income is more than $44,000, up to 85 percent of your Social Security benefits is subject to income tax.

If you are married and file a separate tax return, you probably will pay taxes on your benefits.

Every January individuals receive a Social Security Benefit Statement (Form SSA-1099) showing the amount of benefits received in the previous year. You can use this statement when you complete your federal income tax return to find out if your benefits are subject to tax.

Although you are not required to have federal taxes withheld from your Social Security benefits, you may find it easier than paying quarterly estimated tax payments.

If you have taxable income, two Internal Revenue Service publications may be helpful: IRS Publication 554, "*Tax Benefits for Older Americans;*" and IRS Publication 915, "*Tax Information on Social Security (and Equivalent Railroad Retirement) Benefits.*" These and other publications are available at local Internal Revenue Service offices or at www.irs.gov/formspubs/index.html.

BIGGER BENEFITS BY ADDITIONAL WORK

When an insured person works after he or she is eligible for retirement benefits, the amount he or she can receive in benefits will depend upon how much he or she earns in employment or self-employment. Beginning with the month he or she reaches full retirement age however, a person can receive all benefits due him or her regardless of how much he or she earns or works each year.

If you return to work after you start receiving retirement benefits, your added earnings may result in higher benefits. Social Security will automatically refigure your benefit after the additional earnings are credited to your record. Only earned income is subject to Social Security payment. Retirement pay is not earned income.

In addition, a special credit can mean a larger benefit. For each month past 65 that you delay receiving retirement benefits, a credit increases the amount of your benefit. These increases, known as delayed retirement credits, apply up to the month you are 70.

The rate of delayed retirement credits depends on a person's year of birth, as shown in the following table.

CHART OF DELAYED RETIREMENT CREDIT RATES

Attain Age 65	Monthly Percentage	Yearly Percentage
Prior to 1982	1/12 of 1%	1%
1982-1989	1/4 of 1%	3%
1990-1991	7/24 of 1%	3.5%
1992-1993	1/3 of 1%	4%
1994-1995	3/8 of 1%	4.5%
1996-1997	5/12 of 1%	5%
1998-1999	11/24 of 1%	5.5%
2000-2001	1/2 of 1%	6%
2002-2003	13/24 of 1%	6.5%
2004-2005	7/12 of 1%	7%
2006-2007	5/8 of 1%	7.5%
2008 or later	2/3 of 1%	8%

Survivor benefits are payable to dependent parents at age 62, surviving spouses and surviving divorced spouses at age 60, and disabled surviving spouses and surviving divorced spouses as early as age 50. Benefits to them, however, are reduced because of age under 65. Parents' benefits are not reduced for age.

EARNED INCOME AFTER SOCIAL SECURITY STARTS

If you go back to work and are under Full Retirement Age (FRA), your earnings may affect your Social Security benefits. However, you don't have to stop working completely to get Social Security benefits. You can receive all benefits if your earnings do not exceed the annual exempt amount. Retirement pay is not earned income.

The Senior Citizens' Freedom to Work Act of 2000, eliminated the retirement earnings test for beneficiaries reaching FRA. The following chart shows the earnings test exempt amounts for those under FRA.

Retirement Earnings Test Exempt Amounts:	2008	2009
Under full retirement age **Note:** One dollar in benefits will be withheld for every $2 in earnings above the limit.	$13,560/yr ($1,130/mo)	$14,160/yr ($1,180/mo)
The year an individual reaches full retirement age **Note:** Applies only to earnings for months prior to attaining full retirement age. One dollar in benefits will be withheld for every $3 in earnings above the limit.	$36,120/yr ($3,010/mo)	$37,680/yr ($3,140/mo)
There is no limit on earnings beginning the month an individual attains full retirement age (66 for those born from 1943 to 1954).		

Note: Different rules apply to work performed by people getting benefits because they are disabled. For more information, ask for a copy of the pamphlet, "*If You Become Disabled,*" at any Social Security office.

If you are receiving retirement checks, your earnings may affect your dependent's checks as well as your own. If you get checks as a dependent or survivor, your earnings can affect only your own check.

THE LUMP-SUM DEATH PAYMENT

Upon the death of an insured person, a one-time lump-sum payment may be made. It is paid in addition to any monthly benefits due to survivors. The lump-sum payment is made to the widow or widower if she or he was living in the same household as the insured person at the time of his or her death, or is entitled to or eligible for widow(er)'s benefits for the month of death. Otherwise, the lump sum is paid to a child who was entitled to or eligible for monthly benefits for the month of death. The lump-sum payment is $255.

ESTIMATING RETIREMENT BENEFITS

The following table, although it does not reflect the specific amounts you will receive, gives you some idea of amounts of maximum and average payments as of January 1, 2009.

EXAMPLES OF MONTHLY PAYMENTS

Benefit Category	Amounts
I. Maximum Social Security benefit for workers retiring at full retirement age in 2009 (Age 66)	$2,323
II. Average Social Security benefit	
All retired workers	$1,153
Aged couple, both receiving benefits	$1,876
Widowed mother and two children	$2,399
Aged widow alone	$1,112
Disabled worker, spouse, and children	$1,793
All disabled workers	$1,064
III. Maximum Federal SSI payments	
Individual	$674
Couple	$1,011

SOCIAL SECURITY FINANCING SCHEDULE CALENDAR YEARS 1937-2009[1]

Calendar Year	Maximum annual taxable earning $	Tax Rate: Employer and Employee, Each			Maximum Employee OASDI & HI Tax $
		OASDI* %	HI** %	Total %	
1937-49	3,000	1.0	—	1.0	30.00
1950	3,000	1.5	—	1.5	45.00
1951-53	3,600	1.5	—	1.5	54.00
1954	3,600	2.0	—	2.0	72.00
1955-56	4,200	2.0	—	2.0	84.00
1957-58	4,200	2.25	—	2.25	94.50
1959	4,800	2.5	—	2.5	120.00
1960-61	4,800	3.0	—	3.0	144.00
1962	4,800	3.125	—	3.125	150.00
1963-64	4,800	3.625	—	3.625	174.00
1966	6,600	3.85	—	4.2	277.20
1967	6,600	3.9	.35	4.4	290.40
1968	7,800	3.8	.5	4.4	343.40
1969-70	7,800	4.2	.6	4.8	374.40
1971	7,800	4.6	.6	5.2	405.60
1972	9,000	4.6	.6	5.2	468.00
1973	10,800	4.85	.6	5.85	631.80
1974	13,200	4.95	1.0	5.85	772.20
1975	14,100	4.95	.9	5.85	824.85
1976	15,300	4.95	.9	5.95	895.05
1977	16,500	4.95	.9	5.85	965.25
1978	17,700	5.05	.9	6.05	1,070.85
1979	22,900	5.08	1.0	6.13	1,403.77
1980	25,900	5.08	1.05	6.13	1,587.67
1981	29,700	5.35	1.05	6.65	1,975.05
1982	32,400	5.40	1.30	6.70	2,170.80
1983	35,700	5.40	1.30	6.70	2,391.90
1984	37,800	5.70	1.30	7.0	2,646.00
1985	39,600	5.70	1.30	7.05	2,791.80
1986	42,000	5.70	1.35	7.15	3,003.00
1987	43,800	5.70	1.45	7.15	3,131.70
1988	45,000	6.06	1.45	7.51	3,379.50
1989	48,000	6.06	1.45	7.51	3,604.80
1990	51,300	6.20	1.45	7.65	3,924.45
1991	53,400	6.20	1.45	7.65	4,085.10
1992	55,500	6.20	1.45	7.65	4,245.75
1993	57,600	6.20	1.45	7.65	4,406.40
1994	60,600	6.20	1.45	7.65	4,635.90
1995	61,200	6.20	1.45	7.65	4,681.80
1996	62,700†	6.20	1.45	7.65	4,796.55
1997	65,400†	6.20	1.45	7.65	5,003.10
1998	68,400†	6.20	1.45	7.65	5,232.60
1999	72,600†	6.20	1.45	7.65	5,553.90
2000	76,200†	6.20	1.45	7.65	5,829.30
2001	80,400†	6.20	1.45	7.65	6,150.60
2002	84,900†	6.20	1.45	7.65	6,494.85
2003	87,000†	6.20	1.45	7.65	6,655.50
2004	87,900†	6.20	1.45	7.65	6,724.35
2005	90,000†	6.20	1.45	7.65	6,885.00
2006	94,200†	6.20	1.45	7.65	7,206.30
2007	97,500†	6.20	1.45	7.65	7,458.75
2008	102,000†	6.20	1.45	7.65	7,803.00
2009	106,800†	6.20	1.45	7.65	8,170.20

[1]After 1983 the base will increase per average wage levels.
*Old Age, Survivors and Disability Insurance.
**Hospital Insurance.
†Earnings above this amount will continue to be taxed for Hospital Insurance (Medicare) purposes.

PART VI

ARMED FORCES RETIREMENT HOMES

GULFPORT/WASHINGTON

In 1833, Congress established the U.S. Naval Home, a home for destitute naval officers, sailors and Marines in Philadelphia. In 1851, Congress established the Soldiers' and Airmen's Home, an asylum for "old and disabled soldiers" in Washington, DC.

In 1991, Congress incorporated the U.S. Naval Home, now in Gulfport, MS, and U.S. Soldiers' and Airmen's Home into an independent establishment of the Executive Branch of the Federal Government known as the Armed Forces Retirement Home (AFRH) Agency. In 2002, the names were officially changed to The Armed Forces Retirement Home (Gulfport) and The Armed Forces Retirement Home (Washington).

The two Homes are funded from the Armed Forces Retirement Home Trust Fund. The trust fund receives monthly support of 50 cents withholdings from active duty enlisted personnel, fines and forfeitures, interest from the trust fund, donations and resident fees.

The Armed Forces Retirement Home (AFRH) is home to over 1,200 eligible retirees and veterans. AFRH is a model retirement center, where residents can maintain an independent lifestyle in an environment designed for safety, comfort and personal enrichment. Eligible military veterans from each service branch can live at AFRH. Currently, AFRH (Gulfport) is closed due to Hurricane Katrina and is scheduled to reopen in the fall 2010.

Eligibility. Persons eligible for residency are those honorably discharged members of the Armed Forces, at least one-half of whose service was not active commissioned service (other than as a warrant officer or limited-duty officer), who fit into one of the following categories:

1. Veterans with 20 or more years of active service who: (a) are 60 years of age or over; and (b) were discharged or released from service in the Armed Forces under honorable conditions.
2. Persons who, under rules prescribed by the Chief Operating officer, are determined to be incapable of earning a livelihood because of a service-connected disability incurred in the line of duty in the Armed Forces.
3. Persons who: (a) served in a war theater during a time of war declared by Congress or were eligible for hostile fire special pay under Section 310 of Title 37, United States Code; (b) were discharged or released from service in the Armed Forces under honorable conditions; and (c) are, under rules prescribed by the Chief Operating officer, incapable of earning a livelihood because of injuries, disease, or disability.
4. Persons who: (a) served in a women's component of the Armed Forces before June 12, 1948; and (b) are eligible for admission because of compelling personal circumstances.

Coast Guard veterans who had service during wartime while the Coast Guard was operated as part of the Navy are also eligible for admission.

Active service is defined in part as full-time in the active military service of the United States. Such full time duty includes the annual active duty for training that Reservists must fulfill. It does not, however, include reserve inactive duty training commonly known as Reserve drills, nor does it include years spent as a drilling Reservist; the key is active duty time only.

Resident Fees. The resident fee is 35 percent of the resident's total income for independent living and 40 percent for assisted living. For those who require permanent health care after being admitted into independent living, the monthly fee is 65 percent of income. There are maximum fees, which are adjusted annually for inflation, for each category and campus. The maximum fees for AFRH (Washington) are: Independent Living, $1,238; Assisted Living, $1,856; and Skilled Care, $3,094.

Resident fees are computed on an annual basis. For the purpose of resident fee computation, monthly income is determined by dividing total income by 12.

Total income for computing the monthly fee is:

• Income reportable as Adjusted Gross income (AGI) on the U.S. Individual Income Tax Return and as adjusted by adding tax exempt income received during the same year.

• Tax exempt income includes benefits administered by: the Department of Veterans Affairs, Social Security Administration, disability retired pay, pensions, annuities, interest income, dividends or IRA distributions.

Amenities. The Homes feature all the essentials for retirement living: 24 hour security, community involvement, home style prepared meals, on-site bank, recreational activities, chapels, exchange, private furnished room, movie theatre, wellness center, swimming pool, library, auto and craft shops, barber shop, beauty salon, ballroom, mail room, walking trails, bowling, computer center, boating, fishing, beaches, casinos, golfing, resorts, restaurants etc. In addition, the Homes provide shuttle service to the VA Hospital, local military hospital, medical centers and area recreational functions.

For more information about AFRH, call 1-800-422-9988, visit their website www.AFRH.gov or write to AFRH #1305, 3700 N. Capitol St. N.W., Washington, DC 20011-8400.

RETIREMENT RESIDENCES FOR MILITARY PERSONNEL AND WIDOWS

Each service has established foundations whose major purpose is to provide military retirees, widows, and others who qualify, with housing facilities and other accommodations designed to provide active living, comfort, security, and companionship. The foundations and the residences they maintain are privately endowed and are supported primarily by contributions from wives' clubs, military associations, corporate and foundation gifts, bequests, trusts, and fees.

Knollwood is a military residence located on 16 wooded acres alongside Rock Creek Park in Washington, DC. Owned and operated by the Army Distaff Foundation since 1962, this nonprofit continuing care community is open to retired military officers and their female relatives, including spouses, widows, mothers, daughters, sisters and mothers-in-law. The Knollwood community includes independent apartments, an assisted living wing, a Medicare- and Medicaid-certified skilled nursing center, and a new center specializing in dementia care. Amenities include an indoor pool, cafe, chapel, clinic and rehab, library, full-service dining, private gardens, daily scheduled transportation, and more. Multiple financial options are available. For more information about Knollwood, call 1-800-541-4255, visit www.armydistaff.org or write to 6200 Oregon Ave., N.W., Washington, DC 20015.

Vinson Hall, a continuing care retirement community located in McLean, VA has been serving seniors since 1969. Active independent living at Vinson Hall for military officers and their immediate families consists of 177 apartments ranging from efficiencies through two bedroom units. Amenities include penthouse or casual cafe dining, a state-of-the-art fitness center with indoor pool, whirlpool, and art studios. Fitness trainers are on staff to provide personal training services and monitor residents' fitness progress. There is also a chapel, library, cocktail lounge, game room, and hobby and craft shops, and a wellness center with 24 hours nursing and a on-site physician. Transportation is provided to doctor visits, shopping and community events.

Long-term care is available at both Arleigh Burke Pavilion and The Sylvestery. Arleigh Burke Pavilion provides premier assisted living and nursing care in a luxurious setting. The Sylvestery, an award winning assisted living community devoted to those with memory impairments, provides specialized care in a secure environment. Arleigh Burke and The Sylvestery do not require military affiliation.

For more information about Vinson Hall call (703) 536-4344 or visit their website: www.vinsonhall.org.

Air Force Village, located in San Antonio, TX, is an active continuing care retirement community open to retired and honorably separated commissioned officers of all uniformed services (age 55 and older), their spouses, widow, widowers, and elderly parents or grandparents, with priority to Air Force officers' widows & widowers. The term "officer" refers to Regular, Reserve, Warrant, and National Guard officers and retired commissioned officers of the U.S. Public Health Service (USPHS) and National Oceanic and Atmospheric Administration (NOAA). Air Force Village is a not-for-profit 501c(3) organization managed by Air Force Village, Inc., with two campuses to choose from. Air Force Village I, located at 4917 Ravenswood Dr., San Antonio, TX 78227, has 365 homes, while Air Force Village II, located just seven miles away, offers 403 homes. When needed, both campuses provide a complete continuum of care: life enrichment, home health services, independent living with assistance, full nursing services, rehabilitation services, and Alzheimer's/ dementia care. The Village operates the only registered health care facilities following the Eden Alternative concept in San Antonio. Air Force Village is accredited by the Continuing Care Accreditation Commission (CCAC) in Washington, DC and is an active community with over 65 resident-organized activities and clubs on-site. Experience what it will be like as a resident two nights free of charge in guest rooms on either campus. Additional information regarding costs, application procedures, and other details may be obtained by writing to Air Force Village, Inc., Marketing Department, 5100 John D. Ryan Blvd., San Antonio, TX 78245, by calling 1-800-762-1122, or by visiting their website: www.airforcevillages.com.

The Air Force Enlisted Village (AFEV) is a non-profit organization that provides a home for the surviving spouses of career enlisted servicemembers. Funding for subsidy of dire need Air Force surviving spouses is provided through Air Force Assistance Fund Campaign donations.

The AFEV is the parent organization of Bob Hope Village, with 256 apartments in Shalimar, FL; Teresa Village, a 100 apartment complex in nearby Fort Walton Beach; and Hawthorn House, a 64 apartment assisted living residence on the Bob Hope Village campus. All facilities are only a short distance from Eglin Air Force Base and Hurlburt Field.

Any surviving spouse age 55 or older, whose spouse was a retired enlisted member of the Armed Forces, is eligible to reside in the facilities. Financial support is available only to qualifying Air Force surviving spouses.

Spouses of active duty Air Force members, whose sponsor dies while on active duty, may be admitted for up to one year. Dependent mothers of active duty or retired enlisted Airmen are also eligible for permanent housing.

For more information, please write to the AFEV, 92 Sunset Ln., Shalimar, FL 32579-1000 or call 1-800-258-1413. Email for more information to afef@afev.us, or visit the website at www.afenlistedwidows.org.

STATISTICS, PAY TABLES, AND ORGANIZATIONS

NUMBER OF RETIRED MILITARY PERSONNEL RECEIVING RETIRED PAY

Year	Officers	Enlisted	Total	Year	Officers	Enlisted	Total
1950	66,745	66,083	132,828	1984	439,696	914,848	1,354,544
1955	87,323	93,504	180,827	1985	445,325	928,442	1,373,767
1960	122,373	132,716	255,089	1986	449,313	941,168	1,390,481
1961	136,405	154,683	291,268	1987	453,096	952,823	1,405,919
1962	147,997	178,690	326,687	1988	458,189	968,911	1,427,100
1964	178,818	252,790	431,608	1989	465,009	985,920	1,450,929
1965	193,561	287,002	480,563	1990	471,305	1,000,824	1,472,129
1966	205,496	319,203	524,699	1991	477,145	1,018,901	1,496,046
1967	219,166	365,222	584,388	1992	484,905	1,042,932	1,527,837
1968	232,892	411,963	644,855	1993	491,271	1,064,334	1,555,605
1969	245,801	460,953	706,754	1994	496,257	1,085,718	1,581,975
1970	263,422	501,476	764,898	1995	498,734	1,104,374	1,603,108
1970	280,817	541,818	822,635	1996	500,660	1,121,516	1,622,176
1972	294,214	586,450	880,664	1997	502,855	1,141,269	1,644,124
1973	306,426	628,846	935,572	1998	504,923	1,162,550	1,667,473
1974	318,587	671,247	989,834	1999	506,485	1,180,571	1,687,056
1975	330,024	713,899	1,043,923	2000	507,080	1,194,138	1,701,218
1976	344,940	751,244	1,096,184	2001	506,919	1,206,490	1,713,409
1977	362,708	792,645	1,155,353	2002	504,870	1,208,473	1,713,343
1978	375,839	816,216	1,192,055	2003	504,712	1,217,554	1,722,266
1979	398,808	838,686	1,228,494	2004	508,947	1,282,607	1,791,554
1980	403,562	860,963	1,264,525	2005	511,130	1,300,992	1,812,122
1981	413,862	875,100	1,288,962	2006	513,979	1,318,772	1,832,751
1982	423,748	889,685	1,313,433	2007	519,396	1,340,281	1,859,677
1983	431,762	901,598	1,333,360	2008	521,957	1,358,914	1,880,871

(Figures are as of June 30 for 1950 to 1975 and as of September 30 from 1980 on. Figures represent members receiving retired pay. Enlisted figures include Fleet Reserve.)
NOTE: Figures for 2004 and later include those retirees receiving retired pay under the Concurrent Retirement and Disability Payment (CRDP) and Combat Related Special Compensation (CRSC) Programs.

MILITARY RETIREES

RETIREMENT FROM REGULAR COMPONENTS					
	ARMY	NAVY	MARINE CORPS	AIR FORCE	TOTAL
Nondisability	399,641	381,696	83,357	526,571	1,391,265
Permanent Disability	53,222	32,656	21,057	29,038	135,973
Temporary Disability	3,733	2,518	2,219	1,727	10,197
TOTAL	456,596	416,870	106,633	557,336	1,537,435
RETIREMENT FROM RESERVE COMPONENTS					
20+ Years Active Service	39,772	20,095	1,640	28,281	89,788
20+ Years Reserve Service	273,112	98,262	12,656	154,112	538,142
Permanent Disability	20,772	3,380	1,019	3,884	29,055
Temporary Disability	602	263	243	158	1,266
TOTAL	334,258	122,000	15,558	186,435	658,251
TOTAL MILITARY RETIREES	790,854	538,870	122,191	743,771	2,195,686

As of September 30, 2008

MILITARY RETIREE LIFE EXPECTANCY

NONDISABILITY RETIRED LIFE EXPECTATIONS					DISABILITY RETIRED LIFE EXPECTATIONS		
AGE*	OFFICERS		ENLISTED		AGE*	ALL OFFICERS	ALL ENLISTED
	MALE	FEMALE	MALE	FEMALE			
50	34.20	35.58	30.41	33.57	50	26.67	24.42
51	33.18	34.57	29.42	32.55	51	25.83	23.61
52	32.17	33.57	28.44	31.54	52	25.01	22.79
53	31.17	32.57	27.46	30.54	53	24.20	21.99
54	30.16	31.57	26.50	29.55	54	23.40	21.20
55	29.17	30.58	25.55	28.58	55	22.60	20.41
56	28.18	29.61	24.62	27.63	56	21.82	19.64
57	27.20	28.63	23.70	26.69	57	21.06	18.88
58	26.23	27.67	22.80	25.77	58	20.30	18.14
59	25.27	26.72	21.91	24.86	59	19.55	17.41
60	24.31	25.78	21.03	23.97	60	18.86	16.70
61	23.37	24.85	20.16	23.07	61	18.22	16.01
62	22.44	23.93	19.31	22.18	62	17.58	15.34
63	21.52	23.01	18.48	21.29	63	16.92	14.69
64	20.61	22.10	17.66	20.41	64	16.23	14.05
65	19.72	21.19	16.86	19.56	65	15.53	13.43
66	18.85	20.28	16.08	18.73	66	14.83	12.82
67	17.99	19.39	15.33	17.94	67	14.18	12.24
68	17.14	18.51	14.59	17.19	68	13.56	11.66
69	16.32	17.65	13.88	16.47	69	12.98	11.10
70	15.51	16.81	13.19	15.77	70	12.42	10.55
71	14.72	15.98	12.52	15.09	71	11.87	10.02
72	13.94	15.18	11.88	14.43	72	11.32	9.50
73	13.19	14.40	11.25	13.78	73	10.78	8.99
74	12.46	13.64	10.65	13.15	74	10.25	8.50
75	11.75	12.90	10.07	12.53	75	9.74	8.03
76	11.06	12.17	9.51	11.93	76	9.23	7.57
77	10.39	11.47	8.98	11.34	77	8.71	7.13
78	9.75	10.79	8.47	10.75	78	8.20	6.71
79	9.13	10.12	7.97	10.17	79	7.68	6.30
80	8.54	9.47	7.50	9.60	80	7.19	5.91
81	7.97	8.84	7.05	9.05	81	6.71	5.54
82	7.43	8.23	6.61	8.51	82	6.26	5.19
83	6.92	7.65	6.20	7.99	83	5.82	4.86
84	6.43	7.09	5.80	7.48	84	5.41	4.54
85	5.97	6.57	5.43	6.97	85	5.02	4.23
86	5.53	6.07	5.08	6.46	86	4.66	3.94
87	5.13	5.60	4.74	5.96	87	4.31	3.67
88	4.75	5.16	4.43	5.47	88	3.98	3.40
89	4.39	4.74	4.14	5.01	89	3.66	3.13
90	4.06	4.35	3.87	4.58	90	3.34	2.87
91	3.75	3.98	3.62	4.19	91	3.03	2.63
92	3.47	3.63	3.39	3.85	92	2.75	2.41
93	3.21	3.32	3.17	3.56	93	2.48	2.20

* Age nearest birthday

As of September 30, 2007

RETIREES BY STATE

STATE	ARMY	NAVY	MARINES	AIR FORCE	DoD	COAST GUARD
ALABAMA	29,888	7,306	1,848	16,014	55,056	965
ALASKA	3,223	713	168	5,325	9,429	719
ARIZONA	15,379	9,286	3,641	25,567	53,873	574
ARKANSAS	9,157	4,627	963	10,817	25,564	215
CALIFORNIA	28,961	73,277	17,176	49,233	168,647	3,058
COLORADO	17,106	5,940	1,546	23,573	48,165	326
CONNECTICUT	3,273	4,773	513	2,152	10,711	652
DELAWARE	2,034	1,127	249	4,736	8,146	161
DIST OF COLUMBIA	1,310	677	148	818	2,953	43
FLORIDA	46,572	62,096	8,786	68,128	185,582	4,733
GEORGIA	46,138	13,526	4,150	24,541	88,355	785
GUAM, MI	628	713	84	447	1,872	27
HAWAII	5,767	4,581	1,011	4,473	15,832	332
IDAHO	3,358	2,872	596	5,831	12,657	202
ILLINOIS	10,779	8,700	1,966	13,701	35,146	365
INDIANA	10,270	5,059	1,530	6,906	23,765	225
IOWA	4,745	2,846	651	3,492	11,734	115
KANSAS	10,134	2,909	799	6,673	20,515	177
KENTUCKY	15,934	3,911	1,093	5,386	26,324	294
LOUISIANA	9,613	4,614	1,351	9,992	25,570	628
MAINE	3,390	4,355	508	3,877	12,130	903
MARYLAND	18,639	14,752	2,519	14,529	50,439	1,245
MASSACHUSETTS	7,158	4,790	1,213	5,994	19,155	1,184
MICHIGAN	10,503	6,932	1,761	8,290	27,486	1,040
MINNESOTA	6,868	4,213	849	5,439	17,369	221
MISSISSIPPI	9,096	5,953	963	10,008	26,020	326
MISSOURI	15,099	7,227	2,249	11,861	36,436	486
MONTANA	2,632	1,750	419	3,726	8,527	154
NEBRASKA	2,878	1,948	379	8,516	13,721	84
NEVADA	5,221	6,570	1,595	13,839	27,225	391
NEW HAMPSHIRE	2,994	2,051	440	3,944	9,429	345
NEW JERSEY	8,988	4,728	1,085	5,573	20,374	744
NEW MEXICO	5,779	3,171	859	11,541	21,350	182
NEW YORK	16,196	8,497	2,150	10,637	37,480	1,041
NORTH CAROLINA	37,119	14,326	12,193	20,050	83,688	2,103
NORTH DAKOTA	1,414	473	85	2,766	4,738	25
OHIO	13,207	8,632	2,433	19,693	43,965	630
OKLAHOMA	13,901	4,692	1,164	14,675	34,432	240
OREGON	6,584	6,668	1,367	6,747	21,366	1,065
PENNSYLVANIA	20,046	12,349	3,347	13,095	48,837	872
PUERTO RICO	8,221	477	115	1,084	9,897	140
RHODE ISLAND	1,618	2,658	246	1,030	5,552	209
SOUTH CAROLINA	21,008	12,963	3,249	17,498	54,718	640
SOUTH DAKOTA	2,147	1,105	187	3,586	7,025	49
TENNESSEE	23,000	11,445	2,636	13,415	50,496	522
TEXAS	72,095	27,375	7,659	78,271	185,400	1,949
UTAH	4,549	2,019	509	7,630	14,707	147
VERMONT	1,717	695	152	1,077	3,641	77
VIRGINIA	45,005	56,552	10,368	31,657	143,582	3,254
VIRGIN ISLANDS	269	49	3	43	364	5
WASHINGTON	23,218	24,321	2,354	20,510	70,403	2,510
WEST VIRGINIA	4,378	2,366	706	3,287	10,737	169
WISCONSIN	8,061	4,741	1,145	5,456	19,403	518
WYOMING	1,246	799	159	2,747	4,951	50
OTHER	10,704	6,646	1,196	9,367	27,913	220
TOTAL	709,217	497,841	116,531	679,263	2,002,852	38,336

As of September 30, 2008

RETIRED MILITARY MONTHLY PAY* BY STATE (IN THOUSANDS)

STATE	ARMY	NAVY	MARINES	AIR FORCE	DoD	COAST GUARD
ALABAMA	50,549	12,067	3,219	30,398	96,234	1,669
ALASKA	5,263	1,156	237	9,331	15,988	1,296
ARIZONA	27,073	15,705	6,674	50,206	99,658	968
ARKANSAS	13,185	7,318	1,492	18,832	40,826	331
CALIFORNIA	45,290	132,532	33,175	95,482	306,480	4,860
COLORADO	32,096	10,737	2,844	53,703	99,380	542
CONNECTICUT	4,182	8,539	600	3,367	16,687	1,272
DELAWARE	3,049	1,781	394	8,167	13,391	272
DIST OF COLUMBIA	2,361	1,390	259	1,470	5,479	75
FLORIDA	83,131	116,536	16,893	136,571	353,131	8,314
GEORGIA	81,266	22,269	7,258	43,937	154,729	1,233
GUAM, MI	920	1,170	146	747	2,982	43
HAWAII	10,353	9,299	2,248	8,522	30,421	658
IDAHO	5,046	4,891	1,016	10,287	21,239	299
ILLINOIS	14,856	13,333	2,538	26,486	57,213	578
INDIANA	13,977	7,422	2,116	10,809	34,324	319
IOWA	5,953	4,094	910	5,477	16,434	162
KANSAS	18,537	4,495	1,425	11,825	36,283	291
KENTUCKY	24,805	6,046	1,762	9,235	41,848	463
LOUISIANA	14,224	7,377	2,382	17,443	41,426	1,121
MAINE	4,464	7,544	792	6,199	18,998	1,418
MARYLAND	35,120	31,025	4,781	27,499	98,425	2,501
MASSACHUSETTS	9,443	7,012	1,481	9,957	27,894	1,882
MICHIGAN	13,344	9,634	2,171	12,959	38,108	1,695
MINNESOTA	8,498	6,015	1,137	8,483	24,133	280
MISSISSIPPI	12,300	9,979	1,637	16,959	40,874	546
MISSOURI	22,737	10,810	3,586	19,895	57,029	743
MONTANA	3,951	2,828	758	6,511	14,048	220
NEBRASKA	3,948	3,174	561	17,149	24,833	104
NEVADA	8,947	11,380	3,013	25,868	49,208	621
NEW HAMPSHIRE	4,786	3,639	704	7,719	16,848	607
NEW JERSEY	12,591	6,913	1,342	8,848	29,694	1,156
NEW MEXICO	9,553	5,706	1,499	23,562	40,320	264
NEW YORK	19,469	11,347	2,442	15,904	49,161	1,306
NORTH CAROLINA	63,093	25,524	24,585	36,896	150,098	3,669
NORTH DAKOTA	1,755	644	86	4,447	6,933	40
OHIO	16,997	12,407	3,016	36,692	69,112	978
OKLAHOMA	21,618	7,091	1,854	25,278	55,841	399
OREGON	9,493	10,945	2,115	11,959	34,511	1,776
PENNSYLVANIA	29,754	19,646	4,841	20,853	75,093	1,314
PUERTO RICO	7,537	627	96	1,209	9,469	161
RHODE ISLAND	2,207	5,619	460	1,472	9,758	336
SOUTH CAROLINA	33,633	23,218	6,246	32,076	95,174	1,049
SOUTH DAKOTA	3,063	1,799	301	6,179	11,341	86
TENNESSEE	35,513	20,023	4,613	23,460	83,609	821
TEXAS	129,147	45,850	13,139	160,676	348,811	3,208
UTAH	7,060	3,239	863	14,029	25,191	214
VERMONT	2,306	1,152	223	1,699	5,379	107
VIRGINIA	107,094	126,632	26,922	80,417	341,065	7,139
VIRGIN ISLANDS	356	87	4	71	519	8
WASHINGTON	40,804	43,874	4,184	39,535	128,397	4,638
WEST VIRGINIA	5,914	3,708	1,054	4,976	15,651	301
WISCONSIN	10,240	6,901	1,499	8,483	27,123	738
WYOMING	1,730	1,291	236	4,813	8,070	84
OTHER	20,379	11,458	2,503	18,225	52,565	380
TOTAL	1,174,958	886,897	212,331	1,293,251	3,567,437	65,555

As of September 30, 2008

*The monthly amount of payment (net pay) is before deductions for withholding taxes and allotments, but after deductions for survivor benefits, and waivers to obtain benefits from the VA.

MILITARY PERSONNEL BY LOCATION

Geographic Location	Active Duty			Reserve		
	Enlisted	Officer	Total	Selected Reserve	IRR	Total
UNITED STATES, U.S. TERRITORIES, SPECIAL LOCATIONS						
CONUS	814,696	169,055	983,751	798,753	216,436	1,015,189
ALASKA	17,460	2,207	19,667	4,478	1,233	5,711
GUAM	2,608	446	3,054	2,380	147	2,527
HAWAII	31,699	6,005	37,704	9,249	1,557	10,806
PUERTO RICO	91	31	122	12,962	613	13,575
VIRGIN ISLANDS	3	1	4	824	65	889
SHIPS AFLOAT	141,124	17,364	158,488	787	0	787
EUROPE						
BELGIUM	933	372	1,305	5	0	5
GERMANY	45,901	8,073	53,974	692	6	698
GREECE	367	50	417	1	0	1
GREENLAND	113	25	138	0	0	0
ITALY	10,169	1,674	11,843	93	2	95
NETHERLANDS	421	139	560	1	0	1
PORTUGAL	692	120	812	2	0	2
SPAIN	1,173	211	1,384	2	1	3
TURKEY	1,419	206	1,625	6	1	7
UNITED KINGDOM	8,341	1,266	9,607	56	3	59
EAST ASIA AND PACIFIC						
AUSTRALIA	59	50	109	0	6	6
JAPAN	31,288	4,164	35,452	85	22	107
PHILIPPINES	87	21	108	0	4	4
SINGAPORE	95	68	163	1	0	1
SOUTH KOREA	22,495	3,538	26,033	42	0	42
THAILAND	61	56	117	0	3	3
NORTH AFRICA, NEAR EAST and SOUTH ASIA						
AFGHANISTAN	3,473	368	3,841	294	0	294
BAHRAIN	1,973	366	2,339	42	0	42
DJIBOUTI	82	21	103	0	0	0
EGYPT	202	111	313	1	0	1
IRAQ	18,038	2,225	20,263	413	0	413
KUWAIT	2,494	476	2,970	620	0	620
QATAR	235	187	422	2	0	2
SAUDI ARABIA	107	178	285	0	0	0
UNITED ARAB EMIRATES	63	46	109	0	0	0
WESTERN HEMISPHERE						
CANADA	51	48	99	0	22	22
CUBA	1,638	143	1,781	26	0	26
HAITI	6	29	35	0	0	0
HONDURAS	324	95	419	0	0	0

As of September 30, 2008

ARMED FORCES PAY — Basic Pay Rates, Effective January 1, 2009

Pay Grade	YEARS OF SERVICE																	
	Under 2	Over 2	Over 3	Over 4	Over 6	Over 8	Over 10	Over 12	Over 14	Over 16	Over 18	Over 20	Over 22	Over 24	Over 26	Over 30	Over 34	Over 38
COMMISSIONED OFFICERS																		
O-10												14688.60	14760.30	15067.20	15602.10	16382.10	17201.10	18061.20
O-9												12846.90	13032.00	13299.30	13765.80	14454.60	15177.30	15936.00
O-8	9090.00	9387.60	9585.30	9640.50	9887.10	10299.00	10395.00	10786.20	10898.10	11235.30	11722.50	12172.20	12472.50	12472.50	12472.50	12784.50	13104.30	13104.30
O-7	7553.10	7904.10	8066.40	8195.40	8429.10	8660.10	8926.80	9192.90	9460.20	10299.00	11007.30	11007.30	11007.30	11007.30	11063.10	11284.50	11284.50	11284.50
O-6	5598.30	6150.30	6553.80	6553.80	6578.70	6860.70	6897.90	6897.90	7290.00	7983.30	8390.10	8796.60	9027.90	9262.20	9716.70	9910.80	9910.80	9910.80
O-5	4666.80	5257.20	5621.40	5689.80	5916.60	6052.80	6351.60	6570.60	6853.80	7287.30	7493.40	7697.40	7928.70	7928.70	7928.70	7928.70	7928.70	7928.70
O-4	4026.90	4661.40	4972.20	5041.80	5330.40	5640.00	6025.20	6325.50	6534.30	6654.00	6723.30	6723.30	6723.30	6723.30	6723.30	6723.30	6723.30	6723.30
O-3	3540.30	4013.40	4332.00	4722.90	4948.80	5197.20	5358.00	5622.30	5759.70	5759.70	5759.70	5759.70	5759.70	5759.70	5759.70	5759.70	5759.70	5759.70
O-2	3058.80	3483.90	4012.50	4148.10	4233.30	4233.30	4233.30	4233.30	4233.30	4233.30	4233.30	4233.30	4233.30	4233.30	4233.30	4233.30	4233.30	4233.30
O-1	2655.30	2763.60	3340.50	3340.50	3340.50	3340.50	3340.50	3340.50	3340.50	3340.50	3340.50	3340.50	3340.50	3340.50	3340.50	3340.50	3340.50	3340.50
COMMISSIONED OFFICERS WITH: 1,460 POINTS TOWARDS A RESERVE RETIREMENT; OR 4 YEARS OF ACTIVE SERVICE, AS AN ENLISTED MEMBER AND/OR WARRANT OFFICER																		
O-3E				4722.90	4948.80	5197.20	5358.00	5622.30	5844.90	5972.70	6146.70	6146.70	6146.70	6146.70	6146.70	6146.70	6146.70	6146.70
O-2E				4148.10	4233.30	4368.30	4595.70	4771.50	4902.30	4902.30	4902.30	4902.30	4902.30	4902.30	4902.30	4902.30	4902.30	4902.30
O-1E				3340.50	3567.60	3699.30	3834.30	3966.60	4148.10	4148.10	4148.10	4148.10	4148.10	4148.10	4148.10	4148.10	4148.10	4148.10
WARRANT OFFICERS																		
W-5												6505.50	6835.50	7081.20	7353.60	7721.40	8107.50	8513.10
W-4	3658.50	3935.70	4048.80	4159.80	4351.20	4540.50	4732.20	5021.10	5274.00	5514.60	5711.40	5903.40	6185.70	6417.30	6681.90	6815.40	6815.40	6815.40
W-3	3340.80	3480.30	3622.80	3669.90	3819.60	4114.20	4420.80	4565.10	4731.90	4904.10	5213.10	5422.20	5547.30	5680.20	5860.80	5860.80	5860.80	5860.80
W-2	2956.50	3236.10	3322.20	3381.60	3573.30	3871.20	4018.80	4164.30	4341.90	4480.80	4606.80	4757.10	4856.40	4935.00	4935.00	4935.00	4935.00	4935.00
W-1	2595.30	2874.00	2949.60	3108.30	3296.10	3572.70	3701.70	3882.30	4059.90	4199.40	4328.10	4484.40	4484.40	4484.40	4484.40	4484.40	4484.40	4484.40
ENLISTED																		
E-9							4420.50	4520.70	4646.70	4795.50	4944.90	5185.20	5388.00	5601.90	5928.30	6224.70	6536.10	6863.10
E-8						3618.60	3778.80	3877.80	3996.60	4125.00	4357.20	4474.80	4674.90	4785.90	5059.50	5160.90	5160.90	5160.90
E-7	2515.50	2745.60	2850.60	2990.10	3098.70	3285.30	3390.30	3577.50	3732.60	3838.50	3951.30	3995.40	4142.10	4221.00	4521.00	4521.00	4521.00	4521.00
E-6	2175.60	2394.00	2499.60	2602.20	2709.30	2950.80	3044.70	3226.20	3282.00	3322.50	3369.90	3369.90	3369.90	3369.90	3369.90	3369.90	3369.90	3369.90
E-5	1993.50	2127.00	2229.60	2334.90	2499.00	2670.90	2811.00	2828.40	2828.40	2828.40	2828.40	2828.40	2828.40	2828.40	2828.40	2828.40	2828.40	2828.40
E-4	1827.60	1920.90	2025.00	2127.60	2218.50	2218.50	2218.50	2218.50	2218.50	2218.50	2218.50	2218.50	2218.50	2218.50	2218.50	2218.50	2218.50	2218.50
E-3	1649.70	1753.50	1859.70	1859.70	1859.70	1859.70	1859.70	1859.70	1859.70	1859.70	1859.70	1859.70	1859.70	1859.70	1859.70	1859.70	1859.70	1859.70
E-2	1568.70	1568.70	1568.70	1568.70	1568.70	1568.70	1568.70	1568.70	1568.70	1568.70	1568.70	1568.70	1568.70	1568.70	1568.70	1568.70	1568.70	1568.70
E-1	1399.50	1399.50	1399.50	1399.50	1399.50	1399.50	1399.50	1399.50	1399.50	1399.50	1399.50	1399.50	1399.50	1399.50	1399.50	1399.50	1399.50	1399.50

NOTE: Monthly basic pay for O-7 through O-10 is limited to $14,750.10 corresponding to the Level II Executive Schedule.
Cadet and Midshipmen pay is $929.40 as set by 37 USC, section 203(c).
While serving as the senior enlisted member of the Service, basic pay is $7,143.30 regardless of years of service.

REGULAR MILITARY COMPENSATION*

Average Annual Regular Military Compensation (RMC) (Amounts rounded to nearest dollar)

Pay Grade	YEARS OF SERVICE																	
	Under 2	Over 2	Over 3	Over 4	Over 6	Over 8	Over 10	Over 12	Over 14	Over 16	Over 18	Over 20	Over 22	Over 24	Over 26	Over 28	Over 30	Over 34
0-10																	215,995	215,995
0-9																204,129	212,409	215,964
0-8														188,204	188,204	188,204	192,109	196,091
0-7														169,820	170,515	170,515	173,280	173,280
0-6		108,182		113,641	113,978	117,794	118,297	118,297	123,583	132,548	137,521	142,405	145,180	147,992	153,446	153,446	155,777	155,777
0-5	88,366	95,765	100,579	101,492	104,552	106,393	110,431	113,391	117,219	123,063	125,807	128,477	131,433	131,433	131,433	131,433	131,433	131,433
0-4	78,561	86,183	89,986	90,853	94,498	98,561	103,715	107,759	110,570	112,182	113,115	113,115	113,115	113,115	113,115	113,115	113,115	113,115
0-3	69,633	75,376	79,204	83,894	86,609	89,691	91,715	95,117	96,916	96,916	96,916	96,916	96,916	96,916	96,916			
0-2	60,758	66,260	72,725	74,354	75,377	75,377	75,377	75,377	75,377	75,377								
0-1	52,802	54,243	61,858	61,858	61,858	61,858	61,858	61,858	61,858									
COMMISSIONED OFFICERS WITH: 1,460 POINTS TOWARDS A RESERVE RETIREMENT; OR 4 YEARS OF ACTIVE SERVICE, AS AN ENLISTED MEMBER AND/OR WARRANT OFFICER																		
0-3E				85,216	87,977	91,077	93,124	96,562	99,489	101,178	103,485	103,485	103,485	103,485	103,485	103,485	103,485	
0-2E				76,515	77,538	79,158	81,886	83,996	85,566	85,566	85,566	85,566	85,566	85,566				
0-1E				65,031	68,010	69,637	71,261	72,850	75,030	75,030	75,030	75,030	75,030	75,030	75,030			
WARRANT OFFICERS																		
W-5												105,812	110,283	113,612	117,241	117,241	121,975	126,751
W-4								88,795	91,937	95,008	97,599	100,167	103,973	107,109	110,692	110,692	112,500	112,500
W-3					73,122	76,678	80,360	82,092	84,094	86,160	89,925	92,532	94,102	95,843	98,213	98,213	98,213	98,213
W-2	61,095	64,647	65,731	66,477	68,882	72,511	74,289	76,042	78,174	79,840	81,352	83,156	84,348	85,291	85,291	85,291	85,291	
W-1	54,934	58,526	59,496	61,513	63,894	67,367	68,942	71,133	73,271	74,949	76,494							
ENLISTED MEMBERS																		
E-9									85,281	87,080	88,915	91,916	94,503	97,328	101,705	101,705	105,714	109,930
E-8						71,468	73,419	74,623	76,055	77,602	80,390	81,801	84,203	85,535	88,861	88,861	90,101	
E-7				62,464	63,840	66,200	67,516	69,857	71,750	73,040	74,403	74,934	76,702	77,650	81,250	81,250	81,250	
E-6	50,794	53,657	55,036	56,361	57,744	60,850	62,042	64,347	65,052	65,560	66,156	66,156	66,156	66,156	66,156	66,156		
E-5	45,537	47,309	48,660	50,038	52,185	54,422	56,243	56,469	56,469	56,469	56,469	56,469	56,469	56,469	56,469			
E-4	41,460	42,631	43,931	45,228	46,430	46,430	46,430	46,430	46,430	46,430	46,430							
E-3	38,746	40,018	41,326	41,326	41,326	41,326	41,326	41,326	41,326	41,326	41,326	41,326	41,326					
E-2	37,439	37,439	37,439	37,439	37,439	37,439	37,439	37,439	37,439	37,439	37,439	37,439	37,439					
E-1	35,105	35,105	35,105	35,105	35,105	35,105	35,105	35,105	35,105	35,105	35,105	35,105	35,105					

*Regular Military Compensation (RMC) combines basic pay, basic allowance for housing and basic allowance for subsistence and the federal tax advantage on the tax free allowances. The RMC is at best, a guide to income for any member in a given grade and year of service. (As of January 1, 2009)

TOTAL MILITARY LIFETIME RETIREMENT PAY (LUMP SUM EQUIVALENT) (As of January 1, 2009)

PAY GRADE	YEARS OF SERVICE										
	Over 20	Over 21	Over 22	Over 23	Over 24	Over 25	Over 26	Over 27	Over 28	Over 29	Over 30
COMMISSIONED OFFICERS											
O-10	1,657,061	1,889,733	2,100,875	2,316,809	2,382,412	2,461,445	2,532,524	2,630,875	2,707,721	2,928,858	3,115,663
O-9	1,657,061	1,795,692	1,910,046	2,032,871	2,097,102	2,173,094	2,235,499	2,322,010	2,389,398	2,584,211	2,748,965
O-8	1,657,061	1,761,171	1,840,281	1,933,031	2,000,579	2,065,068	2,110,162	2,151,555	2,188,895	2,341,326	2,431,465
O-7	1,544,437	1,632,015	1,684,042	1,733,351	1,779,428	1,822,496	1,862,230	1,902,201	1,938,464	2,076,690	2,146,171
O-6	1,183,313	1,264,782	1,325,835	1,397,761	1,447,276	1,508,186	1,554,236	1,624,842	1,679,315	1,823,966	1,884,969
O-5	1,064,046	1,121,341	1,167,544	1,224,628	1,269,560	1,312,789	1,341,409	1,367,678	1,391,372	1,488,381	1,507,873
O-4	959,876	996,946	1,028,599	1,058,709	1,086,734	1,113,205	1,137,359	1,159,793	1,179,774	1,262,068	1,278,624
O-3	825,209	853,802	881,154	906,929	930,878	953,538	974,448	993,378	1,010,706	1,081,070	1,095,275
O-2	606,301	627,629	647,528	666,562	684,177	700,917	716,095	730,290	742,840	794,684	804,909
O-1	478,579	495,199	510,929	526,036	539,960	552,974	564,977	576,261	586,189	626,891	635,254
COMMISSIONED OFFICERS WITH: 1,460 POINTS TOWARDS A RESERVE RETIREMENT; OR 4 YEARS OF ACTIVE SERVICE, AS AN ENLISTED MEMBER AND/OR WARRANT OFFICER											
O-3E	872,614	911,238	940,367	967,814	993,618	1,017,740	1,039,997	1,060,159	1,078,597	1,153,832	1,169,071
O-2E	702,319	726,729	750,124	771,885	792,339	811,735	829,364	845,542	860,395	920,269	932,214
O-1E	594,223	614,833	634,630	653,000	670,550	686,681	701,833	715,480	727,782	778,629	788,933
WARRANT OFFICERS											
W-5	805,583	883,562	952,385	1,033,893	1,079,069	1,132,465	1,170,819	1,222,536	1,259,552	1,380,403	1,468,566
W-4	805,583	852,909	889,948	940,690	981,127	1,029,742	1,065,229	1,113,208	1,147,809	1,254,300	1,296,122
W-3	731,003	777,022	812,561	848,641	878,074	911,667	939,069	977,221	1,004,630	1,100,232	1,114,548
W-2	647,063	681,494	711,138	744,761	769,912	800,570	822,508	846,081	860,923	926,225	938,554
W-1	603,583	639,236	667,754	693,109	711,714	728,831	744,892	759,372	772,691	841,810	852,839
ENLISTED MEMBERS											
E-9	685,491	727,779	762,172	806,058	837,218	878,895	908,780	955,692	989,808	1,076,840	1,144,124
E-8	599,621	636,553	662,112	696,971	725,313	758,416	780,171	816,255	844,971	918,997	948,535
E-7	545,914	571,183	590,926	617,687	641,315	669,414	687,509	723,297	752,080	821,034	830,887
E-6	468,011	485,186	500,014	514,224	527,199	539,439	550,380	560,621	569,626	612,080	619,366
E-5	394,239	407,614	419,966	431,847	442,924	453,150	462,511	471,057	478,527	513,616	519,856
E-4	310,433	321,036	330,770	340,192	348,979	356,822	364,257	371,049	376,936	402,876	407,600
E-3	260,268	269,031	277,309	285,086	292,336	299,296	305,412	310,992	316,032	337,734	341,669
E-2	219,546	226,904	233,855	240,665	246,744	252,353	257,483	262,424	266,644	284,869	288,237
E-1	195,939	202,500	208,697	214,518	219,942	225,218	229,791	233,962	237,728	254,052	257,109

Lump sum equivalent values are not adjusted for any reduction in DoD benefits due to offsets.
The lump-sum-equivalent is the amount of money required to be on hand at the time of retirement to pay a lifetime annuity that increases with inflation at 3% annually.
The interest rate used in discounting to the present value is 5.75% resulting in what is commonly referred to as a real interest rate of 2.75%.
Longevity is based on military specific tables that assume an entry age of 23 for commissioned officers and warrant officers and 20 for enlisted.
Warning: these figures should not be used in property settlements for the following reasons: (1) They assume full future cost-of-living adjustments (which have been determined in some courts not to be property); (2) The investment return is low (producing large lump-sums) because of the conservative portfolio of government trust funds; and (3) Only the mortality of the retiree is considered (some courts require the use of the spouse's mortality).

TOTAL MILITARY LIFETIME RETIREMENT PAY (LUMP SUM EQUIVALENT) (As of January 1, 2009)

PAY GRADE	YEARS OF SERVICE									
	Over 31	Over 32	Over 33	Over 34	Over 35	Over 36	Over 37	Over 38	Over 39	Over 40
COMMISSIONED OFFICERS										
O-10	3,150,346	3,178,823	3,201,497	3,378,611	3,389,357	3,393,527	3,391,280	3,551,270	3,534,683	3,511,310
O-9	2,779,629	2,804,786	2,824,851	2,981,127	2,990,742	2,994,328	2,992,281	3,133,388	3,118,767	3,098,180
O-8	2,458,292	2,480,719	2,498,423	2,573,938	2,582,217	2,585,263	2,583,477	2,576,627	2,564,547	2,547,600
O-7	2,169,957	2,189,640	2,205,160	2,216,433	2,223,463	2,226,399	2,224,761	2,218,768	2,208,449	2,193,767
O-6	1,905,691	1,923,060	1,936,770	1,946,745	1,952,765	1,955,224	1,953,860	1,948,666	1,939,669	1,926,642
O-5	1,524,553	1,538,351	1,549,463	1,557,350	1,562,257	1,564,136	1,563,173	1,558,933	1,551,655	1,541,313
O-4	1,292,793	1,304,518	1,313,763	1,320,478	1,324,664	1,326,282	1,325,521	1,321,947	1,315,796	1,307,045
O-3	1,107,435	1,117,500	1,125,439	1,131,211	1,134,815	1,136,218	1,135,400	1,132,358	1,127,108	1,119,630
O-2	813,889	821,327	827,202	831,480	834,165	835,009	834,445	832,245	828,419	822,954
O-1	642,178	648,135	652,617	656,079	658,053	658,976	658,391	656,730	653,582	649,342
COMMISSIONED OFFICERS WITH: 1,460 POINTS TOWARDS A RESERVE RETIREMENT; OR 4 YEARS OF ACTIVE SERVICE, AS AN ENLISTED MEMBER AND/OR WARRANT OFFICER										
O-3E	1,181,876	1,192,695	1,201,243	1,207,241	1,211,160	1,212,726	1,211,704	1,208,525	1,202,984	1,194,868
O-2E	942,672	951,100	957,962	962,742	965,911	967,199	966,379	963,881	959,296	953,017
O-1E	797,512	804,833	810,620	814,610	817,274	818,349	817,607	815,480	811,759	806,429
WARRANT OFFICERS										
W-5	1,484,851	1,498,328	1,508,956	1,592,477	1,597,615	1,599,430	1,598,341	1,673,804	1,666,072	1,655,046
W-4	1,310,411	1,322,468	1,331,766	1,338,734	1,342,906	1,344,477	1,343,638	1,339,954	1,333,861	1,324,931
W-3	1,127,038	1,137,148	1,145,338	1,151,085	1,154,859	1,156,167	1,155,435	1,152,228	1,146,980	1,139,266
W-2	948,875	957,649	964,358	969,213	972,441	973,557	972,774	970,297	965,720	959,433
W-1	862,276	870,083	876,237	880,704	883,485	884,553	884,107	881,712	877,599	871,752
ENLISTED MEMBERS										
E-9	1,155,430	1,164,083	1,170,840	1,234,578	1,237,663	1,238,179	1,236,869	1,295,354	1,289,522	1,281,753
E-8	957,828	965,121	970,646	974,773	977,102	977,576	976,605	974,049	969,599	963,696
E-7	839,028	845,416	850,255	853,870	855,911	856,326	855,476	853,237	849,338	844,355
E-6	625,379	630,087	633,872	636,514	637,984	638,245	637,770	636,053	633,101	629,204
E-5	525,021	528,852	531,951	534,280	535,404	535,730	535,261	533,720	531,342	528,165
E-4	411,730	414,759	417,261	418,934	420,057	420,164	419,860	418,670	416,864	414,240
E-3	345,144	347,659	349,770	351,149	352,103	352,172	351,930	350,912	349,410	347,192
E-2	291,013	293,183	295,047	296,254	296,918	297,020	296,890	296,070	294,676	292,844
E-1	259,636	261,621	263,125	264,251	264,889	265,023	264,766	264,078	262,877	261,281

Lump sum equivalent values are not adjusted for any reduction in DoD benefits due to offsets.

The lump-sum-equivalent is the amount of money required to be on hand at the time of retirement to pay a lifetime annuity that increases with inflation at 3% annually.

The interest rate used in discounting to the present value is 5.75% resulting in what is commonly referred to as a real interest rate of 2.75%.

Longevity is based on military specific tables that assume an entry age of 23 for commissioned officers and warrant officers and 20 for enlisted.

Warning: these figures should not be used in property settlements for the following reasons: (1) They assume full future cost-of-living adjustments (which have been determined in some courts not to be property); (2) The investment return is low (producing large lump-sums) because of the conservative portfolio of government trust funds; and (3) Only the mortality of the retiree is considered (some courts require the use of the spouse's mortality).

MILITARY RETIRED PAY, LUMP SUM EQUIVALENTS* AT RETIREMENT, AND MONTHLY DEPOSITS NECESSARY TO ACCUMULATE THE LUMP SUM EQUIVALENT* FOR NONDISABILITY RETIREMENTS

Retired Pay Grade	Retirement After 20 Years			Retirement After 30 Years			Retirement After 40 Years		
	Monthly Retired Pay	Lump Sum Equivalents	Monthly Deposits	Monthly Retired Pay	Lump Sum Equivalents	Monthly Deposits	Monthly Retired Pay	Lump Sum Equivalents	Monthly Deposits
COMMISSIONED OFFICERS									
O-10	5,488	1,657,061	3,748	12,286	3,115,663	3,336	18,061	3,511,310	1,957
O-9	5,488	1,657,061	3,748	10,840	2,748,965	2,943	15,936	3,098,180	1,726
O-8	5,488	1,657,061	3,748	9,588	2,431,465	2,603	13,104	2,547,600	1,420
O-7	5,115	1,544,437	3,493	8,463	2,146,171	2,298	11,284	2,193,767	1,222
O-6	3,919	1,183,313	2,677	7,433	1,884,969	2,018	9,910	1,926,642	1,074
O-5	3,524	1,064,046	2,407	5,946	1,507,873	1,614	7,928	1,541,313	859
O-4	3,179	959,876	2,171	5,042	1,278,624	1,369	6,723	1,307,045	728
O-3	2,733	825,209	1,867	4,319	1,095,275	1,173	5,759	1,119,630	624
O-2	2,008	606,301	1,371	3,174	804,909	862	4,233	822,954	459
O-1	1,585	478,579	1,083	2,505	635,254	680	3,340	649,342	362
COMMISSIONED OFFICERS WITH: 1,460 POINTS TOWARDS A RESERVE RETIREMENT; OR 4 YEARS OF ACTIVE SERVICE, AS AN ENLISTED MEMBER AND/OR WARRANT OFFICER									
O-3E	2,890	872,614	1,974	4,610	1,169,071	1,252	6,146	1,194,868	666
O-2E	2,326	702,319	1,589	3,676	932,214	998	4,902	953,017	531
O-1E	1,968	594,223	1,344	3,111	788,933	845	4,148	806,429	449
WARRANT OFFICERS									
W-5	2,668	805,583	1,822	5,791	1,468,566	1,572	8,513	1,655,046	922
W-4	2,668	805,583	1,822	5,111	1,296,122	1,388	6,815	1,324,931	738
W-3	2,421	731,003	1,653	4,395	1,114,548	1,193	5,860	1,139,266	635
W-2	2,143	647,063	1,464	3,701	938,554	1,005	4,935	959,433	535
W-1	1,999	603,583	1,365	3,363	852,839	913	4,484	871,752	486
ENLISTED MEMBERS									
E-9	2,323	685,491	1,551	4,668	1,144,124	1,225	6,863	1,281,753	714
E-8	2,032	599,621	1,356	3,870	948,535	1,015	5,160	963,696	537
E-7	1,850	545,914	1,235	3,390	830,887	890	4,521	844,355	471
E-6	1,586	468,011	1,059	2,527	619,366	663	3,369	629,204	351
E-5	1,336	394,239	892	2,121	519,856	557	2,828	528,165	294
E-4	1,052	310,433	702	1,663	407,600	436	2,218	414,240	231
E-3	882	260,268	589	1,394	341,669	366	1,859	347,192	193
E-2	744	219,546	497	1,176	288,237	309	1,568	292,844	163
E-1	664	195,939	443	1,049	257,109	275	1,399	261,281	146

*The lump-sum-equivalent is the amount of money required to be on hand at the time of retirement to pay a lifetime annuity that increases with inflation at 3% annually.

NOTES: Lump sum equivalent values are not adjusted for any reduction in DoD benefits due to offsets.

The interest rate used in discounting to the present value is 5.75% resulting in what is commonly referred to as a real interest rate of 2.75%.

Longevity i*The lump-sum-equivalent is the amount of money required to be on hand at the time of retirement to pay a lifetime annuity that increases with inflation at 3% annually.s based on military specific tables that assume an entry age of 23 for commissioned officers and warrant officers and 20 for enlisted.

Warning: these figures should not be used in property settlements for the following reasons: (1) They assume full future cost-of-living adjustments (which have been determined in some courts not to be property); (2) The investment return is low (producing large lump-sums) because of the conservative portfolio of government trust funds; and (3) Only the mortality of the retiree is considered (some courts require the use of the spouse's mortality).

As of January 1, 2009

NUMBER OF MILITARY PERSONNEL RECEIVING RETIRED PAY AS OF SEPTEMBER 30, 2008 BY RANK AND AVERAGE MONTHLY GROSS/NET (ALL DoD)

Rank/ Grouping	All Retirees Including Reserves			All Retirees Excluding Reserves			Reserve Retired Only		
		Monthly			Monthly			Monthly	
	Number	Avg Net	Avg Gross	Number	Avg Net	Avg Gross	Number	Avg Net	Avg Gross
O-10	231	9,561	10,326	231	9,561	10,326	0	0	0
O-9	816	8,641	9,313	809	8,675	9,348	7	4,706	5,257
O-8	2,980	6,310	6,785	1,925	7,885	8,465	1,055	3,436	3,719
O-7	3,175	4,904	5,278	1,721	6,740	7,241	1,454	2,730	2,955
O-6	105,298	4,308	4,598	70,052	5,456	5,817	35,246	2,026	2,175
O-5	183,564	3,042	3,240	126,358	3,733	3,970	57,206	1,516	1,626
O-4	122,385	2,487	2,652	95,492	2,868	3,057	26,893	1,137	1,212
O-3	38,158	2,130	2,318	31,063	2,398	2,614	7,095	956	1,020
O-2	4,900	1,438	1,631	4,366	1,526	1,737	534	712	762
O-1	1,120	1,096	1,225	1,057	1,125	1,260	63	600	636
All Commission	462,627	3,133	3,346	333,074	3,741	3,993	129,553	1,571	1,684
W-5	2,336	3,452	3,761	1,668	4,060	4,413	668	1,932	2,136
W-4	24,120	2,465	2,659	15,746	3,066	3,302	8,374	1,336	1,450
W-3	20,750	2,008	2,186	18,130	2,161	2,354	2,620	953	1,024
W-2	11,467	1,705	1,867	10,185	1,821	1,995	1,282	788	844
W-1	638	1,236	1,344	488	1,415	1,545	150	654	690
All Warrant	59,311	2,184	2,370	46,217	2,455	2,664	13,094	1,228	1,332
All Officers	521,938	3,026	3,235	379,291	3,584	3,831	142,647	1,540	1,652
E-9	103,991	2,537	2,736	87,242	2,799	3,018	16,749	1,172	1,268
E-8	230,637	1,808	1,972	194,699	1,973	2,154	35,938	914	986
E-7	539,935	1,479	1,632	474,645	1,573	1,739	65,290	791	850
E-6	376,811	1,135	1,270	325,723	1,226	1,376	51,088	552	590
E-5	81,760	837	970	66,208	932	1,089	15,552	434	465
E-4	15,366	491	690	13,997	508	724	1,369	317	341
E-3	6,011	412	568	5,992	412	569	19	305	332
E-2	2,309	384	497	2,306	384	497	3	344	353
E-1	724	402	522	717	403	524	7	248	277
All Enlisted	1,357,544	1,463	1,616	1,171,529	1,577	1,744	186,015	750	807
Unknown	1,389	1,129	1,329	1,387	1,130	1,330	2	529	537
Total	1,880,871	1,897	2,065	1,552,207	2,067	2,254	328,664	1,093	1,174

Figures include retirees receiving payment from DoD and have not been adjusted to DoD budget figures. Those retirees receiving a net dollar amount of zero are not included in the average net, likewise for average gross. "Monthly net retired pay" is the amount chargeable to the appropriation after deducting survivor payments, dual compensation reductions, VA payments and the like, but before making individual deductions such as for income taxes or savings bonds. Figures include amounts paid to retirees under the Concurrent Retirement and Disability Pay and Combat Related Special Compensation Programs.

NUMBER OF MILITARY PERSONNEL RECEIVING RETIRED PAY AS OF SEPTEMBER 30, 2008 BY RANK AND AVERAGE MONTHLY GROSS/NET (ALL DoD)

Rank/ Grouping	Nondisability Including Reserves			Nondisability Excluding Reserves			Disability Only		
		Monthly			Monthly			Monthly	
	Number	Avg Net	Avg Gross	Number	Avg Net	Avg Gross	Number	Avg Net	Avg Gross
O-10	229	9,566	10,331	229	9,566	10,331	2	8,916	9,746
O-9	797	8,633	9,312	790	8,668	9,347	19	8,961	9,366
O-8	2,929	6,275	6,751	1,874	7,873	8,457	51	8,340	8,744
O-7	3,125	4,870	5,241	1,671	6,731	7,230	50	7,048	7,608
O-6	103,604	4,289	4,576	68,358	5,456	5,814	1,694	5,445	5,951
O-5	180,526	3,031	3,223	123,320	3,734	3,964	3,038	3,694	4,218
O-4	119,133	2,488	2,641	92,240	2,882	3,057	3,252	2,471	3,056
O-3	32,720	2,203	2,339	25,625	2,549	2,704	5,438	1,690	2,193
O-2	2,027	1,713	1,821	1,493	2,070	2,199	2,873	1,244	1,497
O-1	185	1,269	1,361	122	1,615	1,735	935	1,061	1,198
All Commission	445,275	3,158	3,362	315,722	3,810	4,050	17,352	2,490	2,951
W-5	2,303	3,447	3,753	1,635	4,066	4,414	33	3,776	4,333
W-4	23,589	2,455	2,644	15,215	3,071	3,301	531	2,913	3,351
W-3	20,199	2,009	2,180	17,579	2,167	2,353	551	1,970	2,381
W-2	10,876	1,724	1,869	9,594	1,849	2,006	591	1,365	1,819
W-1	516	1,275	1,359	366	1,529	1,633	122	1,072	1,280
All Warrant	57,483	2,189	2,367	44,389	2,473	2,673	1,828	2,021	2,443
All Officers	502,758	3,048	3,248	360,111	3,645	3,880	19,180	2,445	2,903
E-9	101,808	2,537	2,732	85,059	2,805	3,020	2,183	2,571	2,908
E-8	225,691	1,808	1,967	189,753	1,977	2,153	4,946	1,837	2,181
E-7	528,231	1,481	1,629	462,941	1,578	1,739	11,704	1,388	1,743
E-6	363,532	1,143	1,269	312,444	1,239	1,380	13,279	929	1,285
E-5	68,479	882	981	52,927	1,014	1,132	13,281	605	917
E-4	3,139	607	662	1,770	831	910	12,227	461	697
E-3	237	666	733	218	697	768	5,774	401	562
E-2	36	566	628	33	587	653	2,273	381	495
E-1	87	497	618	80	518	648	637	389	509
All Enlisted	1,291,240	1,492	1,637	1,105,225	1,617	1,777	66,304	911	1,206
Unknown	1,371	1,137	1,338	1,369	1,137	1,339	18	584	654
Total	1,795,369	1,927	2,088	1,466,705	2,114	2,293	85,502	1,255	1,587

Figures include retirees receiving payment from DoD and have not been adjusted to DoD budget figures. Those retirees receiving a net dollar amount of zero are not included in the average net, likewise for average gross. "Monthly net retired pay" is the amount chargeable to the appropriation after deducting survivor payments, dual compensation reductions, VA payments and the like, but before making individual deductions such as for income taxes or savings bonds. Figures include amounts paid to retirees under the Concurrent Retirement and Disability Pay and Combat Related Special Compensation Programs.

ASSOCIATIONS AND ORGANIZATIONS

ORGANIZATION**	MEMBERSHIP ELIGIBILITY	ANNUAL DUES	MAGAZINE OR NEWSPAPER	REGIONAL OR LOCAL POSTS	LIFE INSURANCE PROGRAMS	HEALTH OR HOSP. INS. PROGRAMS	PARTIAL LISTING OF OTHER PROGRAMS AND/OR SERVICES
Air Force Association (AFA) 1501 Lee Hwy. Arlington, VA 22209 (703) 247-5800 1-800-727-3337 (142,000)	Members/former members of Armed Forces, current Reserve or Guard, spouses, widow(er)s, lineal descendants or patrons	$36.00 life membership available	"AIR FORCE" magazine monthly	Yes	Yes	Yes	National and local seminars and symposia on national defense; AF activities and aerospace development; 47 state organizations and almost 250 local chapters; insurance programs; MasterCard; rental car; medical care discounts and resume assistance. (www.afa.org)
Air Force Sergeants Association (AFSA) P.O. Box 50 Temple Hills, MD 20757-0500 (301) 899-3500 (130,000)	Active duty, retired, or veteran enlisted members of the USAF, Air National Guard, Air Force Reserve Command, and their families	$25.00 life membership available	"SERGEANTS Magazine" monthly and AFSA Bulletin Newsletter weekly	Yes	Yes	Yes	Professional Airmen's Conference and Annual Convention, over 200 local and international chapters, discount car rentals, auto insurance, scholarships and grant programs, legislative programs, association credit card, discount wireless service, discount computer hardware and accessories, online banking center, online employment service. (www.hqafsa.org)
American Ex-Prisoners of War 3201 E. Pioneer Pkwy., Ste. 40 Arlington, TX 76010-5396 (817) 649-2979	Ex-POW or family member	$40.00/ $50.00	"Ex-POW Bulletin" monthly	Yes	No	No	Medical research; National Service Officer Program; Veterans Administration Voluntary Service. (axpow@sbcglobal.net)
The American Legion 700 N. Pennsylvania St. Indianapolis, IN 46206 (317) 630-1200 1-800-433-3318 (2.7 million)	Vets with honorable service during WWI, II, Korea, Vietnam, Lebanon, Grenada, Panama, Desert Storm, and Iraq	$25.00 to $40.00 depending on posts	"The American Legion" magazine quarterly	Yes	Yes	Yes	Vets counseling and assistance; scholarship guidance and assistance; assistance to needy children of vets; military family support network; National Youth Baseball and Scout programs; national security, foreign relations and civil defense programs; Heroes to Hometowns program and American Legion scholarship program. (www.legion.org)
American Military Retirees Association (AMRA) 5436 Peru St., Ste.1 Plattsburgh, NY 12901 1-800-424-2969	Military retirees, active duty, Guardmembers, Reservists, 100% service-connected disabled, and surviving spouse of the above.	$18.00 life membership available	"AMRA News Report" quarterly	Yes	Yes	Yes	Legislative programs; scholarship awards; rental car discount; auto/home insurance; TRICARE/CHAMPVA supplemental insurance; hotel/motel discounts; and state to state moving discounts. (www.amra1973.org)
American Military Society P.O. Box 98186 Washington, DC 20090-8186 1-800-808-4517 (48,000)	Open to all grades, ranks, and components of the uniformed services; active duty, National Guard, Reserve, retired, and veterans and their spouses, widow(er)s, and former spouses	$20.00 life membership available	"The Advocate" bi-monthly	Yes	Yes	Yes	Legislative program; TRICARE/Medicare supplemental insurance; hospital income plan; accidental death/dismemberment; term life; dental plan; legal services plan; auto and homeowners insurance; car rentals; hotel/motel discounts; VISA card. (www.amsmilitary.org)
American Retirees Association P.O. Box 2333 Redlands, CA 92373 (909) 557-0107	All uniformed, military, Coast Guard, Public Health, and National Oceanic and Atmospheric Administration service personnel	$25.00	"ARA Newsletter" quarterly	No	No	No	Purpose is to help Uniformed Service personnel avoid being targeted by the Uniformed Services Former Spouses Protection Act, (USFSPA), Public Law 97-252; 10 USC 1408. Also, provides information and group counseling on (USFSPA) impact. (www.americanretireesassociation.org)

ORGANIZATION**	MEMBERSHIP ELIGIBILITY	ANNUAL DUES	MAGAZINE OR NEWSPAPER	REGIONAL OR LOCAL POSTS	LIFE INSURANCE PROGRAMS	HEALTH OR HOSP. INS. PROGRAMS	PARTIAL LISTING OF OTHER PROGRAMS AND/OR SERVICES
AMVETS (American Veterans) 4647 Forbes Blvd. Lanham, MD 20706-4380 (301) 459-9600 (250,000)	All who are currently serving or have served honorably in the U.S. Armed Forces (active, Reserve or National Guard) after Sept. 15, 1940	$30.00 (varies with post)	"American Veteran" magazine quarterly	Yes	No	No	Counseling and assistance in obtaining benefits and claims; VA hospital volunteers; legislative program; scholarship program; ROTC program; blood donor program; prescription card. (www.amvets.org)
Armed Forces Services Corporation 2800 Shirlington Rd., Ste. 350 Arlington, VA 22206 1-888-237-2872	All branches of the uniformed services (active, Reserve, retired, veterans, and their widow(er)s)	Varies based on age	"AFSC Benefactor" annually	No	No	No	Advising on federal benefits, life insurance matters and financial counseling; central repository service for important documents of all members. (www.afsc-usa.com)
Army and Air Force Mutual Aid Association (AAFMAA) 102 Sheridan Ave. Fort Myer, VA 22211-1110 1-800-336-4538 (77,800)	All members of the Army/AF, Army/AF Guard and Reserve (active duty/retired under age 66). Army/AF Academy cadets, ROTC contract and scholarship cadets. Honorably discharged veterans residing in VA, NC, or SC.	None	"AAFMAA Advantage" quarterly	No	Yes	No	Life insurance and survivor assistance services. (www.aafmaa.com)
Association of the United States Army (AUSA) 2425 Wilson Blvd. Arlington, VA 22201 (703) 841-4300 (100,000)	Any individual, military or civilian who subscribe to AUSA's aims and objectives	Based on grade. Discount for E1-E4, life membership on request	"ARMY" magazine and "AUSA News" both monthly	Yes	Yes	Yes	Educational programs; national and regional seminars; AUSA Position Papers; several discount programs; auto insurance. (www.ausa.org)
Catholic War Veterans of the United States of America 441 N. Lee St. Alexandria, VA 22314-2301 (703) 549-3622 (15,000)	Catholic active duty members or veterans of the U.S. Armed Forces, or Guard and Reserve. Merchant Marines having served 90 days AD between 12/7/41 and 12/31/46	$25.00 (varies with post)	"Catholic War Veteran" quarterly	Yes	Yes	No	www.cwv.org
Chief Warrant Officers Association, USCG 200 V St. S.W. Washington, DC 20024 (202) 554-7753	Chief warrant officer or former chief warrant officer serving in a higher grade on the U.S. Coast Guard active duty, Reserve, or retired list	$60.00	"CWO News" bi-monthly	Yes	No	No	Scholarship awards (www.cwoauscg.org)
Coast Guard Chief Petty Officer Association 5520G Hempstead Way Springfield, VA 22151 (703) 941-0395 (11,000)	Any E-7, E-8, E-9 in any service (active, retired or Reserve)	$24.00 check or allotment	"The Chief" magazine quarterly	Yes	Yes	No	Scholarship awards; legislative program. (www.uscgcpoa.org) (cgcpoa@aol.com)

ORGANIZATION**	MEMBERSHIP ELIGIBILITY	ANNUAL DUES	MAGAZINE OR NEWSPAPER	REGIONAL OR LOCAL POSTS	LIFE INSURANCE PROGRAMS	HEALTH OR HOSP. INS. PROGRAMS	PARTIAL LISTING OF OTHER PROGRAMS AND/OR SERVICES
Coast Guard Enlisted Association 5520G Hempstead Way Springfield, VA 22151 (703) 941-0395 (2,000)	All services, enlisted E-6 or below (active, retired or Reserve)	$24.00 check or allotment	"The Chief" magazine quarterly	Yes	Yes	No	Scholarship awards; legislative representation. (www.uscgcpoa.org)(cgcpoa@aol.com)
The Commissioned Officers Association of the U.S. Public Health Service (COA) 8201 Corporate Dr., Ste. 200 Landover, MD 20785 (301) 731-9080 1-866-366-9593 (7,000)	For active duty, Reserve, and former Public Health Service officers, as well as anyone who is a supporter of the Public Health Service Commissioned Corps	$50.00-$135.00 depending on grade and/or category of membership	"Frontline" 10 issues annually	Yes	Yes	Yes	Capitol Hill representation; local representation; annual meeting; financial planning seminars; scholarship program; ribbon; and Hertz rental discount. (www.coausphs.org)
Disabled American Veterans (DAV) P.O. Box 14301 Cincinnati, OH 45256 1-877-426-2838 (1.3 million)	All war disabled veterans of the U.S. Armed Forces	Varies based on age	"DAV Magazine" monthly	Yes	No	No	Provides counseling and assistance in obtaining veterans benefits and claims; employment assistance; VA hospital volunteers; scholarship programs; vocational rehab and education assistance; legislative programs. (www.dav.org) (feedback@davmail.org)
Enlisted Association of the National Guard of the U.S. 3133 Mount Vernon Ave. Alexandria, VA 22305 (703) 519-3846 1-800-234-3264 (85,000)	All enlisted ARNG and ANG (grades E-1 thru E-9)	$9.00 plus state association add-ons	"The New Patriot" quarterly	Yes	Yes	Yes	National scholarship; student loans; Visa card; travel service; eye care; employment referral service; state and national legislative programs; state and national conferences; family education and support programs; auxiliary; and accidental death and dismemberment. (www.eangus.org) (eangus@eangus.org)
FRA 125 N. West St. Alexandria, VA 22314-2754 (703) 683-1400 1-800-FRA-1924 (100,000)	Active duty, Reserve, retired members and veterans of the U.S. Navy, Marine Corps and Coast Guard who have served at least one day of enlisted duty	$30.00	"FRA Today" magazine monthly "On Watch" AD newspaper quarterly	Yes	Yes	Yes	Career counseling and assistance; legislative programs; veterans claims and grievances representation with the VA; assistance with corrections of military records and disability claims; TRICARE/Medicare supplemental insurance term life. (www.fra.org)
Jewish War Veterans of the U.S. (JWV) 1811 R St. N.W. Washington, DC 20009 (202) 265-6280 (100,000)	Veterans of wartime service	$40.00	"The Jewish Veteran" 5 issues annually	Yes	Yes	Yes	Legislative program; assistance to Veterans Services Officers; services to homeless; annual convention; VA hospital volunteer service; scholarship program for JWV dependents; blood donor post level programs. (www.jwv.org) (jwv@jwv.org)
Marine Corps Association P.O. Box 1775 Quantico, VA 22134 (703) 640-6161 1-800-336-0291	All U.S. Marines, Reserve Marines, retired Marines, Marine veterans, and dependents of Marines	Varies for print and online magazine/membership	"Leatherneck" and "Marine Corps Gazette" monthly	No	Yes	Yes	The Marine Shop at Quantico, VA, the MCA Bookstore at Camp Lejeune, NC and online store. All purchases help support MCA's award program. (www.mca-marines.org)

ORGANIZATION**	MEMBERSHIP ELIGIBILITY	ANNUAL DUES	MAGAZINE OR NEWSPAPER	REGIONAL OR LOCAL POSTS	LIFE INSURANCE PROGRAMS	HEALTH OR HOSP. INS. PROGRAMS	PARTIAL LISTING OF OTHER PROGRAMS AND/OR SERVICES
Marine Corps League P.O. Box 3070 Merrifield, VA 22116-3070 1-800-MCL-1775 (703) 207-9588 (62,000)	All active duty, Reserve, or Marine veterans and FMF Corpsmen with at least 90 days active duty or 90 Reserve retirement points. If discharged, an honorable discharge is required	$20.00-$30.00 depending on local chapters	"Semper Fi" bi-monthly	Yes	No	Yes	Auto quote and purchase plan; discount lodging program; service group insurance; TRICARE/Medicare supplement plan; veteran service program, claims and grief assistance. (www.mcleague.org)
Marine Corps Mustang Association, Inc. Bunker 127 Mountain City, GA 30562-0127 1-866-937-6262 (1,400)	Marines who, after having served on active duty in the enlisted ranks of the Marine Corps or Marine Corps Reserve, have risen to the officer ranks and served as a commissioned or warrant officer in the U.S. Marine Corps	$40.00 first year $35.00 thereafter	"MESTENGO" quarterly	No	No	No	www.marinecorpsmustang.org
Marine Corps Reserve Association (MCRA) 8626 Lee Hwy., Ste. 205 Fairfax, VA 22031-2135 (703) 289-1204 (877) 289-8780 (6,000)	Any Reserve or regular member of the USMC on active duty, inactive duty, or retired; associate membership for others with USMC service; friends of MCRA members	$15.00-$100.00 depends on grade. Life membership available	"The Word" quarterly	Yes	No	No	www.usmcra.org
Military Officers Association of America (MOAA) 201 N. Washington St. Alexandria, VA 22314-2539 (703) 549-2311 (370,000)	All active duty, retired, National Guard, Reserve, former commissioned officers, and warrant officers of the uniformed services and their surviving spouses	$31.00 (surviving spouses $27.00)	"Military Officer" magazine monthly; "Today's Officer" and "Today's Officer Online" quarterly	Yes	Yes	Yes	Career transistion assistance; legislative programs; insurance and support; counseling regarding entitlements and benefits; travel program; survivor assistance; discount legal services; educational assistance. (www.moaa.org) (msc@moaa.org)
Military Order of the Purple Heart 5413-B Backlick Rd. Springfield, VA 22151-3960 (703) 642-5360 FAX (703) 642-2054	All persons who have been awarded the Purple Heart medal by the U.S. Government	$50.00 life membership	"Purple Heart" bi-monthly	Yes	No	No	Veteran Service Program; scholarship for Purple Heart veterans or direct descendant of MOPH member or veteran KIA. (www.purpleheart.org) (info@purpleheart.org)
Military Order of the World Wars 435 N. Lee St. Alexandria, VA 22314 (703) 683-4911 1-877-320-3774	Any federally recognized commissioned or warrant officer, retired or active duty, of the uniformed services	$40.00	"Officer Review" 10 issues annually	Yes	No	No	Patriotic, nonpartisan organization established in 1919 to promote welfare of nation. Promotes national security, patriotism, good citizenship and service to country; conducts youth leadership conferences throughout the country. (www.militaryorder.net)

ORGANIZATION**	MEMBERSHIP ELIGIBILITY	ANNUAL DUES	MAGAZINE OR NEWSPAPER	REGIONAL OR LOCAL POSTS	LIFE INSURANCE PROGRAMS	HEALTH OR HOSP. INS. PROGRAMS	PARTIAL LISTING OF OTHER PROGRAMS AND/OR SERVICES
National Association for Uniformed Services® (NAUS) 5535 Hempstead Way Springfield, VA 22151-4094 (703) 750-1342 (160,000)	Open to all grades, ranks and components of the uniformed services; active duty, National Guard, Reserve, retired, and veterans and their spouses, widow(er)s, and former spouses	$15.00 (widows $12.00)	"USJ" bi-monthly	Yes	Yes	Yes	Legislative program; TRICARE/Medicare supplemental insurance; hospital income plan; accidental death/dismemberment; term life; dental plan; legal services plan; auto and homeowners insurance; Pentagon Federal Credit Union; car rentals; hotel/motel discounts; VISA card. (www.naus.org)
National Chief Petty Officers Association 106 Waring-Welfare Rd. Boerne, TX 78006-7925 (830) 537-4899 (1,500)	Served as a Chief Petty Officer for at least 30 days in the U.S. Navy or U.S. Coast Guard	$15.00 life membership available	"The Chiefs" quarterly	Yes	No	No	www.goatlocker.org/ncpoa
National Guard Association of the United States (NGAUS) One Massachusetts Ave. N.W. Washington, DC 20001 (202) 789-0031 (45,000)	Any officer or warrant officer who is serving or ever served honorably in the National Guard	$26-$153 depending on grade. Life membership available	"National Guard" magazine monthly	Yes, each state has an association	Yes	Yes	Recognition and awards programs; legislative programs; insurance program; TRICARE/ Medicare supplements; Edward L. Martin Library; National Guard Memorial Museum. (www.ngaus.org)
Naval Enlisted Reserve Association (NERA) 6703 Farragut Ave. Falls Church, VA 22042 (703) 534-1329 (12,000)	All enlisted or veterans who have served in the Naval Reserve, Marine Corps Reserve, or Coast Guard Reserve	$30.00	"THE MARINER" quarterly	Yes	Yes	Yes	Assistance with filing for Naval Reserve retirement; legislative program; record reviews; group insurance programs; recognition, scholarship and awards program. (www.nera.org)
Naval Reserve Association (NRA) 1619 King St. Alexandria, VA 22314-2793 (703) 548-5800 (22,000)	Commissioned, Reserve, retired, or WOs having served honorably in the Armed Forces and their spouses & widow(er)s	$40.00 ($105 for three years)	"NRA News" monthly	Yes	Yes	Yes	Professional counseling service; TRICARE/ Medical insurance supplement; record review service; promotional and service jacket services; legislative program. (www.navy-reserve.org)
Navy League of the U.S. 2300 Wilson Blvd., Ste. 200 Arlington, VA 22201 (703) 528-1775 1-800-356-5760 (60,000)	Citizen of the U.S. not on active duty with the Armed Forces	$40.00 life membership available	"Sea Power" magazine monthly	Yes	No	No	Education programs; support of Sea Service through seminars and awards programs; legislative programs; youth program; Sea Cadets. (www.navyleague.org) (mail@navyleague.org)
Navy Mutual Aid Association Henderson Hall 29 Carpenter Rd. Arlington, VA 22212 1-800-628-6011 (100,000)	Active duty, Reserve or retired Sea Service members (USN, USMC, USCG, NOAA, and USPHS)	None	Semi-annual member newsletter	No	Yes	No	Advising on federal benefits, life insurance matters and financial counseling; central repository service for members' important documents. (www.navymutual.org) (info@navymutual.org)

ORGANIZATION**	MEMBERSHIP ELIGIBILITY	ANNUAL DUES	MAGAZINE OR NEWSPAPER	REGIONAL OR LOCAL POSTS	LIFE INSURANCE PROGRAMS	HEALTH OR HOSP. INS. PROGRAMS	PARTIAL LISTING OF OTHER PROGRAMS AND/OR SERVICES
Non Commissioned Officers Association (NCOA) P.O. Box 33790 San Antonio, TX 78265 1-800-662-2620 (60,000)	Active, retired, Reserve or National Guard NCO or POs, honorably separated veterans regardless of rank. All branches of Armed Forces	$30.00	"NCOA Journal" magazine quarterly	Yes	Yes	Yes	Congressionally chartered; legislative program; scholarships; medical fund; disaster relief fund; employment assistance; recognition/awards program; national discount programs; social activites through local chapters. (www.ncoausa.org)
Paralyzed Veterans of America 801 18th St. N.W. Washington, DC 20006 (202) 872-1300 1-800-424-8200 (19,700)	Honorably discharged member who served in the Armed Forces and suffered a spinal cord injury as a result of trauma or disease	None	"PN/Paraplegia News" monthly and "Sports 'N Spokes" bi-monthly	Yes	No	No	www.pva.org
Reserve Enlisted Association (REA) 1 Constitution Ave. N.E. Washington, DC 20002 (202) 646-7715	Enlisted members from all branches of the United States military Reserve Components	$20-$40 based on grade	"The Officer" magazine monthly	Yes	No	No	Regular membership available to those serving or retired from a Reserve Component, with dues based on grade and length of membership. Associate membership also available. (www.reaus.org)
Reserve Officers Association (ROA) 1 Constitution Ave. N.E. Washington, DC 20002 (202) 479-2200 1-800-809-9448 (70,000)	Any officer–active, Reserve or retired–of the uniformed services, and their spouses and widow(er)s	$40.00	"The Officer" magazine monthly	Yes	Yes	Yes	Congressionally chartered; Reserve Component focus for national security issues via legislative advocacy and education; Membership benefits include: discounts, insurance, scholarships, e-commerce mall. (www.roa.org)
The Retired Enlisted Association (TREA) 1111 S. Abilene Ct. Aurora, CO 80012 1-800-338-9337 (100,000)	Any enlisted, retired from the Armed Forces of the U.S. for length of service or permanently medically retired. Any active duty enlisted with 10 or more years of service	$20.00	"The VOICE" bi-monthly (odd months); e-magazine (even months)	Yes	Yes	Yes	Various programs; legislative; veterans counseling and assistance; scholarship; TRICARE/Medicare supplement; car rental; prescription drug; counsels members on retirement rights, benefits and privileges; discount eye care program. (www.trea.org)
Society of Military Widows (SMW) 5535 Hempstead Way Springfield, VA 22151 (703) 750-1342 1-800-842-3451 (6,000)	Widows of former servicemen from all services	$12.00	"Uniformed Services Journal" bi-monthly	Yes	Yes	Yes	Legislative program; TRICARE/Medicare supplemental insurance; term life; dental plan; legal services plan; auto and homeowners insurance; Pentagon Federal Credit Union; car rentals; hotel/motel discounts; VISA card. Affiliated with NAUS. (www.militarywidows.org)

ORGANIZATION**	MEMBERSHIP ELIGIBILITY	ANNUAL DUES	MAGAZINE OR NEWSPAPER	REGIONAL OR LOCAL POSTS	LIFE INSURANCE PROGRAMS	HEALTH OR HOSP. INS. PROGRAMS	PARTIAL LISTING OF OTHER PROGRAMS AND/OR SERVICES
U.S. Army Warrant Officers Association 462 Herndon Pkwy., Ste. 207 Herndon, VA 20170-5235 (703) 742-7727 (6,000)	Army warrant officers, active duty, National Guard, Reserve and retired warrant officers	Retired $30.00 others $45.00 life membership available	"Newsliner" magazine monthly	Yes	No	No	Membership service to include legislative programs; professional development seminars; awards; distinctive WO memorabilia; TRICARE/Medicare supplement; rental discounts; discounted multi-year "dues" rates available. (www.usawoa.org) (usawoahq@verizon.net)
Veterans of Foreign Wars of the U.S. (VFW) 406 W. 34th St. Kansas City, MO 64111 (816) 756-3390 (2 million)	Any AD or honorably discharged veteran who has overseas service recognized by campaign medal or ribbon; or 30 consecutive days or 60 days on the Korean Peninsula or territorial waters, or received hostile fire or imminent danger pay	$35.00 (average)	"VFW" magazine monthly	Yes	Yes	Yes	Assistance for any veteran, without regard to membership or cost, who has a claim or appeal with VA or DoD; maintains service officers in chapters to provide counseling services to all veterans; scholarship programs; maintains home for children of members. (www.vfw.org)
Vietnam Veterans of America 8605 Cameron St., #400 Silver Spring, MD 20910 1-800-882-1316 (57,000)	Vietnam era veterans	$20.00	"The VVA Veteran" newspaper bi-monthly	Yes	Yes	No	Leading advocate for recognition of health issues affecting Vietnam era veterans and family members; provide representation in VA benefits cases; expanding economic opportunities; promote youth outreach, community service and scholarship programs; active nationwide in homeless veteran projects. (www.vva.org)
Women's Army Corps Veterans Association P.O. Box 5577 Ft. McClellan, AL 36205 (256) 820-6824 (3,500)	WAC & WAAC; Women in or former members of Army, Army Reserve & National Guard	$20.00	"The Channel" 5 issues annually	Yes	No	No	Provide services and support to veterans through VA hospital volunteer program; annual award to outstanding female ROTC graduate in four regions; annual scholarship award; promotes general welfare of all veterans, especially women. (www.armywomen.org)

**Figures in ( ) reflect organization membership.

INTERNET ADDRESSES

Note: Many organizations frequently change their website addresses. All Internet addresses were accessed on January 15, 2009 and verified to link to a valid site. We apologize for any inconvenience caused by addresses that are not valid when you attempt to access one of the sites listed.

SERVICE LINKS

Department of Defense: www.defenselink.mil
Assistant Secretary of Defense for Reserve Affairs: www.defenselink.mil/ra/
U.S. Army Homepage: www.army.mil
U.S. Air Force Homepage: www.af.mil
U.S. Navy Homepage: www.navy.mil
Army Reserve Homepage: www.armyreserve.army.mil
Air National Guard Homepage: www.ang.af.mil
Air Force Reserve Command Homepage: www.afreserve.com
Army National Guard: www.1800goguard.com
Coast Guard Reserve: www.uscg.mil/Reserve/
Marine Reserve: www.marforres.usmc.mil
Navy Reserve: www.navyreserve.com/

VETERANS' LINKS

Veterans News & Information Service: www.vnis.com
Department of Veterans Affairs: www.va.gov
American Legion: www.legion.org
Veterans of Foreign Wars: www.vfw.org
Disabled American Veterans: www.dav.org

FAMILY READINESS AND MILITARY SUPPORT LINKS

DEERS: https://www.dmdc.osd.mil/appj/address/index.jsp
Armed Services YMCA: www.asymca.org
Army Well-Being: www.armywell-being.org
Lifelines: www.lifelines.navy.mil
American Red Cross: www.redcross.org
Fisher House: www.fisherhouse.org
Military Women: www.militarywoman.org
Consumer Information Center: www.pueblo.gsa.gov
Transition Information: www.TurboTap.org
Military Spouse's Career Network: www.mscn.org
Civilian Personnel Office: http://cpol.army.mil

MILITARY RECREATION & TRAVEL

Morale Welfare Recreation (MWR) Army Gateway: www.armymwr.com
USO, Morale, Welfare and Recreation-Type Services: www.uso.org
Mountaineering and Recreation Information: www.hillwalking.org.uk
National Parks Information: www.nps.gov

LEGAL LINKS

LAWLINKS-Legal Information: www.kent.ac.uk/lawlinks/
Better Business Bureau: http://welcome.bbb.org/
Findlaw Website: www.findlaw.com
Hieros Gamos-The Comprehensive Law & Government Site: www.hg.org
Internet Legal Resource Group: www.ilrg.com
Widener Law Library: www.law.widener.edu/Law-Library

EDUCATION/INFORMATION

Library of Congress: www.loc.gov
Government Research Center: http://grc.ntis.gov
US Department of Education: www.ed.gov
Military Child Education Coalition: www.militarychild.org
Dictionary: http://dictionary.reference.com

MILITARY MOVES/DEPLOYMENT

Foreign Languages for Travelers: www.travlang.com/languages
Military Assistance Program: http://www.militaryhomefront.dod.mil/

VOLUNTEER LINKS

Energize: www.energizeinc.com
Make a Difference Day: www.makeadifferenceday.com
Points of Life Institute: http://pointsoflight.org
Volunteer Today: www.volunteertoday.com

OTHER LINKS

AF Publications: www.e-publishing.af.mil
Air Force Links: www.af.mil/sites
Airman Magazine: www.airmanonline.af.mil/
Early Bird: http://ebird.osd.mil/
FedWorld Home Page: www.fedworld.gov
Defense Commissary Agency: www.commissaries.com
Army Reserve Association: www.armyreserve.org
Defense Finance and Accounting Service (DFAS): www.dfas.mil
Army Retirement Services: www.armyg1.army.mil/rso
Nifty Ideas on Anything: www.tutorials.com
Kelly Blue Book: http://kbb.com
Weather: www.weather.com
All Kinds of Free Stuff: www.thefreesite.com
Flights Airline Travel: www.travelocity.com

STAY INFORMED

The **UNIFORMED SERVICES ALMANAC,** for active duty personnel, provides accurate, timely information on pay, allowances, taxes, health care and TRICARE benefits, SBP, retirement, travel and transportation, Space-A, Veterans' and Social Security Benefits and many more subjects of vital concern to active duty personnel and their families. The almanac also contains detailed installation listings and important statistical data.

The **RESERVE FORCES ALMANAC** is specifically prepared for members of the Army, Navy, Air Force, Marine Corps and Coast Guard Reserve. This annual 256 page edition contains detailed drill pay tables for daily, weekend and annual drill periods, plus information on military retirement, tax information and tips, federal benefits, Reserve Component SBP, VA and Social Security coverage, promotion information and many more subjects of importance for all Reservists and their families.

The **NATIONAL GUARD ALMANAC** is published for members of the Army and Air National Guard and their families. This 256 page edition contains detailed drill and annual pay tables, retirement point and pay computation tables, and other important information relative to state and federal benefits available to Guardmembers. This is the most comprehensive book on Guard related material available.

The **RETIRED MILITARY ALMANAC** is published, not only for the already retired members, but for all military personnel considering retirement. This 256 page edition contains information on benefits, entitlements, privileges and restrictions as well as detailed coverage of retired pay, military health care, TRICARE, VA and Social Security, SBP, legislation, installation listings and many other subjects. Also includes information on federal and state taxes.